Environmental Earth Sciences

Series Editor

James W. LaMoreaux, Tuscaloosa, USA

Environmental Earth Sciences encompass multidisciplinary studies of the Earth's atmosphere, biosphere, hydrosphere, lithosphere and pedosphere and humanity's interaction with them. This book series provides a forum for this diverse range of studies, reporting on the very latest results and documenting our emerging understanding of the Earth's system and our place in it. The type of material accepted for publication includes:

- proceedings that are peer-reviewed and published in association with a conference
- post-proceedings consisting of thoroughly revised final papers
- research monographs that may be based on individual research projects
- tutorials or collections of lectures for advanced courses
- contemporary surveys that offer an objective summary of a current topic of interest
- emerging areas of research directed at a broad community of practitioners.

Xiaodong He · Peiyue Li

Geochemical Processes in Tight Gas Hydraulic Fracturing Stimulation

Springer

Xiaodong He
School of Water and Environment
Chang'an University
Xi'an, China

Peiyue Li
School of Water and Environment
Chang'an University
Xi'an, China

ISSN 2199-9155 ISSN 2199-9163 (electronic)
Environmental Earth Sciences
ISBN 978-3-032-15179-7 ISBN 978-3-032-15180-3 (eBook)
https://doi.org/10.1007/978-3-032-15180-3

This Springer imprint is published by the registered company Springer Nature Switzerland AG
The registered company address is: Gewerbestrasse 11, 6330 Cham, Switzerland

Preface

Hydraulic fracturing has sparked a global revolution in unconventional oil and gas extraction, enabling the exploitation of unconventional resources that comprise approximately 80% of the world's total energy reserves. With the continuous advancement and application of hydraulic fracturing technology, unconventional hydrocarbon resources represented by tight oil/gas, shale oil/gas, coalbed methane, and oil shale have developed rapidly. However, while significantly increasing unconventional hydrocarbon production, hydraulic fracturing has also induced a series of engineering and environmental issues that warrant serious attention. During the fracturing process, substantial volumes of fluid containing diverse chemical additives are injected into subsurface formations. This process inevitably disrupts the original water-rock-gas equilibrium of the formation, inducing a series of chemical reactions that alter the quality of the flowback water and cause various forms of reservoir damage such as secondary mineral precipitation, clay swelling, and particle migration, thereby reducing fracturing efficiency and gas well productivity.

In response to these challenges, this book takes the Sulige Gas Field (the largest natural gas field in China) as the study area, systematically investigates the geochemical processes associated with hydraulic fracturing in tight sandstone gas reservoirs through integrated field observations, laboratory experiments, and model simulations. By characterizing the mineralogy, reservoir properties, and post-fracturing water chemistry, this research elucidates the geochemical processes active during hydraulic fracturing of tight sandstones, thereby providing theoretical and technical foundations for fluid optimization and process enhancement. Meanwhile, regional water environmental issues induced by hydraulic fracturing were further explored, focusing on the identification of potential groundwater contamination in the Sulige Gas Field and discussion on the management and resource utilization of flowback water.

This publication comprises ten chapters. Chapter 1 outlines the significance and objectives of this study, introducing fundamental concepts of hydraulic fracturing and unconventional hydrocarbon systems, including resource distribution, fracturing procedures, fracturing fluid and additives, water-rock interactions, and related engineering and environmental challenges. This introductory chapter delineates the publication's research framework and thematic boundaries, while systematically mapping its organizational architecture to equip readers with foundational context for navigating subsequent analytical content.

Chapter 2 provides an overview of the geographical location, hydrometeorology, regional geology, reservoir conditions, hydrogeology, and development status of the Sulige Gas Field, aiming to give readers a comprehensive understanding and background for later discussions. The Sulige Gas Field is located in the central-northern part of the Ordos Basin, China's second largest petroleum basin. The gas reservoirs are mainly developed in the Upper Paleozoic Permian Shihezi (He 8) and Shanxi (Shan 1) formations, consisting primarily of fluvial sandstone bodies forming tight sandstone lithologic gas reservoirs. The Sulige area is characterized by a temperate arid to semi-arid climate, with poorly developed surface drainage and scarce surface water resources. Thus, groundwater serves as the principal source of water supply.

Chapter 3 presents a study on the lithology, reservoir properties, and gas reservoir characteristics of tight sandstones in the Sulige Gas Field. Core samples from five tight sandstone wells were collected from the Sulige Gas Field and analyzed using X-ray fluorescence (XRF), X-ray diffraction (XRD), scanning electron microscopy (SEM), and casting thin sections to obtain a comprehensive understanding of mineral composition, occurrence, and pore distribution characteristics. Given the intrinsic complexity of pore distribution in tight sandstone reservoirs, the diagenetic processes were further investigated to elucidate the factors controlling pore structure formation and evolution. The tight sandstone in the Lower He 8 Member of the Sulige region mainly consists of quartz (45.9–81.6%) and clay minerals (16.5–47.4%), classified as lithic sandstone and lithic quartz sandstone. Illite (3.9–27.3%), kaolinite, and chlorite (10.9–31.8%) are abundant, while feldspars are generally absent. During diagenesis, the tight sandstones underwent significant compaction, cementation, and dissolution, resulting in the development of intergranular, intragranular, and intercrystalline pores. These alterations yielded reservoir properties with 7.7–12.6% of porosity and 0.16–1.42 mD of permeability, characteristic of low-permeability tight reservoirs.

Chapter 4 presents hydraulic fracturing experiments conducted on six tight sandstone gas wells with different well types (horizontal and directional wells) and fracturing fluid systems (guar, variable-viscosity slickwater, slickwater, and biopolymer). Flowback fluid samples were collected and analyzed after fracturing to identify geochemical and water quality characteristics and to determine the sources of solutes. Furthermore, by compiling and analyzing global geochemical data of produced waters, comparative studies were performed on the geochemical characteristics of flowback waters from tight gas, shale gas, and coalbed methane wells, offering site-scale evidence for understanding the geochemical mechanisms. Flowback water from the Sulige Gas Field are characterized by high ionic concentrations (Cl^-, Na^+, Ca^{2+}, K^+, Sr^{2+}, Ba^{2+}, Fe, etc.) and enriched heavy isotopes (δ^2H, $\delta^{18}O$, $^{87}Sr/^{86}Sr$). These fluids are of poor quality and possess high contamination potential.

Based on field-scale hydraulic fracturing experiments, Chap. 5 employs core samples of tight sandstone to perform laboratory-scale studies of water-rock interactions, thereby investigating the microscale geochemical processes occurring during hydraulic fracturing. To evaluate the effects of individual fracturing fluid additives on water-rock reactions, separate experiments were conducted between each additive and tight sandstone, with deionized water used as the control. Additional experiments were conducted under varying pH, temperature, and solid–liquid ratios to analyze the effects of these environmental factors on the interactions between fracturing fluids and tight sandstone. Within the guar-based fracturing system, the presence of biocide, surfactant, and clay stabilizer exerted little effect on the inorganic composition of the fluid, with their water-rock interaction effects similar to those of deionized water. Conversely, pH control and crosslinking are the key determinants of the fracturing fluid's chemical characteristics and exert significant control over its geochemical interactions with tight sandstone.

Based on the results from field-scale flowback fluids and laboratory water-rock interaction experiments, Chap. 6 characterizes and elucidates the geochemical processes involved in the interactions among fracturing fluid, tight sandstone, and formation water during hydraulic fracturing. Given the water-bearing nature of tight sandstone reservoirs, the chapter first provides an in-depth investigation of the reactions between fracturing fluids and formation water. Integrating laboratory findings, the study identifies key geochemical mechanisms during fracturing, including mineral dissolution and precipitation, redox reactions, and ion exchange. During hydraulic fracturing, mixing, mineral dissolution and precipitation, redox reactions, and ion exchange processes are interwoven and coupled. The flowback fluid is thus a complex mixture derived from the combined effects of fracturing fluid-formation water mixing and fracturing fluid-sandstone interactions.

Geochemical processes occurring in fracturing stimulation influence not only the chemical characteristics of the flowback fluids but also trigger a range of impacts on the reservoir itself. Chapter 7 investigates the reservoir impacts and damage resulting from fracturing fluid-tight sandstone interactions during hydraulic fracturing, focusing on clay mineral instability and pore blockage caused by scaling. The high clay mineral content and their distribution patterns in tight sandstone can easily induce instability of clay minerals during hydraulic fracturing. Illite minerals tend to swell during hydraulic fracturing, while kaolinite is highly susceptible to crystal fragmentation, detachment, dispersion, and migration under alkaline conditions, leading to pore blockage and formation damage.

In response to the significant environmental risks associated with hydraulic fracturing, Chap. 8 discusses the fracturing activities in the Sulige region, along with groundwater hydrochemistry, potential contamination sources, and migration pathways. Building on these findings, a flowback fluid contamination identification model based on hydrochemical and isotopic indicators was constructed, employing strontium isotopes to trace potential groundwater contamination in the study area. In the northern desert plateau of Sulige, $^{87}Sr/^{86}Sr$ is sensitive enough to detect contamination events when as little as 0.49–2.15% of flowback fluid mixes with groundwater. With the anticipated intensification of hydraulic fracturing in Sulige, establishing robust groundwater protection and pollution control strategies becomes crucial. It is vital both to promote efficient treatment and reuse of flowback fluids and to enhance groundwater monitoring, establishing a regional water-quality baseline and a comprehensive geochemical database for the Sulige area.

Chapter 9 focuses on the management strategies and resource recovery potential of flowback water. Understanding the temporal variation of flowback water chemistry enables stage-wise storage and targeted treatment or resource recovery. The flowback water from tight sandstone reservoirs contains high-temperature formation water, representing a potential geothermal resource. The chapter highlights the lithium extraction potential of flowback water. Through comprehensive geochemical understanding, flowback fluids can be valorized via multiple pathways such as fracturing fluid reuse, elements recovery, geothermal exploitation, deep reinjection, road deicing, ecological irrigation, and soil remediation.

Chapter 10 provides a summary of the entire book, outlining the major findings and conclusions. Additionally, it highlights research gaps and future perspectives in three frontier areas: (1) Biogeochemical processes during hydraulic fracturing, (2) lithium enrichment and resource evaluation of flowback waters, and (3) migration and transformation of key contaminants in coupled soil-groundwater systems.

This book provides theoretical foundations and technical support for the green and sustainable development of unconventional energy and significantly contribute to the advancement of geochemistry in deep subsurface environments. The book bridges fundamental geochemistry and practical engineering, offering actionable insights for reservoir engineers, geochemists, and policymakers to navigate environmental and engineering issues.

Xi'an, China

Dr. Xiaodong He
Prof. Peiyue Li

Acknowledgments This book synthesizes Dr. Xiaodong He's contributions during his doctoral and postdoctoral research, with focused investigations into geochemical processes associated with hydraulic fracturing in unconventional hydrocarbon reservoirs. Dr. Xiaodong He hereby expresses his profound gratitude to his doctoral supervisor, Prof. Peiyue Li, and his postdoctoral mentor, Prof. Hui Qian, for their invaluable academic guidance. The successful realization of this scholarly work stems from the synergistic collaboration between publisher, editorial teams, and reviewers. First and foremost, we would like to express gratitude to Prof. Wanfang Zhou for his assistance during the book's publication process. Thanks to his suggestions and help, this book was able to be presented to everyone. We extend our sincere gratitude to Annett Buettner and Ragavendar Mohan, our editorial contacts at Springer, for their unwavering support and prompt guidance throughout the publication process.

The authors gratefully acknowledge the invaluable contributions provided by the Prof. Wei Wang, Prof. Xiuhua Liu, Prof. Jifeng Guo, Prof. Yaoguo Wu, Prof. Jinting Huang, and Prof. Jianhua Wu. Without their comments, the book would have never become so attractive. Their expert comments and profound insights significantly enhanced the academic rigor and presentation quality of this book. At the same time, we would like to express our gratitude to Yanan Guo, Haotian Liang, Hanghang Zhao, Ningning Yang, and Jing Ning for their assistance in completing the laboratory experiments. We would like to express our gratitude to Engineer Hua Shi, Senior Engineer Yuanxiang Xiao, and Senior Engineer Xiaoyong Wen from the Oil and Gas Technology Research Institute of Changqing Oilfield Company, PetroChina for their technical support and guidance during the indoor and outdoor tests. We are also deeply grateful to Dr. Yongchun Liang, Xiaohui Gao, Wei Zhao, Nan Liu, Dawei Mu, Jiajia Kong, Menyu Gong, Xinshuo Yuan, Jiayao Li, Le Wang, and Tiechuan Jing for their invaluable assistance in preparing the manuscripts.

Finally, we acknowledge the research grants from the National Natural Science Foundation of China (42472316 and 42272302), the National Key Research and Development Program of China (2023YFC3706901), the Postdoctoral Fellowship Program (Grade C) of China Postdoctoral Science Foundation (GZC20241452), the China Postdoctoral Science Foundation (2024M752738), the Natural Science Basic Research Program of Shaanxi Province (2025JC-YBQN-375, 2024JC-YBQN-0303), Special Program for Serving Local Areas of the Shaanxi Provincial Department of Education (24JE002), and the Fundamental Research Funds for the Central Universities, CHD (300102294101 and 300102295201).

Xi'an, China

Dr. Xiaodong He
Prof. Peiyue Li

Competing Interests The authors have no competing interests to declare that are relevant to the content of this manuscript.

Contents

About the Authors

Dr. Xiaodong He is an assistant professor at Chang'an University, China. He holds a Ph.D. in hydraulic engineering from Chang'an University and possesses extensive expertise in hydrogeochemistry, water-rock interactions, groundwater quality, and groundwater pollution. Dr. He has published over 30 scientific papers in renowned international journals. He has edited and co-edited 3 textbooks and academic monographs. In recent years, focusing on deep Earth geochemistry, he has led five research projects including "*Formation and Evolution Mechanisms of Fracturing Flowback Fluids in Tight Gas Reservoirs Based on Isotopic Fingerprinting, and Their Application in Pollution Tracing*", "*Formation and Evolution Mechanisms of Fracturing Flowback Fluids in Tight Gas Reservoirs Driven by Mixing Processes and Water-Rock Interactions*", and "*Geochemical Processes and Mechanisms in Hydraulic Fracturing of Tight Gas Reservoirs*", with six peer-reviewed papers published in *Journal of Hydrology*, *Geoenergy Science and Engineering*, *Chemosphere*, and *Environmental Earth Sciences*. In recognition of his research excellence, he was selected as one of the Top 2% Top Scientists Worldwide by Stanford University in 2022 and 2025.

Prof. Peiyue Li is a full professor in the field of hydrogeology and environmental sciences at Chang'an University, China. He serves as the vice dean of School of Water and Environment at this university. He holds a Ph.D. in Groundwater Science and Engineering (2014) and has wide experience in groundwater quality assessment, hydrogeochemistry, and groundwater modeling. Peiyue Li has been selected for the National Ten Thousand Talents Program as a Young Talent and recognized as a Young Changjiang Scholar by the Ministry of Education. He has edited and co-edited 14 textbooks and academic monographs, and published over 220 articles in refereed journals. He at present serves as Associate Editors for *Exposure and Health, Mine Water and the Environment, Archives of Environmental Contamination and Toxicology, Environmental Monitoring and Assessment, Human and Ecological Risk Assessment, Applied Water Science, Water Practice and Technology, and Discover Water*. He is also book series editors for *Springer Hydrogeology, Cave and Karst Systems of the World, Advances in Karst Science, and GeoHealth: The Impact of Geology on Environment and Human Health*. He has been awarded one of the most highly cited researchers by Clarivate, Elsevier, and other organizations.

Chapter 1
Introduction

As a pivotal technology for extracting unconventional hydrocarbon, hydraulic fracturing stimulation alters the native balance between aqueous phases, mineral matrices, gaseous components, and microbial communities in deep geological systems, fostering a novel ecosystem that drives multifaceted geochemical processes. These processes not only lead to significant deterioration in the quality of fracturing flowback fluid but also cause reservoir damage, thereby impacting oil and gas production. Hydraulic fracturing activities implicate a broad range of critical issues spanning geology, hydrogeology, geochemistry, and environmental science. This chapter provides an overview of hydraulic fracturing and introduces foundational concepts such as unconventional hydrocarbon, hydraulic fracturing steps, fracturing fluids and additives, water–rock interactions, and engineering and environmental challenges associated with hydraulic fracturing. This chapter delineates the book's aims and thematic boundaries while elucidating its organizational framework, equipping readers with foundational knowledge to navigate the detailed discussions in later sections.

1.1 Significance of This Study

Climate change driven by the extensive use of fossil fuels has emerged as a global issue threatening human survival and sustainable societal development, making the achievement of net-zero carbon emissions imperative in the context of global warming (Burke et al. 2018). The global energy structure is rapidly shifting toward green and low-carbon pathways. Renewable energy represents an inevitable trend in the carbon–neutral era. Although countries worldwide have implemented various policies to promote the development of nuclear energy and renewable energy (e.g., solar, wind, hydro), technological limitations mean that widespread adoption of renewable energy will still require time, and these new energy sources remain unable

X. He and P. Li, *Geochemical Processes in Tight Gas Hydraulic Fracturing Stimulation*, Environmental Earth Sciences, https://doi.org/10.1007/978-3-032-15180-3_1

to fully replace fossil fuels. Compared with coal and oil, natural gas emits less carbon dioxide, sulfur dioxide, and particulate matter during combustion, and is widely recognized as a cleaner, lower-carbon energy source. The advent of hydraulic fracturing has triggered an unconventional hydrocarbon revolution worldwide, enabling the exploitation of unconventional hydrocarbon resources that account for 80% of the world's total energy reserves (Zou et al. 2015). As the continuous advancement of hydraulic fracturing, unconventional natural gas represented by shale gas, tight sandstone gas, and coalbed methane, has emerged as a cornerstone of the global energy transition, serving as a key substitute for coal and oil in reducing CO_2 emissions (He et al. 2026).

As the extremely low porosity and permeability of tight reservoirs, conventional development techniques are generally insufficient to achieve commercial production, necessitating stimulation measures to enhance reservoir permeability (Zou 2017). The introduction of hydraulic fracturing and horizontal drilling has made the commercial development of unconventional oil and gas resources economically feasible (Zou 2017; Belyadi et al. 2019; He et al. 2026). Hydraulic fracturing utilizes high-pressure fluids to generate artificial fractures and connecting natural fractures in the tight reservoirs, thereby providing pathways for hydrocarbon flow. However, the injection of large volumes of external fluids into the formation leads to interactions with reservoir rocks and fluids (including water and gas), inducing various water–rock processes (Jew et al. 2022). The type and progression of water–rock interactions directly determine the extent of fracturing fluid impact on the reservoir, thereby decreasing (or increasing) porosity and permeability, and ultimately affecting natural gas production. "Positive" water–rock interactions increase reservoir porosity and permeability, facilitating hydrocarbon flow and production, whereas "negative" interactions cause formation damage, block flow pathways, and ultimately reduce hydrocarbon yield (Civan 2015; He et al. 2022a). Thus, investigating and characterizing water–rock interactions in hydraulic fracturing offers essential information for improving fracturing operations and boosting oil and gas yields, which holds substantial practical significance. Furthermore, the hydraulic fracturing process opens a new frontier for water–rock interaction studies, contributing fresh perspectives to the advancement of water–rock interaction theory and offering theoretical insights into deep formation (reservoir) geochemical processes influenced by human activities, with notable theoretical value.

Hydraulic fracturing has significantly influenced the global energy structure, enabling the shale gas revolution in the North America and positioning it as the leading natural gas producer worldwide (He 2023). However, while increasing oil/gas output, hydraulic fracturing may also trigger a series of regional environmental issues. As hydraulic fracturing continues to evolve and expand, its associated environmental, public health, and geological risks have become more evident, including high water consumption, induced earthquakes, and pollution of soil, groundwater, and air, among which water pollution has garnered widespread concern (Vidic et al. 2013; He et al. 2026). During fracturing stimulation, large volumes of fracturing fluids containing various chemical additives are injected into the formation, where they undergo numerous chemical reactions with reservoir rocks and fluids, simultaneously producing large volumes of highly contaminated wastewater. Such wastewater may migrate into overlying freshwater formations via multiple routes, leading to groundwater contamination (He et al. 2022b). Currently, in regions with mature unconventional hydrocarbon exploitation such as North America, extensive hydraulic fracturing operations have resulted in various groundwater-related environmental problems (Warner et al. 2012; Vidic et al. 2013; Bondu et al. 2021). In China, groundwater issues induced by hydraulic fracturing have received limited attention, primarily in the shale gas areas of the Sichuan Basin (Zou et al. 2018; Huang et al. 2020a). Relevant studies remain scarce in other hydraulic fracturing regions, particularly in arid and semi-arid areas. Therefore, it is necessary to investigate the impacts of unconventional hydrocarbon development on groundwater environments in ecologically fragile arid and semi-arid regions.

China's energy resource is characterized by being "rich in coal, poor in oil, and scarce in gas". Despite a recent steady rise in the share of natural gas in China's energy mix, its overall consumption structure remains heavily carbon-intensive, and the proportion of natural gas is still far below the global average (Fig. 1.1). As a clean fossil fuel, natural gas will play a pivotal role during the transition from high-carbon fossil fuels to renewable resources. The discovery of numerous large tight gas fields, exemplified by the Sulige Gas Field, has established tight gas as the largest unconventional gas resource in China. China's total natural gas production reached 2463.7×10^8 m^3 in 2024, with unconventional gas contributing 40%. Within the unconventional category, shale gas production was 254×10^8 m^3, coalbed methane was 166.6×10^8 m^3, and tight gas was 620×10^8 m^3 which accounts for over 25% of total national production and nearly 60% of unconventional output (Fig. 1.1). This makes China the world leader in tight gas production and a core growth driver for domestic natural gas supply expansion. The Ordos and Sichuan basins are the principal production areas, together accounting for over 98% of China's tight gas output. Sulige Gas Field, situated in the northern-central region of Ordos Basin, is the largest onshore integrated gas field discovered in China, hosting abundant resources within Upper Paleozoic tight sandstones. Since its discovery in 2020, it has progressed through initial evaluation, large-scale development, and rapid production growth, and is now in a phase of stable long-term output. From 2022 to 2024, annual production remained above 300×10^8 m^3 for three consecutive years, representing roughly half of China's tight gas yield. Enhancing gas well recovery rates is critical to maintaining stable production, and understanding water–rock interactions is essential to improving gas extraction efficiency. Moreover, the Sulige Gas Field lies in the Yellow River Basin in an arid to semi-arid region with a fragile ecology and scarce water resources, where groundwater is particularly valuable. Long-term hydraulic fracturing poses substantial pressure and risks to the local environment.

Given this context, this book investigates geochemical processes during tight gas fracturing stimulation in the Sulige Gas Field through data compilation, literature analysis, field experiments, laboratory tests, and modeling. By determining reservoir mineral composition, physical properties, and the hydrochemistry of post-fracturing

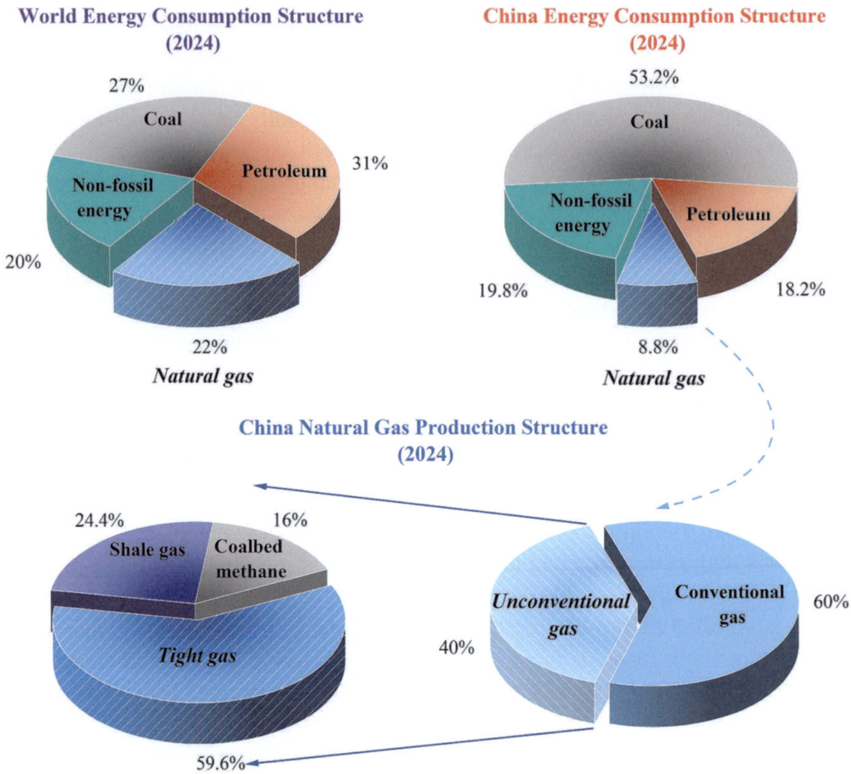

Fig. 1.1 Energy structure of China and world in 2024

flowback fluids, this study reveals the processes and mechanisms governing water–rock interactions, offering essential theoretical and technical guidance for optimizing fracturing fluids and improving hydraulic fracturing technology. Additionally, it addresses regional water environmental impacts and the potential for resource utilization of flowback water, with a focus on identifying groundwater contamination risks in the field's hydraulic fracturing zones. The findings not only contribute a theoretical basis for achieving China's dual-carbon strategy but also support unconventional hydrocarbon development and ecological protection in the Yellow River Basin, offering both significant theoretical and practical value.

1.2 Unconventional Oil and Gas

1.2.1 Distribution of Unconventional Oil and Gas

Conventional oil and gas resources are generally concentrated in favorable structural or trap formations, characterized by reservoirs with good quality, high permeability, and sufficient hydrocarbon mobility, thus typically requiring no large-scale stimulation. In contrast to conventional resources, unconventional hydrocarbon resources occur in tight formations with poor reservoir quality, making it difficult to extract hydrocarbons using traditional techniques (Ma and Holditch 2016; Zou 2017). However, substantial unconventional petroleum and natural gas resources are stored globally. Globally, unconventional petroleum resources are approximately equal in volume to conventional petroleum, including shale oil, tight sandstone oil, tight limestone oil, oil shale, oil sands, and heavy oil. In terms of natural gas, resources represented by shale gas, tight sandstone gas, and coalbed methane are approximately eight times greater than conventional natural gas resources (Zou et al. 2015). Unconventional hydrocarbon resources generally occur as large-scale continuous accumulations, primarily distributed in basin centers and slope regions within negative structural units. Wang et al. (2016) and Tong et al. (2018) conducted evaluations of global unconventional hydrocarbon resources, revealing that they are mainly distributed across 363 basins in over 60 countries. Among these, North America is the most resource-rich region, with recoverable unconventional hydrocarbon accounting for approximately 33.8% of the global total. Unconventional natural gas resources with commercial potential are chiefly located in 147 basins distributed across 37 countries (Wang et al. 2016; Tong et al. 2018). Figure 1.2 illustrates the global distribution of major unconventional natural gas-rich basins. Currently, tight sandstone gas, shale gas, and coalbed methane are the main categories of unconventional natural gas in commercial production globally. CNPC's 2020 statistics (excluding China) indicate that shale gas accounts for the largest share of recoverable unconventional natural gas resources, about 223.8×10^{12} m^3 (83.6%), with coalbed methane and tight sandstone gas representing 13.9% and 2.5%, respectively (Dou et al. 2022).

The rapid development of shale gas and tight gas in the North America and China has driven the steady growth of unconventional gas production worldwide. By 2023, global unconventional natural gas output amounted to 1.13×10^{12} m^3, representing 27.8% of total natural gas production. Since 2020, the proportion of unconventional natural gas in global production has remained largely stable, establishing it as a mainstay of global gas supply. In 2024, the United States produced 1.07×10^{12} m^3 of natural gas, with shale gas contributing 8578×10^8 m^3 (80%). China's production totaled 2463.7×10^8 m^3, of which shale gas and tight gas accounted for 254×10^8 and 620×10^8 m^3, respectively, jointly exceeding 30% of total domestic output. The combined unconventional gas output of China and the United States accounts for over 80% of the global total, making them the world's leading producers (Jia et al. 2025). For China, numerous basins and advantageous geological settings are conducive to

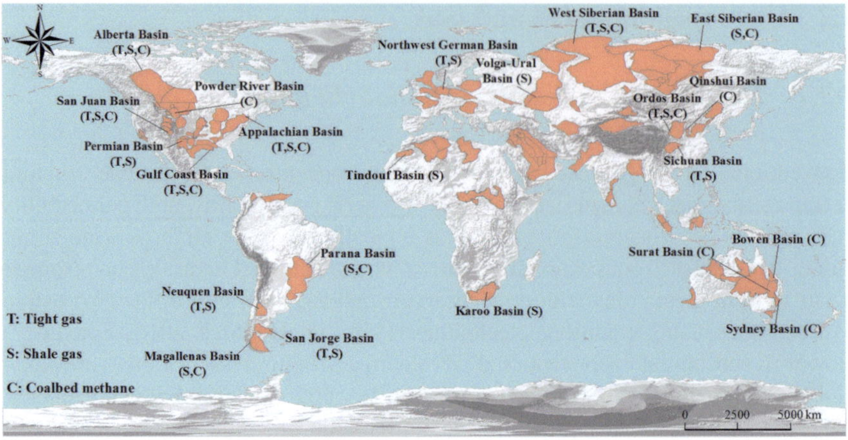

Fig. 1.2 Sedimentary basins with commercial accumulations of tight gas, shale gas, and coalbed methane (Data from Zou et al. 2015; Ma and Holditch 2016; EIA 2021)

the development and preservation of unconventional natural gas resources. Technically recoverable unconventional natural gas resources in China are distributed as 19.9% coalbed methane, 39.4% shale gas, and 40.7% tight gas (Liu et al. 2025). China possesses abundant tight sandstone gas resources, accounting for about 10% of the global total. With the discovery of two major tight gas regions: (1) the Shihezi and Shanxi formations in the Ordos Basin, and (2) the Xujiahe Formation in the Sichuan Basin, alongside the emergence of several high-yield fields in the Songliao, Bohai Bay, Tuha, and Tarim basins, China has become the global leader in tight gas (Jia et al. 2025).

1.2.2 Unconventional Oil and Gas Reservoirs

Unlike conventional hydrocarbons, unconventional reservoirs are significantly tighter, characterized by lower porosity, lower permeability, and stronger heterogeneity. Typically, unconventional reservoirs have nano-porous throat structures (< 1 μm), porosity below 10%, and permeability less than 0.1 mD (Nelson 2009; Zou 2017). Reservoir properties and quality in unconventional systems directly control hydrocarbon potential, recoverable reserves, and productivity. Hence, many researchers worldwide have conducted studies on unconventional reservoir properties. Chabalala et al. (2020) compared the petrological characteristics of shales from the Appalachian Basin (USA) and the Karoo Basin (South Africa), providing deeper insights into the shale gas potential of the Karoo Basin. Jia et al. (2021) applied X-ray diffraction (XRD), scanning electron microscopy (SEM), tensile strength, and fracture toughness analyses to investigate the mechanical and petrophysical properties of the Longmaxi Formation shales. Findings indicated that quartz is the primary mineral in Longmaxi shales, contributing to high brittleness and fracturability, while well-developed microfractures suggest considerable gas storage potential and favorable conditions for hydraulic fracturing. Based on mineral composition and microstructural analysis, Wu et al. (2021) examined the formation and diagenetic features of tight sandstones in the Shihezi Formation within closed to semi-closed systems. Shan et al. (2022) studied the diagenesis of shale oil reservoirs in the Mafengcheng Formation of the Junggar Basin, revealing that diagenesis primarily controls pore types and distribution. Many investigations indicate that while compaction and cementation impede pore development, dissolution and recrystallization generate secondary pore, thereby increasing permeability (Lai et al. 2018; Cui et al. 2024; Yang et al. 2025).

Pore size, porosity, and permeability are the three most critical physical attributes of tight reservoirs, affecting reservoir quality and hydrocarbon distribution (Li et al. 2023). Nelson (2009) systematically summarized pore-throat size distributions in conventional sandstones, tight sandstones, and shales, establishing a relatively complete distribution spectrum. Chalmers et al. (2012a) conducted detailed characterization of porosity, pore-size distributions, surface area, organic geochemistry, and mineralogical attributes of Barnett, Woodford, Haynesville, Marcellus, and Doig shales. Mukherjee et al. (2020) investigated how coal composition and fracture intensity affect initial permeability in the Walloon coal measures of the Surat Basin, Australia. Kang et al. (2017) reviewed permeability characteristics of coal reservoirs in China and concluded that permeability is mainly controlled by cleat/fracture networks, coal structure, and in-situ stress. Across regions, the key controls on coal reservoir permeability vary. Selecting targeted development strategies is essential for coalbed methane exploitation. Examination of pore-size features across major hydrocarbon basins (Fig. 1.3) shows shale reservoirs with pore sizes mostly < 0.1 μm, leading to generally < 10% porosity (Fig. 1.4a) and < 0.1 mD permeability (Fig. 1.4b). Compared with shales, tight sandstones show broader pore-size ranges (0.01–1 μm), porosity typically < 15%, and permeability clustered between 0.01 and 1 mD (Fig. 1.3). Coal seams possess a dual pore system of pores and fractures, with wide pore-size variation strongly linked to coal rank: micropores dominate in high-rank coals, whereas macropores prevail in low-rank coals (Flores 2014). Overall, coal seams exhibit higher porosity and permeability than both shales and tight sandstones (Fig. 1.4).

1.3 Hydraulic Fracturing

1.3.1 Hydraulic Fracturing Operation

Unconventional hydrocarbon resources stored in low-permeability reservoirs require large-scale fracturing stimulation to achieve commercial development. The integration of hydraulic fracturing and horizontal drilling has rendered the development of

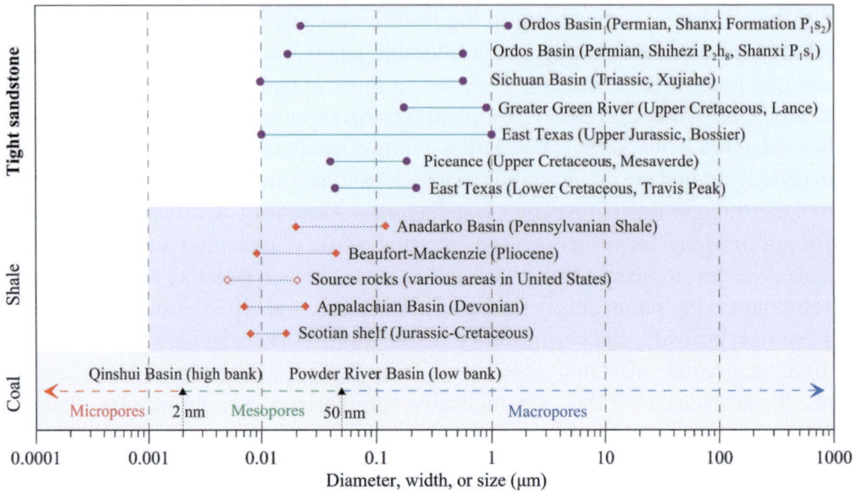

Fig. 1.3 Pore throat sizes observed in low-permeability formations across representative sedimentary basins (Data from Nelson 2009; Flores 2014; Zou 2017)

unconventional resources economically feasible, where fracturing serves as the critical method for improving fluid flow in tight reservoirs. In tight reservoirs, fracturing fluid efficiency critically determines hydraulic stimulation outcomes. Hydraulic fracturing was first implemented experimentally in 1947, when the first fracturing test was conducted in Texas, USA, followed by the first small-scale shale gas fracturing in 1965. However, hydraulic fracturing was not widely commercialized until the late twentieth century, which drove the shale gas revolution in the North America and profoundly reshaped the global energy market (Belyadi et al. 2019; Zhao et al. 2021).

Unlike conventional sandstone formations, tight sandstone gas reservoirs require artificial stimulation to achieve industrial gas flow. Vertically, tight gas formations are featured by multiple strata and overlapping sand bodies, which makes them well-suited for multi-layer fracturing (Kadkhodaie and Kadkhodaie 2022). Consequently, vertical (directional) wells were predominantly employed in the initial development of tight sandstone gas. Since single vertical wells often failed to meet production expectations, multi-stage fracturing of horizontal well has become an important technique for enhancing tight sandstone gas production. Shale reservoirs, being denser than tight sandstones, can develop extensive fracture networks through hydraulic fracturing, thereby markedly boosting shale gas productivity. At present, widely applied techniques in shale reservoirs include multi-stage fracturing of a horizontal well and multi-well simultaneous fracturing (Belyadi et al. 2019). In coalbed methane (CBM) production, water drainage from coal seams decreases reservoir pressure, causing adsorbed methane to desorb and diffuse into coal pores and fractures (Flores 2014). Nevertheless, as the majority of CBM reservoirs exhibit low permeability, hydraulic fracturing has become essential for enhancing methane recovery. In practice, vertical well completions are typically employed in high-permeability coal

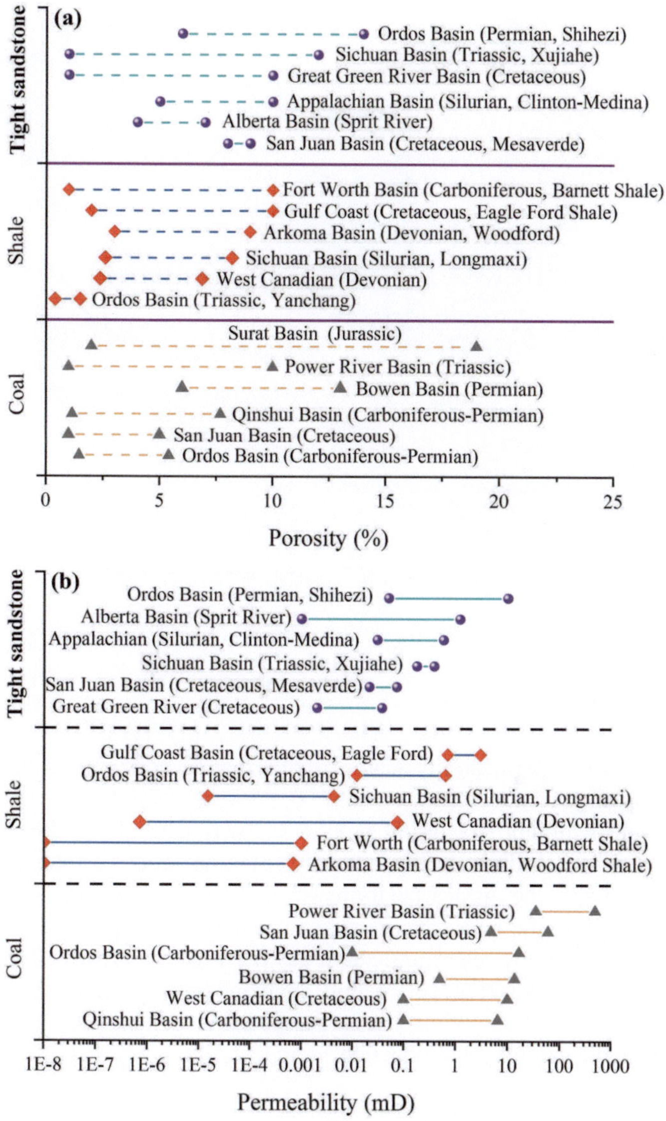

Fig. 1.4 Porosity (**a**) and permeability (**b**) observed in low-permeability formations across representative sedimentary basins (Data from Byrnes et al. 2008; Chalmers et al. 2012b; Flores 2014; Zou 2017; Sun et al. 2020)

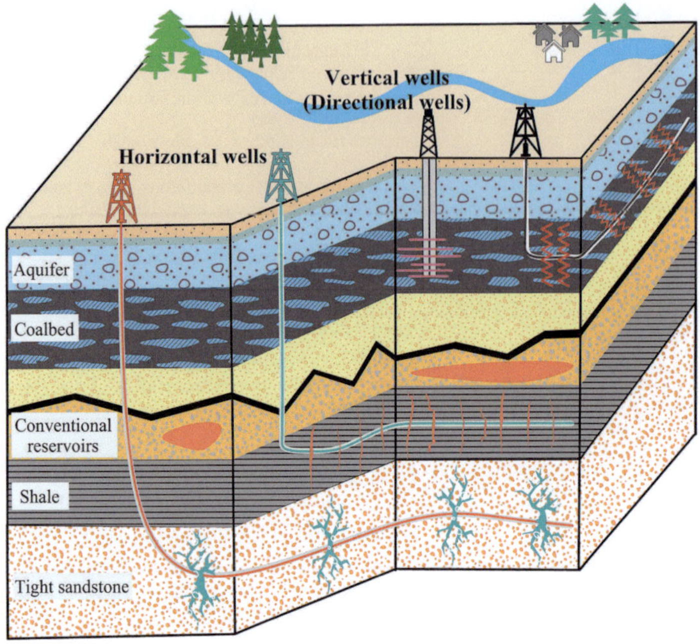

Fig. 1.5 Illustration of hydraulic fracturing in tight reservoirs

seams, whereas low-permeability seams usually require horizontal well fracturing modifications (Fig. 1.5).

A complete hydraulic fracturing process consists of multiple steps: (1) well construction, where drilling, casing, and cementing operations establish a connection channel between the surface and the reservoir; (2) fracturing, in which fracturing fluid is pumped into the reservoirs to generate artificial fractures and interconnect natural fractures, thereby releasing natural gas from the rock matrix. This process mainly involves well cleaning, perforation, acidizing, pre-pad fluid injection, proppant slurry, and flush fluid. Ultimately, part of the injected fluid is flowed back to the gas well to enable natural gas production.

1.3.2 Fracturing Fluids and Additives

Fracturing fluid choice significantly impacts hydraulic stimulation outcomes. These fluids are categorized by primary composition into water-based, oil-based, acid-based, alcohol-based, and foam-based types (Barati and Liang 2014; Abdelaal et al. 2021). Owing to their low cost and operational simplicity, water-based fluids are extensively employed worldwide in the fracturing of low-permeability reservoirs.

Three dominant aqueous fracturing fluid systems prevail in field operations: guar-based, slickwater, and viscoelastic surfactant fluids (Yang et al. 2018; He et al. 2026). Slickwater, consisting primarily of water, friction reducers, proppants, and additives such as scale inhibitors and biocides, are extensively applied in shale reservoirs (Belyadi et al. 2019). As one of the earliest water-based fracturing fluids, guar gum systems consist of guar gum, water, and additives including crosslinkers and breakers. Their advantages, such as high viscosity and excellent proppant transport, ensure their continued widespread application in tight reservoir stimulation (Al-Muntasheri 2014). Similarly, these high-performance water-based fluids are also extensively used in CBM production. In addition, active water, composed of water and surfactants, has the advantages of low cost and minimal damage to coal reservoirs, and has been widely applied in coalbed methane development (Li et al. 2022a).

In recent years, the drawbacks of water-based fluids in terms of freshwater consumption and environmental pollution have become increasingly evident. Consequently, foam fracturing fluids, with advantages such as high viscosity, strong leak-off resistance, excellent proppant transport, easy flowback, and minimal reservoir damage, have attracted widespread attention (Yekeen et al. 2018; Abdelaal et al. 2021). Foam-based fluids, made up of gaseous CO_2 or N_2, aqueous or gel phases, and surfactants, can effectively reduce freshwater consumption during hydraulic fracturing. Furthermore, waterless fracturing technologies, represented by supercritical CO_2 and liquid nitrogen (LN_2) fracturing fluids, have developed rapidly and hold promise as alternatives to conventional water-based fluids (Middleton et al. 2015; Huang et al. 2020c). Nonetheless, widespread field implementation is limited by complex operational procedures, transportation and storage requirements, and high costs.

In tight reservoirs, various fracturing fluid systems present unique strengths and limitations. In recent years, hybrid fracturing technology has been increasingly applied. Depending on reservoir properties such as temperature, pressure, and mineral composition, tailored use of different fracturing fluid systems in specific layers or stages can improve gas recovery and even achieve combined production of multiple gases (Belyadi et al. 2019; Reynolds 2020). Application of the same fracturing fluid and hydraulic fracturing method across different reservoirs frequently results in substantial variations in gas output. Even within the same area, variations in fracturing fluid types and fracturing techniques can lead to significant differences in gas yield. Thus, selecting fracturing fluid types and technologies compatible with reservoir characteristics is essential. A comprehensive understanding of reservoir characteristics and overall cost, combined with field and laboratory data, is crucial for designing appropriate fracturing fluid systems for low-permeability reservoirs. The optimization of fracturing is a highly complex system problem involving multimodality, high dimensionality, large scale, and fine spatiotemporal dynamics. Intelligent fracturing integrates multi-source fused data through geological big-data visualization, cloud computing, and AI algorithms. This approach supports the full lifecycle of reservoir sweet spot selection, real-time parameter optimization, risk prediction, fracture monitoring, and online regulation (Jia et al. 2024).

Engineered chemical agents serve as indispensable functional elements in fracturing fluid systems, enabling their application in complex reservoir environments. In the case of water-based fracturing fluids, over 98% is composed of water and proppants, while chemical additives account for less than 2% by mass (Reynolds 2020). Despite their small proportion, additives govern the performance of fracturing fluids and play a decisive role in hydraulic fracturing success and production efficiency. Typical additives used in fracturing fluids, such as biocides, crosslinkers, and breakers, are shown in Table 1.1. These additives contain numerous chemical substances, with a wide range of organic and inorganic substances. They perform key roles, including bacterial suppression, enhancing viscosity for proppant suspension, and aiding post-fracturing flowback, making their optimal selection pivotal for effective reservoir stimulation (Barbati et al. 2016). Designing effective fracturing fluid systems requires thorough knowledge of reservoir attributes such as mineralogy, pressure, and temperature and cost considerations, informed by field and laboratory studies (Reynolds 2020; He et al. 2026).

Hydraulic fracturing fluids and their additives also have strongly environmental impacts. Currently, the environmental toxicity of many additives substances is still poorly characterized (He et al. 2026). A functional taxonomy of 81 common additives by Stringfellow et al. (2014) revealed organic compounds dominate (55 substances), with 17 displaying elevated theoretical oxygen depletion potential. Given the complexity of subsurface environments, additional evaluation of fracturing additive behaviors and effectiveness is necessary (Faber et al. 2023). Advancing green chemistry innovation has emerged as a central focus of hydraulic fracturing (He et al. 2026).

1.4 Water–Rock Interactions During Fracturing Stimulation

"Interaction" is a universal principle in nature. Acting as a major geological force, water governs and drives a wide array of geological and ecological-environmental processes, including diagenesis, mineralization, weathering, and karst processes, extending from Earth's surface to the subsurface (He 2023). Across both temporal and spatial scales, water–rock interaction is an omnipresent natural phenomenon. Studies on water–rock interactions can be categorized into two major branches: one emphasizes hydrochemistry, focusing on water itself, including groundwater genesis and evolution, geochemical cycling of water within the crust, hydrogeochemical processes governing water chemistry, groundwater and soil contamination and remediation, as well as hydrogeochemical modeling (He 2023; Chemeri et al. 2025). The second branch centers on hydrodynamics, concentrating on geotechnical media, with research topics including the impacts of water–rock interactions on geomechanical behavior, their relevance to geological hazards, and their influence on rock properties during diagenetic processes.

Table 1.1 Common chemical agents and substances employed in fracturing operations (Khan et al. 2021; He et al. 2026)

Chemical agents	Substance	Proportion (%)	Role
Gelling agents	Guar gum, cellulosic compounds	0.056	Enhance fluid viscosity in order to effective proppant suspension
Biocide	Sodium hypochlorite, glutaraldehyde	0.001	Control bacteria that degrade fracturing chemicals and contribute to corrosion of well tubing, casings, and equipment
Surfactant	Ethoxylated alcohols, isopropanol, sodium lauryl sulfate, and dimethyl dihydrogenated tallow ammonium chloride	0.08	Reduce surface tension of the fluid and assist fluid recovery
pH adjustment agent	Acetic acid, $(K/Na)_2CO_3$, (K/Na)OH	0.011	Maintain the effectiveness of components, such as crosslinker
Crosslinker	Borate salts, titanium and zirconium compounds, monoethylamine	0.007	Link up specialized additives to maintain optimal fluid viscosity
Breaker	Ammonium persulfate, sodium bromate, enzyme complexes	0.009	Remove the polymer gel inside the fracture
Clay stabilizer	Choline chloride, tetramethyl ammonium chloride, KCl, NaCl	0.06	Prevent clay from detaching, migrating, and expanding
Scale inhibitor	Ethylene glycol, phosphonic acid salts, sodium polycarboxylate, copolymers of acrylamide, sodium acrylate	0.04	Prevent scale deposition
Iron control	Thioglycolic acid, ethylene glycol, acetic acid, sodium erythorbate	0.004	Prevent precipitation of iron
Friction reducer	Polyacrylamide, kerosene	0.08	Minimize friction of the fluid
Proppant	Sand, ceramsite	/	Prevent the crack from closing
Corrosion inhibitor	N,N-dimethyl formamide	0.001	Inhibit steel tubing corrosion of the tools, well casings, and tanks by acids, salts, and corrosive gasses

Recently, as green, low-carbon, and high-quality development has become a dominant paradigm, new fields have emerged, including unconventional oil and gas development, geothermal utilization, CO_2 geological sequestration, deep reinjection of wastewater, and subsurface energy storage. These fields involve fracturing fluid-reservoir rock interactions, reservoir formation processes in deep geothermal systems (e.g., hot dry rock), and mechanisms of gas–water-rock reactions in deep reservoirs involving CO_2 (Zheng et al. 2015; Herz-Thyhsen et al. 2019; Osselin et al. 2019). The study of deep water–rock interactions has created new frontiers and research directions, among which the exploration of water–rock processes in fracturing stimulation is a pivotal aspect.

During fracturing stimulation, a large volume fluid is pumped into tight reservoirs, generating a complicated fracture network that enhances oil/gas flow capacity. However, the injected fracturing fluid, as an external fluid, inevitably disrupts the original water–rock equilibrium of the reservoir, triggering a series of interactions with the rock, which subsequently influence fracturing effectiveness. Hence, clarifying the potential interactions between fracturing fluids, rocks, and reservoir fluids during hydraulic fracturing is crucial, as it facilitates technological improvements and production enhancement, while also providing deeper insights into the mechanisms of geochemical processes in deep formations. At present, research on water–rock interactions in hydraulic fracturing stimulation mainly focuses on the analysis of post-fracturing flowback fluids and laboratory experiments (He et al. 2026).

Driven by the shale gas revolution, interactions between shale and fracturing fluids during shale gas hydraulic fracturing have attracted extensive attention and research. Numerous scholars and institutions have examined the water quality of post-fracturing flowback water to infer and elucidate the types of fracturing fluid-shale interactions (Rowan et al. 2015; Cui et al. 2020; Gusa et al. 2020; Huang et al. 2020b; Phan et al. 2020). Osselin et al. (2019) investigated the hydrochemical composition and sulfate isotope evolution of post-fracturing flowback water from the Montney shale reservoir in Canada. The findings indicated that the geochemistry of the flowback water is governed by mixing processes between formation brines and the injected fracturing fluid. Through hydrochemical analysis of flowback fluids from the Jurassic shale reservoirs in the Qaidam Basin, Cui et al. (2020) revealed that boron and strontium are predominantly affected by water–rock processes involving feldspar and carbonate dissolution. A study by Zhong et al. (2019) revealed that reusing produced water facilitates the proliferation of halotolerant anaerobic bacteria, and microbial activity may adversely impact shale gas exploitation. Birkle and Makechnie (2022) analyzed geochemical properties of flowback fluids to evaluate fracture behavior and fracturing effectiveness in low-permeability sandstone formations. Nevertheless, studies concerning post-fracturing flowback fluids in tight sandstone reservoirs are still limited.

Laboratory experiments are also an important approach to investigate water–rock interactions. While lab-generated results may not perfectly align with actual well fracturing data, controlled experiments allow researchers to replicate diverse fracturing scenarios and thoroughly examine liquid–solid phase compositions, which the insights are frequently unattainable through field trials alone (Zolfaghari et al. 2016; He 2023). Using X-ray fluorescence (XRF), XRD, SEM, and isotope analyses, laboratory studies have provided detailed insights into processes such as pyrite oxidation, ion exchange, and mineral dissolution–precipitation during shale gas hydraulic fracturing stimulation (Zolfaghari et al. 2016; Jew et al. 2017; Huang et al. 2022). Building on previous findings in shale reservoirs, Khan et al. (2021) and Jew et al.

(2022) reviewed the chemical and reactive transport processes and their impacts on shale reservoirs, providing a relatively systematic and comprehensive summary of fracturing fluid-shale interactions. Herz-Thyhsen et al. (2019) demonstrated through controlled experiments that calcite dissolution markedly enhances sandstone porosity when fracturing fluids interact with tight sandstone. Compared with shale reservoirs, laboratory studies of water–rock interactions in tight gas fracturing stimulation are also scarce. Given the complex nature of fracturing fluid compositions and tight reservoir properties, neither field-derived flowback data nor isolated lab tests alone can comprehensively capture the full subsurface geochemical dynamics, making in situ field experiments the most ideal and reliable approach. Nevertheless, the depth of tight reservoirs, coupled with cost and technological challenges, often hampers the feasibility of conducting field-scale water–rock interaction experiments (Jew et al. 2022). At present, combining flowback fluid testing and analysis with laboratory experiments remains a key approach for studying water–rock processes in hydraulic fracturing.

Microorganisms, as highly active agents in hydrocarbon development, participate across the entire life cycle of petroleum formation, pipeline corrosion, oilfield wastewater treatment, and pollution prevention, significantly influencing multiple stages of hydrocarbon production (Li et al. 2022b; Hernandez-Becerra et al. 2023; Tang et al. 2024). In recent years, microbial activity during hydraulic fracturing has attracted widespread attention (Mouser et al. 2016; Hernandez-Becerra et al. 2023). Hydraulic fracturing not only provides spaces and energy sources for indigenous microorganisms such as methanogens and sulfate-reducing bacteria but may also introduce surface microorganisms into reservoirs through fracturing and drilling fluids, thereby triggering a series of complex biochemical reactions (Mouser et al. 2016; He et al. 2026). Analyzing microbial communities of flowback water has become a key pathway to infer biochemical processes in fracturing stimulation. Studies have revealed that microbial communities in shale flowback fluids are dominated by halotolerant and thermophilic species, such as *Halanaerobium*, *Marinobacter*, *Methanohalophilus*, and *Candidatus* Frackibacter, with community structures varying significantly across different flowback stages (Cluff et al. 2014; Daly et al. 2016). Thus, improvements in analytical methods have enabled progressively deeper insights into flowback fluids, extending from ionic composition and isotopic features to microbial community profiling. The hydraulic fracturing process is characterized by multiscale and multiprocess biogeochemical coupling, spanning nanopores to large fractures and physical transport to microbial metabolism (He et al. 2026). Consequently, identifying the multiprocess-driven biogeochemical mechanisms of hydraulic fracturing is essential to improve fracturing efficiency and mitigate ecological impacts linked to flowback fluids.

1.5 Engineering and Environmental Challenges Associated with Hydraulic Fracturing

The fluid-rock interactions between injected fracturing fluid and reservoir rocks not only modify the chemical composition of the flowback fluid but also strongly affect the properties of tight reservoirs. Because fracturing fluid originates externally and is usually incompatible with the formation, injecting large volumes disturbs the reservoir's environmental balance, induces formation damage, decreases permeability, and consequently reduces or nullifies the effectiveness of hydraulic fracturing, leading to lower hydrocarbon production. Moreover, during high-pressure injection, subsequent flowback, or surface storage and transport, fracturing fluid may migrate into overlying groundwater aquifers, posing contamination risks (Bondu et al. 2021). Given the potential environmental risks of fracturing, including freshwater consumption, groundwater pollution, soil contamination, and induced earthquakes, countries such as France and Germany, as well as U.S. states including New York, Vermont, and Maryland, have passed laws prohibiting hydraulic fracturing (Chailleux et al. 2018; Mohammad-Pajooh et al. 2018).

1.5.1 Formation Damage

Reservoir damage has remained a central concern across both conventional and unconventional hydrocarbon development (Bennion 1999; Civan 2015; Ngata et al. 2022). Formation damage refers to adverse processes that impair the permeability of hydrocarbon formations, thereby reducing well productivity. Formation damage occurs throughout the entire oil and gas development cycle, arising from multiple factors, and may occur during drilling, completion, fracturing, and workover operations (Civan 2015). To systematically study damage in reservoir stimulation, Civan (2015) wrote *Reservoir Formation Damage Fundamentals, Modeling, Assessment, and Mitigation*, covering mechanisms, models, evaluation, prediction, and mitigation, currently updated to its third edition. Formation damage arises from physical, chemical, biological, and thermal interactions between external fluids (e.g., drilling, fracturing fluids) and formation constituents (fluids and rocks). Additionally, mechanical interactions under stress also influence reservoir properties (Reinicke et al. 2010; Kang et al. 2014). Classic damage comprises incompatibility between fracturing fluids and reservoir fluids, incompatibility of fracturing fluids and formation rocks, invasion of solids from drilling or fracturing fluids, chemical adsorption with wettability reversal, fines migration, and bacterial activity that clogs pores and reduces permeability (Fig. 1.6).

Among various operations, hydraulic fracturing induces the most severe reservoir damage, particularly in tight reservoirs (Khan et al. 2021). Clay minerals rank as the predominant constituents in tight formations, second only to quartz in abundance. Their geochemical stability directly governs fracturing effectiveness. Clay

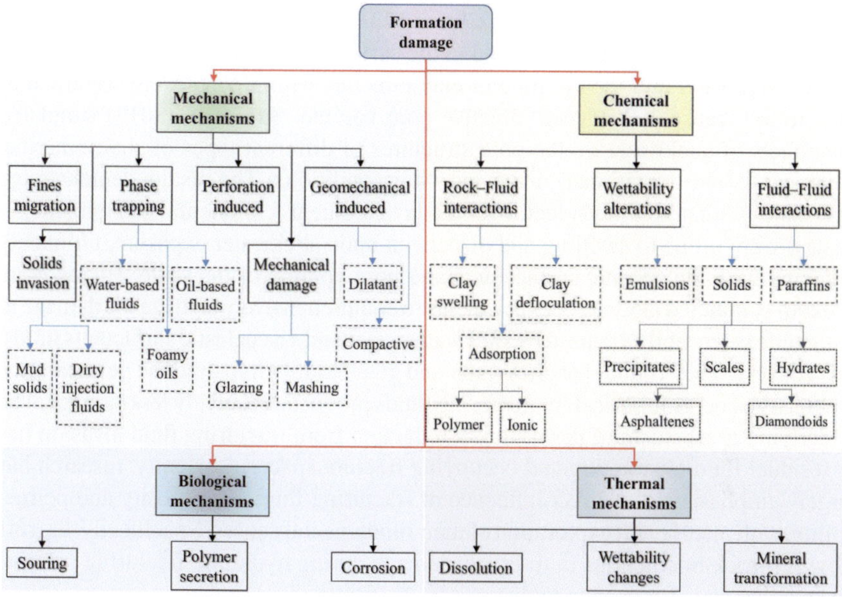

Fig. 1.6 Types of formation damage (Modified from Civan 2015)

destabilization may trigger particle migration, resulting in pore throat blockage and significant permeability reduction. Based on clay mineral sensitivity to fracturing fluid, formation damage is classified as water sensitivity (montmorillonite), acid sensitivity (chlorite), alkali sensitivity (kaolinite), velocity sensitivity (illite), and salt sensitivity (Zhao et al. 2022; He et al. 2026). Radwan et al. (2022) investigated damage types, mechanisms, and mitigation strategies of the sandstone reservoirs in the Suez Rift Basin, and identified four key mechanisms of reservoir damage: water blocking, solid intrusion, kaolinite dispersion and migration leading to pore throat clogging, and sulfate mineral precipitation induced by mixing between fracturing fluids and reservoir fluids. According to two-phase gas–water flow simulations in shale formations, Eveline et al. (2017) demonstrated that water invasion triggers swelling of clay minerals, reducing permeability, while prolonged shale-water interactions during production exacerbate reservoir damage. Proppant performance is also vital for the success of hydraulic fracturing. Variations in reservoir stress during fracturing may cause proppant crushing, embedment, and chemical reactions, all of which contribute to reservoir damage. Coal seams, due to their relatively softer surfaces, are more prone to proppant embedment than shale and sandstone, which causes fracture closure and induces reservoir damage (Ahamed et al. 2019).

In China, research on unconventional reservoir damage has primarily focused on shale and tight sandstone reservoirs in the Ordos Basin and the Sichuan Basin. Recently, with the gradual development of coalbed methane exploitation, research on the damage caused by hydraulic fracturing to coal seams has also increased. Qi et al. (2021) investigated the impact of slickwater fracturing fluids on the microscopic pore structure of tight sandstones in Chang 7 Member of the Ordos Basin, revealing that fracture damage mainly occurred in the early stage of fluid invasion. The retention of slickwater and the swelling of clay minerals within pores were identified as the primary causes of reduced effective pore volume. Sun et al. (2018) simulated the effects of slickwater on the pore structures of different types of shale (marine, continental, and transitional) under reservoir conditions. The results demonstrated that shales with high illite/smectite mixed-layer content (continental and transitional shales) were prone to swelling and dispersion upon slickwater exposure, ultimately reducing both the specific surface area and pore volume of the shale. Huang et al. (2020d) conducted laboratory experiments to comprehensively analyze the damage to anthracite permeability caused by slickwater, guar, and viscoelastic surfactants fluids, respectively. They found that slickwater and guar notably impaired the permeability, whereas viscoelastic surfactant fracturing fluids produced relatively less damage. The underlying mechanism of permeability reduction from fracturing fluid invasion lies in residual fluid obstructing and occupying fracture spaces. Currently, research has largely emphasized the direct influence of fracturing fluids on porosity and permeability, with insufficient exploration of their fundamental causes. Specifically, the role of water–rock interactions in reservoir damage during hydraulic fracturing remains an area that calls for deeper study.

1.5.2 Groundwater Contamination

Additives usually account for a small fraction of fracturing fluids. However, they encompass a wide range of chemicals (e.g., biocides, crosslinkers, gel breakers, and surfactants), among which many are environmentally hazardous compounds, such as benzene, toluene, and formaldehyde (Zhou et al. 2022). Once injected into the reservoir, fracturing fluids can mobilize chemical elements such as As, Mn, and Ni from the rocks and mix with formation fluids, generating flowback water that is discharged to the surface. Consequently, flowback water generated after hydraulic fracturing usually contains high concentrations of organic compounds, metals, salts, and even radioactive elements (Vidic et al. 2013). Owing to the large volume of fracturing fluids and the mixing of formation water, substantial quantities of flowback water are returned to the surface. For example, a nine-well pad can generate as much as 9000 m^3 of flowback within 28 days in the Sulige Gas Field (He et al. 2015). US shale gas wells typically generate between 1.7 and 14.3 million liters of produced water within their initial 5–10 year production period (Kondash et al. 2017). As the first nation to explore and exploit unconventional hydrocarbon resources, the United States has experienced frequent pollution incidents associated with hydraulic fracturing during its long history of development (Peterman et al. 2012; Warner et al. 2012; Llewellyn 2014). Several key shale gas regions have already experienced cases of groundwater contamination (Drollette et al. 2015; Harkness et al. 2017; Kondash et al. 2017). A study by Darrah et al. (2014) systematically examined 133 drinking water samples from shallow aquifers in the Marcellus and Barnett shale regions, identifying eight discrete gas contamination clusters and observing an increasing trend in pollution over time. Llewellyn et al. (2015) detected methane and foaming in some household drinking water supplies in Pennsylvania. In addition, Nelson et al. (2015) monitored radionuclides (natural uranium, ^{210}Pb, and ^{210}Po) in groundwater drinking sources near unconventional drilling areas, while Harkness et al. (2015) monitored chloride, bromide, iodide, and ammonium in shale gas flowback fluids. The both studies concluded that the discharge and accidental leakage of flowback water pose substantial hazards for ecosystems and public health.

In China, groundwater environmental issues caused by hydraulic fracturing have not yet received sufficient attention, and related studies are relatively limited, with existing research primarily focused on the more mature shale developments in the Sichuan Basin. Huang et al. (2020a) traced the contamination of shallow groundwater by formation water during shale gas exploitation through sensitive inorganic monitoring indicators (Ba, Cl, Br, $^{87}Sr/^{86}Sr$, $\delta^{11}B$) in the Fuling Gas Field. The results indicated that 0.2–0.9% of formation water may have already infiltrated shallow aquifers, leading to groundwater contamination. Similarly, Ni et al. (2022) investigated the hydrochemistry of shale gas wastewater and its effects on surface water in the Fuling area. They found that even after treatment at wastewater plants, highly polluted flowback water still contained some ions (Cl^-, B, Ba^{2+}, Cu, and NO_3^-) at concentrations exceeding drinking water standards. However, after discharge into local rivers and dilution, all indicators were below drinking and ecological thresholds. For emerging unconventional hydrocarbon fields in China, the potential groundwater contamination risks of hydraulic fracturing need to be taken seriously. The Sulige Gas Field is sited in the middle-upper reaches of the Yellow River Basin, spanning the northern Loess Plateau and the Ordos Desert Plateau, where water resources are scarce and groundwater is the essential source of supply. Therefore, investigating the groundwater groundwater consequences of hydraulic fracturing activities in tight sandstone gas development is essential not only for the sustainable use and protection of groundwater in the Sulige region but also advancing ecological conservation throughout the Yellow River Basin.

1.6 Objectives and Scope of This Book

1.6.1 Purpose of the Book

A synthesis of global and domestic research on unconventional hydrocarbon exploitation, water–rock interactions, and the engineering and environmental issues related to hydraulic fracturing indicates that research on the processes of fracturing fluid-tight sandstone interactions remains scarce. Existing work primarily focuses on the direct damage fracturing fluids may inflict, such as decreased porosity and permeability. Investigating geochemical processes in fracturing stimulation is crucial for elucidating reservoir damage mechanisms and enhancing natural gas recovery. Current research on water–rock interaction mechanisms in tight reservoirs during hydraulic fracturing faces the following main challenges:

(1) Previous studies have predominantly focused on shale reservoirs, with very limited work on tight sandstones. The interaction types and mechanisms of water–rock processes in tight sandstone hydraulic fracturing are poorly understood. Hence, systematic investigation is required.
(2) Water–rock interactions between fracturing fluids and reservoir rocks during hydraulic fracturing directly or indirectly influence tight reservoirs and may even cause damage. However, studies addressing the impacts and damage mechanisms of these interactions are relatively lacking.
(3) The Sulige Gas Field, currently the largest natural gas field in China, undergoes frequent hydraulic fracturing activities. However, detailed reports on the hydrochemical and water-quality characteristics of flowback fluids in this field remain unavailable. Moreover, investigations into the environmental impacts of fracturing, especially groundwater contamination, are still insufficient.

To address these gaps in understanding water–rock interactions in tight gas fracturing stimulation, this book takes the Sulige Gas Field as the research area. By integrating literature review, field fracturing trials, laboratory experiments, isotopic tracers, and modeling approaches, it systematically examines the geochemical processes in tight sandstone reservoirs subjected to hydraulic fracturing. The primary research objectives are as follows:

(1) Investigation of water–rock chemical interaction mechanisms in tight gas fracturing stimulation.

    After determining the petrographic and tight sandstones properties of in the Sulige region, this study employs field fracturing tests and laboratory experiments, combined with XRF, XRD, SEM, isotopic tracers, hydrogeochemical modeling, and end-member mixing approaches, to depict the geochemical processes of water–rock interactions at multiple scales and to elucidate their mechanisms.

(2) Study on the impacts and damage mechanisms of water–rock chemical interactions on tight sandstone reservoirs.

    Drawing from the results of water–rock interaction studies, this work investigates the influence and damage mechanisms of chemical interactions on tight reservoirs. Focusing on the impacts of fracturing fluids on microstructural morphology, particular attention is given to clay mineral destabilization and scaling damage, providing critical guidance for fracturing fluid optimization and technological improvement.

(3) Research on identifying potential groundwater pollution in hydraulic fracturing areas.

    Considering the groundwater pollution risks potentially induced by frequent hydraulic fracturing in the Sulige region, this study focuses on identifying groundwater contamination caused by fracturing flowback fluids. By analyzing potential pollution sources and pathways, and by constructing contamination identification models. a strontium isotope-based identification curve is established. This method aids in recognizing, assessing, and predicting groundwater pollution in fracturing zones, while offering practical recommendations for prevention and control.

1.6.2 Scientific Issues Addressed in This Book

This book primarily addresses hydrogeological scientific issues associated with tight gas fracturing stimulation, as well as the consequent engineering and environmental challenges.

- What are the characteristics of tight sandstone reservoirs, and how do they influence the effectiveness of hydraulic fracturing?
- Which types of water–rock interactions take place in tight gas fracturing stimulation, and what impacts do they exert?
- How is the water quality of flowback fluids characterized, and which factors govern its changes?
- How does hydraulic fracturing affect reservoirs, and in turn, how do these effects alter natural gas recovery?
- How might flowback wastewater constituents affect aquifer systems, and how can contamination be identified?
- Given the high pollution characteristics of flowback fluids, how can their resource utilization be achieved?

1.6.3 Overview of the Book's Structure

A comprehensive review of previous studies on water–rock interactions in tight gas fracturing stimulation has been conducted. The book begins with an Introduction (Chapter 1), outlining the background and significance of this study, reviewing existing research, and defining the book's purpose, key questions, and structure. Building on this foundation, the Sulige Gas Field, China's largest natural gas field and its principal tight sandstone production base, was chosen as the research site to examine fracturing fluid-tight sandstone interactions. Chapter 2 offers a detailed overview of the study area, covering geography, climate, topography, geology, hydrogeology, water resources, and energy development conditions. Tight sandstone reservoir cores were collected, and methods including XRF, XRD, and SEM were applied to characterize their lithological and petrophysical attributes. Chapter 3 presents a systematic analysis of tight gas reservoirs, focusing on mineral composition, physical reservoir properties, gas reservoir characteristics, and diagenetic evolution.

The core of this book is to determine the water quality of flowback fluids and the geochemical processes during hydraulic fracturing. Chapter 4 identifies the hydrochemical characteristics of flowback fluids and their water quality evolution. Chapter 5 investigated the microscale geochemical processes occurring during hydraulic fracturing by conducting water–rock reaction experiments involving fracturing fluids, deionized water, and major additives in the laboratory. Based on the field fracturing and water–rock reaction experiments, Chapter 6 reveals the processes and mechanisms of fracturing fluid-tight sandstone interactions.

The last part of this book discusses the engineering and environmental impacts of hydraulic fracturing activities. Chapter 7 examines how water–rock interactions during hydraulic fracturing induce reservoir damage. Chapter 8 discusses the potential impacts of flowback fluids on groundwater environments. Chapter 9 examines strategies for managing flowback fluids, with emphasis on new pathways for their resource recovery and utilization. The concluding chapter (Chapter 10) provides a summary of the entire book and highlights research gaps and future perspectives.

By integrating geology, geochemistry, engineering, environment, and resources analysis, this book provides a multidisciplinary perspective on tight gas hydraulic fracturing stimulation. The book bridges fundamental geochemistry and practical engineering, offering a vital reference for researchers, reservoir engineers, geochemists, policymakers, and environmental managers.

References

Abdelaal A, Aljawad MS, Alyousef Z et al (2021) A review of foam-based fracturing fluids applications: from lab studies to field implementations. J Nat Gas Sci Eng 95:104236. https://doi.org/10.1016/j.jngse.2021.104236

Ahamed MAA, Perera MSA, Li D et al (2019) Proppant damage mechanisms in coal seam reservoirs during the hydraulic fracturing process: a review. Fuel 253:615–629. https://doi.org/10.1016/j.fuel.2019.04.166

Al-Muntasheri G (2014) A critical review of hydraulic fracturing fluids over the last decade. In: SPE Western North American and Rocky Mountain joint meeting, SPE-169552-MS. https://doi.org/10.2118/169552-MS

Barati R, Liang J (2014) A review of fracturing fluid systems used for hydraulic fracturing of oil and gas wells. J Appl Polym Sci 131(16):1–11. https://doi.org/10.1002/app.40735

Barbati AC, Desroches J, Robisson A et al (2016) Complex fluids and hydraulic fracturing. Annu Rev Chem Biomol Eng 7:415–453. https://doi.org/10.1146/annurev-chembioeng-080615-033630

Belyadi H, Fathi E, Belyad F (2019) Hydraulic fracturing in unconventional reservoirs: theories, operations, and economic analysis. Elsevier, Cambridge. https://doi.org/10.1016/C2018-0-01643-6

Bennion B (1999) Formation damage-the impairment of the invisible, by the inevitable and uncontrollable, resulting in an indeterminate reduction of the unquantifiable! J Can Pet Technol 38(2):11–17. https://doi.org/10.2118/99-02-DA

Birkle P, Makechnie GK (2022) Geochemical cycle of hydraulic fracturing fluids: Implications for fracture functionality and frac job efficiency in tight sandstone. J Petrol Sci Eng 208:109596. https://doi.org/10.1016/j.petrol.2021.109596

Bondu R, Kloppmann W, Naumenko-D'ezes MO et al (2021) Potential impacts of shale gas development on inorganic groundwater chemistry: implications for environmental baseline assessment in shallow aquifers. Environ Sci Technol 55(14):9657–9671. https://doi.org/10.1021/acs.est.1c01172

Burke M, Davis WM, Diffenbaugh NS et al (2018) Large potential reduction in economic damages under UN mitigation targets. Nature 557:549–553. https://doi.org/10.1038/s41586-018-0071-9

Byrnes AP, Cluff R, Webb J et al (2008) Analysis of critical permeability, capillary pressure and electrical properties for Mesaverde tight gas sandstones from Western U.S. Basins. https://doi.org/10.2172/971248

Chabalala VP, Wagnerb N, Malumbazoc N et al (2020) Geochemistry and organic petrology of the Permian Whitehill formation, Karoo Basin (RSA) and the Devonian/Carboniferous shale of the Appalachian Basin (USA). Int J Coal Geol 232:103612. https://doi.org/10.1016/j.coal.2020.103612

Chailleux S, Merlin J, Gunzburger Y (2018) Unconventional oil and gas in France: from popular distrust to politicization of the underground. Extr Ind Soc 5:682–690. https://doi.org/10.1016/j.exis.2018.05.007

Chalmers GR, Bustin RM, Power IM (2012a) Characterization of gas shale pore systems by porosimetry, pycnometry, surface area, and field emission scanning electron microscopy/transmission electron microscopy image analyses: Examples from the Barnett, Woodford, Haynesville, Marcellus, and Doig units. AAPG Bull 96(6):1099–1119. https://doi.org/10.1306/10171111052

Chalmers GR, Ross DJK, Bustin RM (2012b) Geological controls on matrix permeability of Devonian Gas Shales in the Horn River and Liard basins, northeastern British Columbia, Canada. Int J Coal Geol 103:120–131. https://doi.org/10.1016/j.coal.2012.05.006

Chemeri L, Taussi M, Cabassi J et al (2025) Water-rock interaction processes and hydrogeological pathways in seismically active areas as revealed by a multi-isotopic (C, S, O, H, B, Sr) approach. J Hydrol 661:133533. https://doi.org/10.1016/j.jhydrol.2025.133533

Civan F (2015) Reservoir formation damage: fundamentals, modeling, assessment, and mitigation, 3rd edn. Gulf Professional Publishing, Elsevier. https://doi.org/10.1016/C2014-0-01087-8

Cluff MA, Hartsock A, MacRae JD et al (2014) Temporal changes in microbial ecology and geochemistry in produced water from hydraulically fractured Marcellus Shale gas wells. Environ Sci Technol 48:6508–6517. https://doi.org/10.1021/es501173p

Cui X, Zheng Z, Zhang H et al (2020) Impact of water-rock interactions on indicators of hydraulic fracturing flowback fluids produced from the Jurassic shale of Qaidam Basin, NW China. J Hydrol 590:125541. https://doi.org/10.1016/j.jhydrol.2020.125541

Cui G, Xu S, Wei Q et al (2024) The influence of diagenesis on the quality of tight sandstone reservoirs in Longdong, Ordos Basin. J Pet Explor Prod Technol 14:3331–3348. https://doi.org/10.1007/s13202-024-01883-8

Daly RA, Borton MA, Wilkins MJ et al (2016) Microbial metabolisms in a 2.5-km-deep ecosystem created by hydraulic fracturing in shales. Nat Microbiol 1(10):16146. https://doi.org/10.1038/nmicrobiol.2016.146

Darrah TH, Vengosh A, Jackson RB et al (2014) Noble gases identify the mechanisms of fugitive gas contamination in drinking-water wells overlying the Marcellus and Barnett Shales. PNAS 111(39):14076–14081. https://doi.org/10.1073/pnas.1322107111

Dou L, Li D, Wen Z et al (2022) History and outlook of global oil and gas resources evaluation 43(8):1035–1048 (in Chinese with English abstract)

Drollette BD, Hoelzer K, Warner NR et al (2015) Elevated levels of diesel range organic compounds in groundwater near Marcellus gas operations are derived from surface activities. PNAS 112(43):13184–13189. https://doi.org/10.1073/pnas.1511474112

EIA (U.S. Energy Information Administration) (2021) International Energy Outlook 2021. https://www.eia.gov/outlooks/ieo

Eveline VF, Akkutlu IY, Moridis GJ (2017) Numerical simulation of hydraulic fracturing water effects on shale gas permeability alteration. Transp Porous Media 116:727–752. https://doi.org/10.1007/s11242-016-0798-4

Faber A, Brunner AM, Schimmel M et al (2023) Effects of high pressure and temperature conditions on the chemical fate of flowback water related chemicals. Sci Total Environ 888:163888. https://doi.org/10.1016/j.scitotenv.2023.163888

Flores RM (2014) Coal and coalbed gas fueling the future. Elsevier, Waltham. https://doi.org/10.1016/C2011-0-06861-8

Gusa AV, Tomani A, Zhang Z et al (2020) Sulfate precipitation in produced water from Marcellus Shale for the control of naturally occurring radioactive material. Water Res 177:115765. https://doi.org/10.1016/j.watres.2020.115765

Harkness JS, Dwyer GS, Warner NR et al (2015) Iodide, bromide, and ammonium in hydraulic fracturing and oil and gas wastewaters: environmental implications. Environ Sci Technol 49:1955–1963. https://doi.org/10.1021/es504654n

Harkness JS, Darrah TH, Warner NR et al (2017) The geochemistry of naturally occurring methane and saline groundwater in an area of unconventional shale gas development. Geochim Cosmochim Acta 208:302–334. https://doi.org/10.1016/j.gca.2017.03.039

He X (2023) Mechanism of chemical interactions between water and rocks during hydraulic fracturing in tight sandstone reservoirs. Ph.D. thesis, Chang'an University (in Chinese with English abstract)

He M, Lai X, Li N et al (2015) Recovery and treatment of fracturing flowback fluids in the Sulige Gasfield, Ordos Basin. Nat Gas Ind B 2:467–472. https://doi.org/10.1016/j.ngib.2015.09.024

He X, Li P, Ning J et al (2022a) Geochemical processes during hydraulic fracturing in a tight sandstone reservoir revealed by field and laboratory experiments. J Hydrol 612:128292. https://doi.org/10.1016/j.jhydrol.2022a.128292

He X, Li P, Shi H et al (2022b) Identifying strontium sources of flowback fluid and groundwater pollution using $^{87}Sr/^{86}Sr$ and geochemical model in Sulige gasfield, China. Chemosphere 135594. https://doi.org/10.1016/j.chemosphere.2022.135594

He X, Li P, Qian H et al (2026) Biogeochemistry during hydraulic fracturing: a critical review of reservoirs, fluids, processes, and implications. Geoenergy Sci Eng 256:214143. https://doi.org/10.1016/j.geoen.2025.214143

Hernandez-Becerra N, Clife L, Xiu W et al (2023) New microbiological insights from the Bowland shale highlight heterogeneity of the hydraulically fractured shale microbiome. Environ Microbiome 18:14. https://doi.org/10.1186/s40793-023-00465-1

Herz-Thyhsen RJ, Kaszuba JP, Dewey JC (2019) Dissolution of minerals and precipitation of an aluminosilicate phase during experimentally simulated hydraulic fracturing of a mudstone and a tight sandstone in the Powder River Basin, WY. Energy Fuels 33:3947–3956. https://doi.org/10.1021/acs.energyfuels.8b04443

Huang T, Pang Z, Li Z et al (2020a) A framework to determine sensitive inorganic monitoring indicators for tracing groundwater contamination by produced formation water from shale gas development in the Fuling Gasfield, SW China. J Hydrol 581:124403. https://doi.org/10.1016/j.jhydrol.2019.124403

Huang T, Li Z, Mayer B et al (2020b) Identification of geochemical processes during hydraulic fracturing of a shale gas reservoir: a controlled field and laboratory water-rock interaction experiment. Geophys Res Lett 47:e2020GL090420. https://doi.org/10.1029/2020GL090420

Huang Z, Zhang S, Yang Y et al (2020c) A review of liquid nitrogen fracturing technology. Fuel 266:117040. https://doi.org/10.1016/j.fuel.2020.117040

Huang Q, Liu S, Cheng W et al (2020d) Fracture permeability damage and recovery behaviors with fracturing fluid treatment of coal: an experimental study. Fuel 282:118809. https://doi.org/10.1016/j.fuel.2020.118809

Huang T, Li Z, Long Y et al (2022) Role of desorption-adsorption and ion exchange in isotopic and chemical (Li, B, and Sr) evolution of water following water-rock interaction. J Hydrol 610:127800. https://doi.org/10.1016/j.jhydrol.2022.127800

Jew AD, Dustin MK, Harrison AL et al (2017) Impact of organics and carbonates on the oxidation and precipitation of iron during hydraulic fracturing of shale. Energy Fuels 31(4):3643–3658. https://doi.org/10.1021/acs.energyfuels.6b03220

Jew AD, Druhan JL, Ihme M et al (2022) Chemical and reactive transport processes associated with hydraulic fracturing of unconventional oil/gas shales. Chem Rev 122(9):9198–9263. https://doi.org/10.1021/acs.chemrev.1c00504

Jia Y, Tang J, Lu Y et al (2021) Laboratory geomechanical and petrophysical characterization of Longmaxi shale properties in Lower Silurian Formation, China. Mar Pet Geol 124:104800. https://doi.org/10.1016/j.marpetgeo.2020.104800

Jia J, Fan Q, Jing J et al (2024) Intelligent hydraulic fracturing under industry 4.0—a survey and future directions. J Pet Explor Prod Technol 14:3161–3181. https://doi.org/10.1007/s13202-024-01877-6

Jia A, Wang G, Li Y (2025) Natural gas development in China: present situation and prospect. Nat Gas Ind 45(5):31–42 (in Chinese with English abstract)

Kadkhodaie A, Kadkhodaie R (2022) Reservoir characterization of tight gas sandstones: exploration and development. Elsevier, Cambridge. https://doi.org/10.1016/C2020-0-02860-4

Kang Y, Xu C, You L et al (2014) Comprehensive evaluation of formation damage induced by working fluid loss in fractured tight gas reservoir. J Nat Gas Sci Eng 18:353–359. https://doi.org/10.1016/j.jngse.2014.03.016

Kang Y, Sun L, Zhang B et al (2017) The controlling factors of coalbed reservoir permeability and CBM development strategy in China. Geol Rev 63(5):1401–1418 (in Chinese with English abstract)

Khan HJ, Spielman-Sun E, Jew AD et al (2021) A critical review of the physicochemical impacts of water chemistry on shale in hydraulic fracturing systems. Environ Sci Technol 55:1377–1394. https://doi.org/10.1021/acs.est.0c04901

Kondash AJ, Albright E, Vengosh A et al (2017) Quantity of flowback and produced waters from unconventional oil and gas exploration. Sci Total Environ 574:314–321. https://doi.org/10.1016/j.scitotenv.2016.09.069

Lai J, Wang G, Wang S et al (2018) Review of diagenetic facies in tight sandstones: diagenesis, diagenetic minerals, and prediction via well logs. Earth Sci Rev 185:234–258. https://doi.org/10.1016/j.earscirev.2018.06.009

Li J, Huang Q, Wang G et al (2022a) Influence of active water on gas sorption and pore structure of coal. Fuel 310:122400. https://doi.org/10.1016/j.fuel.2021.122400

Li Y, Zheng B, Yang Y et al (2022b) Soil microbial ecological effect of shale gas oil-based drilling cuttings pyrolysis residue used as soil covering material. J Hazard Mater 436:129231. https://doi.org/10.1016/j.jhazmat.2022.129231

Li P, He X, Zhou C et al (2023) Mineral compositions and microstructural characteristics of the tight sandstone reservoir in the Sulige area and their potential influence on hydraulic fracturing. Hydrogeol Eng Geol 50(3):1–11 (in Chinese with English abstract)

Liu C, Wang X, Che C et al (2025) Development prospects of unconventional oil and gas and renewable energy in China. World Pet Ind 32(3):12–21 (in Chinese with English abstract)

Llewellyn GT (2014) Evidence and mechanisms for Appalachian Basin brine migration into shallow aquifers in NE Pennsylvania, USA. Hydrogeol J 22:1055–1066. https://doi.org/10.1007/s10040-014-1125-1

Llewellyn GT, Dormanb F, Westlandb JL et al (2015) Evaluating a groundwater supply contamination incident attributed to Marcellus Shale gas development. PNAS 112(20):6325–6330. https://doi.org/10.1073/pnas.1420279112

Ma YZ, Holditch SA (2016) Unconventional oil and gas resources handbook: evaluation and development. Elsevier, Waltham. https://doi.org/10.1016/C2014-0-01377-9

Middleton RS, Carey JW, Currier RP et al (2015) Shale gas and non-aqueous fracturing fluids: opportunities and challenges for supercritical CO_2. Appl Energy 147:500–509. https://doi.org/10.1016/j.apenergy.2015.03.023

Mohammad-Pajooh E, Weichgrebe D, Cuff G et al (2018) On-site treatment of flowback and produced water from shale gas hydraulic fracturing: a review and economic evaluation. Chemosphere 212:898–914. https://doi.org/10.1016/j.chemosphere.2018.08.145

Mouser PJ, Borton M, Darrah TH et al (2016) Hydraulic fracturing offers view of microbial life in the deep terrestrial subsurface. FEMS Microbiol Ecol 92(11):fiw166. https://doi.org/10.1093/femsec/fiw166

Mukherjee S, Rajabi M, Esterle J et al (2020) Subsurface fractures, in-situ stress and permeability variations in the Walloon Coal Measures, eastern Surat Basin, Queensland, Australia. Int J Coal Geol 222:103449. https://doi.org/10.1016/j.coal.2020.103449

Nelson PH (2009) Pore-throat sizes in sandstones, tight sandstones, and shales. AAPG Bull 93:329–340. https://doi.org/10.1306/10240808059

Nelson AW, Knight AW, Eitrheim ES et al (2015) Monitoring radionuclides in subsurface drinking water sources near unconventional drilling operations: a pilot study. J Environ Radioact 142:24–28. https://doi.org/10.1016/j.jenvrad.2015.01.004

Ngata MR, Yang B, Aminu MD et al (2022) Review of developments in nanotechnology application for formation damage control. Energy Fuels 36(1):80–97. https://doi.org/10.1021/acs.energyfuels.1c03223

Ni Y, Yao L, Sui J et al (2022) Shale gas wastewater geochemistry and impact on the quality of surface water in Sichuan Basin. Sci Total Environ 851(2):158371. https://doi.org/10.1016/j.scitotenv.2022.158371

Osselin F, Saada S, Nightingalea M et al (2019) Geochemical and sulfate isotopic evolution of flowback and produced waters reveals water-rock interactions following hydraulic fracturing of a tight hydrocarbon reservoir. Sci Total Environ 687:1389–1400. https://doi.org/10.1016/j.scitotenv.2019.07.066

Peterman ZE, Thamke J, Futa K et al (2012) Strontium isotope systematics of mixing groundwater and oil-field brine at Goose Lake in northeastern Montana, USA. Appl Geochem 27:2403–2408. https://doi.org/10.1016/j.apgeochem.2012.08.004

Phan TT, Hakala JA, Sharma S (2020) Application of isotopic and geochemical signals in unconventional oil and gas reservoir produced waters toward characterizing in situ geochemical fluid-shale reactions. Sci Total Environ 714:136867. https://doi.org/10.1016/j.scitotenv.2020.136867

Qi A, Qian P, Huang H et al (2021) Research on the characteristics of microscopic damage to different pore types by slippery water fracturing fluid in the Chang Member tight sandstone reservoir, Ordos Basin, NW China. Geofluids 1–8. https://doi.org/10.1155/2021/6652877

Radwan AE, Wood DA, Abudeif AM et al (2022) Reservoir formation damage; reasons and mitigation: a case study of the Cambrian-Ordovician Nubian 'C' sandstone gas and oil reservoir from the Gulf of Suez Rift Basin. Arab J Sci Eng 47:11279–11296. https://doi.org/10.1007/s13369-021-06005-8

Reinicke A, Rybacki E, Stanchits S et al (2010) Hydraulic fracturing stimulation techniques and formation damage mechanisms-Implications from laboratory testing of tight sandstone-proppant systems. Chem Erde 70(S3):107–117. https://doi.org/10.1016/j.chemer.2010.05.016

Reynolds MA (2020) A technical playbook for chemicals and additives used in the hydraulic fracturing of shales. Energy Fuels 34:15106–15125. https://doi.org/10.1021/acs.energyfuels.0c02527

Rowan EL, Engle MA, Kraemer TF et al (2015) Geochemical and isotopic evolution of water produced from Middle Devonian Marcellus shale gas wells, Appalachian basin, Pennsylvania. AAPG Bull 99(2):181–206. https://doi.org/10.1306/07071413146

Shan X, He W, Guo H et al (2022) Pore space characteristics and diagenesis of shale oil reservoir of the Permian Fengcheng formation in Mahu Sag Junggar Basin. Mar Orig Pet Geol 27(3):325–336 (in Chinese with English abstract)

Stringfellow WT, Domen JK, Camarillo MK et al (2014) Physical, chemical, and biological characteristics of compounds used in hydraulic fracturing. J Hazard Mater 275:37–54. https://doi.org/10.1016/j.jhazmat.2014.04.040

Sun Z, Wang Y, Wu B et al (2018) Chemical reactions and effects of slick water fracturing fluid on the pore structures of shale reservoirs in different deposition environments. J Univ Chin Acad Sci 35(5):712–719 (in Chinese with English abstract)

Sun M, Zhao J, Pan Z et al (2020) Pore characterization of shales: a review of small angle scattering technique. J Nat Gas Sci Eng 78:103294. https://doi.org/10.1016/j.jngse.2020.103294

Tang P, Zhong C, Zhang D et al (2024) Microbial insights into shale gas wastewater from the Sichuan Basin: diversity, potential functions, and implications. Energy Fuels 38:2992–3001. https://doi.org/10.1021/acs.energyfuels.3c04325

Tong X, Zhang G, Wang Z et al (2018) Distribution and potential of global oil and gas resources. Pet Explor Dev 45(4):779–789. https://doi.org/10.1016/S1876-3804(18)30081-8

Vidic RD, Brantley SL, Vandenbossche JM et al (2013) Impact of shale gas development on regional water quality. Science 340(6134):1235009. https://doi.org/10.1126/science.1235009

Wang H, Ma F, Tong X et al (2016) Assessment of global unconventional oil and gas resources. Pet Explor Dev 43(6):925–940. https://doi.org/10.1016/S1876-3804(16)30111-2

Warner NR, Jackson RB, Darrah TH et al (2012) Geochemical evidence for possible natural migration of Marcellus formation brine to shallow aquifers in Pennsylvania. PNAS 109(30):11961–11966. https://doi.org/10.1073/pnas.112118110

Wu H, Zhao J, Wu W et al (2021) Formation and diagenetic characteristics of tight sandstones in closed to semi-closed systems: typical example from the Permian Sulige gas field. J Petrol Sci Eng 199:108248. https://doi.org/10.1016/j.petrol.2020.108248

Yang F, Ge Z, Zheng J et al (2018) Viscoelastic surfactant fracturing fluid for underground hydraulic fracturing in soft coal seams. J Petrol Sci Eng 169:646–653. https://doi.org/10.1016/j.petrol.2018.06.015

Yang Z, Chen D, Tang X et al (2025) Pore structure formation mechanism of lacustrine fine-grained sedimentary system in Fengcheng formation, Junggar Basin, China. Energy Fuels 39(15):7235–7253. https://doi.org/10.1021/acs.energyfuels.4c06392

Yekeen N, Padmanabhan E, Idris AK (2018) A review of recent advances in foam-based fracturing fluid application in unconventional reservoirs. J Ind Eng Chem 66:45–71. https://doi.org/10.1016/j.jiec.2018.05.039

Zhao J, Ren L, Jiang X et al (2021) Ten years of gas shale fracturing in China: review and prospect. Nat Gas Ind 41(8):121–142 (in Chinese with English abstract)

Zhao S, Wang Y, Li Y et al (2022) Tight sandstone petrophysics and flow rate/water/alkali sensitivities variations: a cross section of Permian in the northeastern of Ordos Basin, China. J Petrol Sci Eng 208:109443. https://doi.org/10.1016/j.petrol.2021.109443

Zheng H, Feng XT, Pan PZ (2015) Experimental investigation of sandstone properties under CO_2–NaCl solution-rock interactions. Int J Greenhouse Gas Control 37:451–470. https://doi.org/10.1016/j.ijggc.2015.04.005

Zhong C, Li J, Flynn SL et al (2019) Temporal changes in microbial community composition and geochemistry in flowback and produced water from the Duvernay formation. ACS Earth Space Chem 3:1047–1057. https://doi.org/10.1021/acsearthspacechem.9b00037

Zhou S, Peng S, Li Z et al (2022) Risk assessment of pollutants in flowback and produced waters and sludge in impoundments. Sci Total Environ 811:152250. https://doi.org/10.1016/j.scitotenv.2021.152250

Zolfaghari A, Dehghanpour H, Noel M, Bearinger D (2016) Laboratory and field analysis of flowback water from gas shales. J Unconventional Oil Gas Resour 14:113–127. https://doi.org/10.1016/j.juogr.2016.03.004

Zou C, Zhai G, Zhang G et al (2015) Formation, distribution, potential and prediction of global conventional and unconventional hydrocarbon resources. Pet Explor Dev 42(1):14–28. https://doi.org/10.1016/S1876-3804(15)60002-7

Zou C, Ni Y, Li J et al (2018) The water footprint of hydraulic fracturing in Sichuan Basin, China. Sci Total Environ 630:349–356. https://doi.org/10.1016/j.scitotenv.2018.02.219

Zou C (2017) Unconventional petroleum geology, 2nd edn. Elsevier. https://www.sciencedirect.com/book/9780128122341/unconventional-petroleum-geology

Chapter 2
Overview of the Study Area

This chapter provides an overview of the Sulige Gas Field, covering its geographical location, hydrometeorology, regional geology, reservoir environment, hydrogeology, and the current status of gas field development. The aim is to enable readers to gain a comprehensive understanding of the Sulige Gas Field and to establish a fundamental background for subsequent chapters.

2.1 Geographical Location

The Ordos Basin, located in the eastern part of Northwest China, is the country's second-largest sedimentary basin and also the second-largest basin rich in oil and gas resources. Abundant natural gas, petroleum, coal, and various other mineral resources are hosted within the basin. The basin covers an area of approximately 370,000 km^2. The study area, the Sulige Gas Field, lies in the central-northern part of the Ordos Basin, spanning Ordos City in Inner Mongolia to the north and Shaanxi Province to the south (Fig. 2.1). Geographically, it extends between 36°30′–39°30′ N and 107°00′–110°00′ E. The southern part lies on the northern margin of the Loess Plateau, characterized by loess tablelands at elevations of 1100–1400 m. The northern part lies within the Mu Us Desert, with aeolian sandy landforms, flat terrain, and elevations of 1200–1350 m, with relative relief of 30–80 m. Therefore, the gas field is located in the central core area of the Yellow River Basin. Its flat topography and convenient transport make the northern area the primary development zone of the gas field. Surrounding the field are other large gas fields, including Yulin, Shenmu, and Jingbian. The location of the Ordos Basin and the Sulige Gas Field is shown in Fig. 2.1.

X. He and P. Li, *Geochemical Processes in Tight Gas Hydraulic Fracturing Stimulation*, Environmental Earth Sciences, https://doi.org/10.1007/978-3-032-15180-3_2

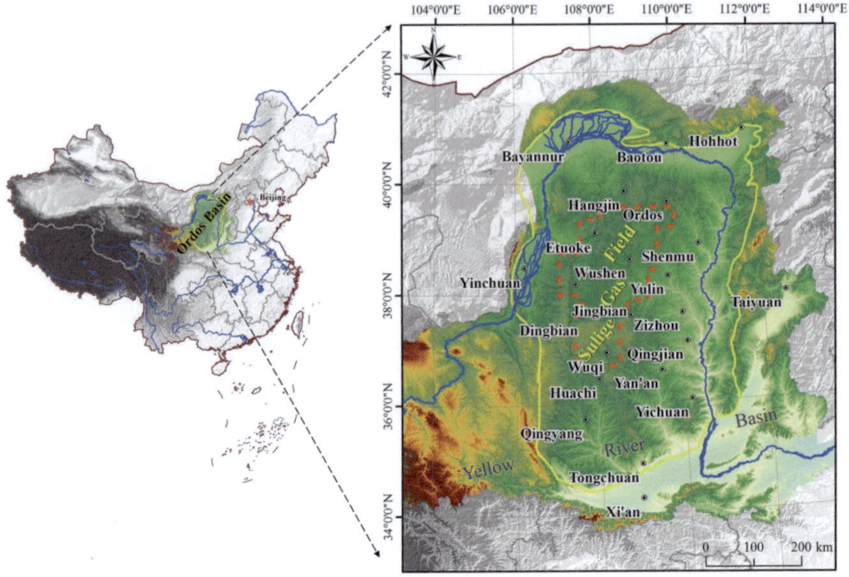

Fig. 2.1 Geographical location of the gas field

2.2 Meteorological and Hydrological Conditions

The Sulige region has a typical temperate arid to semi-arid climate, characterized by low rainfall, strong evaporation, frequent winds, abundant sunshine, short hot summers, and long cold winters. According to over 60 years (1956–2020) of records from eight meteorological stations around the field, the region's mean annual precipitation is 378.37 mm, of which 73.71% falls between June and September. The mean annual temperature is 8.09 °C, and the mean annual evaporation is 1906.03 mm. The mean wind speed, sunshine duration, relative humidity, and 0 cm ground temperature are 2.57 m/s, 2795.84 h, 52.24%, and 10.41 °C, respectively (Table 2.1). Over the past six decades, annual precipitation has fluctuated but shows no clear increasing or decreasing trend, remaining overall stable. Under global warming, the regional mean annual temperature shows a upward trend, with a mean increase of 0.03 °C per year. However, the temperature rise has not led to increased evaporation. Instead, mean annual evaporation shows a marked downward trend (Fig. 2.2). Declining sunshine duration and wind speed are two critical factors driving the reduction in evaporation (Shen and Sheng 2008; Zhao et al. 2023).

The river and lake systems in the Sulige area are poorly developed, with scarce surface water resources, which can generally be categorized into endorheic and exorheic drainage systems. The exorheic systems are tributaries of the Yellow River, mainly including the Dusitu River and the upper reaches of the Wuding and Beiluo Rivers. The Wuding River originates from the northern foothills of Baiyu Mountains within the study area, flowing northeast from Shaanxi into Wushen County, Inner

Table 2.1 Multi-year average monthly climatic elements in the study area

Month	Precipitation (mm)	Temperature (°C)	Evaporation (mm)	Wind speed (m/s)	Sunshine (h)	Relative humidity (%)	0 cm ground temperature (°C)
1	2.25	– 8.55	47.15	2.16	210.60	50.79	– 8.69
2	3.81	– 4.59	65.68	2.40	198.74	48.10	– 3.82
3	9.18	2.52	139.07	2.81	231.79	43.16	4.48
4	18.95	10.15	230.01	3.20	245.97	39.49	13.46
5	31.55	16.27	277.41	3.08	276.68	41.79	20.72
6	42.11	20.75	285.93	2.79	275.10	47.21	25.82
7	80.34	22.59	255.47	2.60	263.01	59.64	27.29
8	102.46	20.60	198.63	2.61	240.53	65.43	24.40
9	54.00	15.11	153.21	2.40	217.10	64.70	17.90
10	23.48	8.47	125.43	2.20	223.53	59.31	9.87
11	8.49	0.39	77.60	2.38	207.25	54.65	0.58
12	1.76	– 6.65	50.44	2.27	205.54	52.61	– 7.02
Mean	378.37	8.09	1,906.03	2.57	2,795.84	52.24	10.41

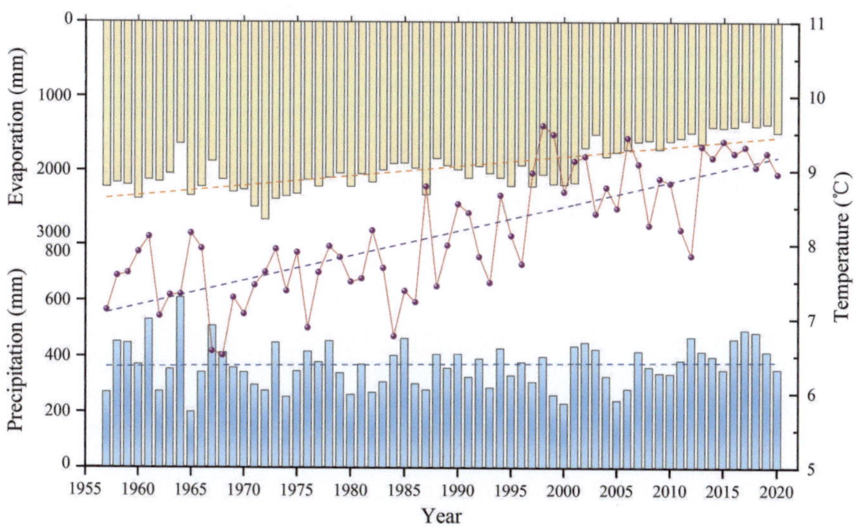

Fig. 2.2 Interannual variations of temperature, precipitation, and evaporation in the study area

Mongolia. After receiving two tributaries, the Nalin River and Hailiutu River, it exits the study area and flows into Yulin City. Hydrological records (1957–2010) from the Liujiamao station on the Hailiutu River indicate a mean annual runoff of 0.91×10^8 m^3 and a mean annual sediment load of 3×10^5 tons (Li 2019). The Beiluo River rises from the southern foothills of Baiyu Mountains, flowing southeastward and leaving the study area through Wuqi County. Hydrological data (1959–2018) from the Liujiahe station on the upper Beiluo River report an average annual runoff of 2.21×10^8 m^3 and an average sediment load of 7×10^5 tons (Wu et al. 2021a). The Dusitu River originates near Etuoke County within the study area, flows westward, and ultimately joins the Yellow River at the border of Inner Mongolia and Ningxia. The river has a total length of about 166 km, with a mean annual runoff of 13.9 million m^3 and a sediment concentration of 10 kg/m^3. Within the Mu Us Desert, the endorheic drainage is dominated by lakes of different scales, including Subei, Wulan, and Bayan. These lakes serve as convergence and discharge centers for surface water and groundwater, but due to intense evaporation, their water quality is generally poor.

2.3 Regional Geological Environment

The Ordos Basin is a large sedimentary basin characterized by multiple tectonic systems, polycyclic evolution, and diverse sedimentary types. The basin contains thick sedimentary strata, thinning in the north and thickening toward the south, with the maximum thickness reaching up to 9000 m in the western thrust-fault zone, while the Yishan Slope in the central basin generally exceeds 5000 m. The Ordos Basin hosts a nearly complete stratigraphic sequence, spanning from the Archean to the Quaternary, including the Cambrian, Ordovician, Carboniferous, Permian, Triassic, Jurassic, Cretaceous, Paleogene, Neogene, and Quaternary systems (Yang and Van Loon 2021). The basin is largely overlain by Cenozoic sediments, though Paleozoic strata crop out locally in the eastern and southeastern parts (Fig. 2.3). As a cratonic basin, the Ordos Basin is structurally simple with gently dipping strata. Structural deformation is largely concentrated at the basin margins, which are divided into six tectonic units: the northern Yimeng Uplift, southern Weibei Uplift, central Yishan Slope, eastern Jinxi Fold Belt, western Tianhuan Depression, and western margin thrust zone (Yang et al. 2016). Overall, faulting in the basin is not strongly developed, consisting mainly of basement faults, normal and reverse faults, folds (anticlines and synclines), as well as some inferred faults of uncertain nature. Folding is most prominent in the western margin thrust belt and the southern Weibei Uplift. The central Yishan Slope, characterized by structural stability, serves as the principal reservoir for the basin's prolific oil and natural gas resources (Ju et al. 2025).

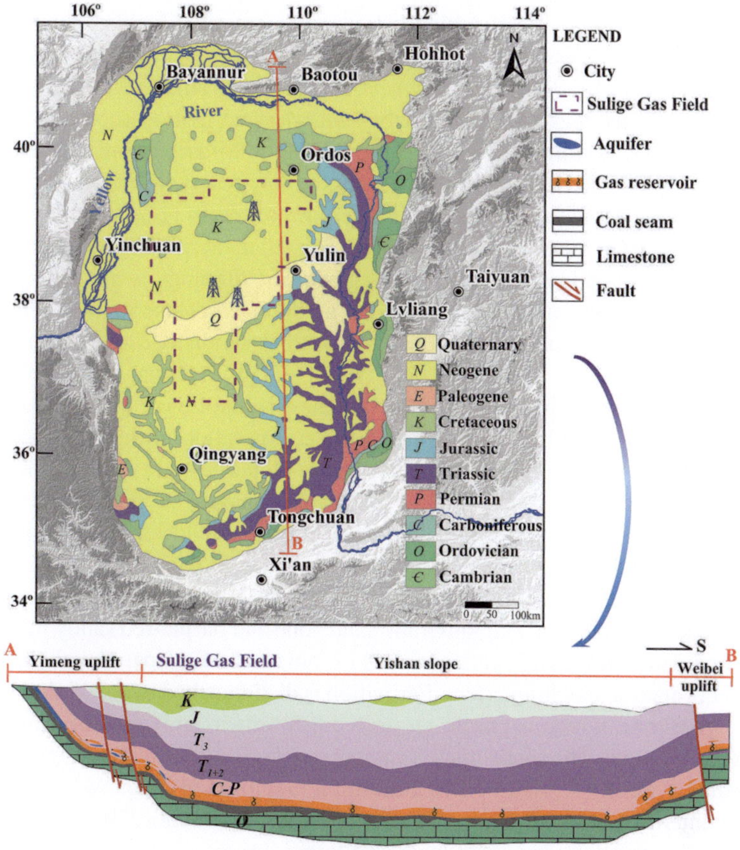

Fig. 2.3 Geological structure of the Ordos Basin

2.3.1 Formation

In the Sulige area, the Lower Paleozoic is dominated by the Ordovician Majiagou Formation, comprising limestones and dolomites, which are rich in carbonate gas reservoirs. The Upper Paleozoic succession is represented by the Carboniferous and Permian, with the Silurian and Devonian missing from the sequence (Fig. 2.4). The Benxi Formation (C_2b) of the Carboniferous is mainly present in the central–eastern study area, with a thickness ranging from 10 to 50 m. At the top of the formation, thick coal seams (No. 8 and No. 9) are developed, marking a significant boundary between the Carboniferous and Permian. The Lower Permian is represented by the Taiyuan Formation (P_1t) and Shanxi Formation (P_1s). The Taiyuan Formation, 40–60 m thick, is mainly composed of dark gray mudstone, limestone, light gray sandstone, and coal. The Shanxi Formation, 80–100 m thick, primarily consists of gray to dark-gray or gray-brown medium- to fine-grained sandy mudstones, with intercalated coal seams and coal streaks. Specifically, the P_1s_2 represents coal-bearing deltaic facies, whereas the P_1s_1 is characterized by channel sandstones deposited in distributary systems (Yang et al. 2016). The extensive coal-bearing strata of the Taiyuan and Shanxi Formations constitute high-quality coal-derived hydrocarbon source rocks for Paleozoic gas accumulations in the region. In the Middle Permian, the Shihezi Formation (P_2s) is developed and serves as the key reservoir-bearing unit of the Upper Paleozoic gas system in the basin. The Lower Shihezi Formation is composed of fluvial-lacustrine facies of a nearshore setting, characterized by sandstones with large-scale cross-bedding and low mud content. The lithology is mainly light-gray conglomeratic coarse sandstone, gray-white medium- to coarse-grained sandstone, and gray-green lithic quartz sandstone. It is further subdivided into five reservoir intervals: Upper He 8, Lower He 8, He 7, He 6, and He 5, among which the Lower He 8 is the most gas-rich (Fig. 2.4). The thickness of the Lower Shihezi Formation ranges from 120 to 160 m. The Upper Shihezi Formation represents deposits of an arid lacustrine environment, and can be subdivided upward into four reservoir groups: He 4, He 3, He 2, and He 1. The lithology mainly consists of interbedded red mudstone and sandy mudstone, with occasional thin sandstones, generally 100–120 m thick. The Upper Permian is dominated by the Shiqianfeng Formation (P_3s), which represents the uppermost strata of the Upper Paleozoic. It is characterized by interbedded purple-red conglomeratic sandstone and purple-red sandy mudstone, with occasional calcareous nodules of marl in some areas. Compared with the Upper Shihezi Formation, the Shiqianfeng Formation is distinguished by its bright-colored, impure mudstone, which is purple-red to reddish-brown, and contains gravels and calcareous material. The sandstones comprise quartz along with lithic fragments and K-feldspar, typically forming feldspathic lithic quartz sandstones, with an overall higher content of epidote. With a stable thickness of around 250 m, this unit is extensively distributed, deposited in an arid lacustrine environment, and constitutes a favorable sealing layer (Yang et al. 2016; Zhang et al. 2019).

The Mesozoic is mainly characterized by the development of Triassic, Jurassic, and Cretaceous strata. The Lower Triassic is represented by the Liujiagou Formation, which consists of purplish-red sandy mudstone. The Jurassic strata mainly consist of continental coal-bearing sediments, containing significant coal resources. The Cretaceous succession, from the base upwards, is primarily composed of the Luohe, Huanhe, and Luohandong Formations, forming the principal aquifer system of the study region. The Cenozoic strata include the Paleogene, Neogene, and Quaternary (He et al. 2022). The Quaternary is characterized by complexly derived continental clastic deposits, mainly loess and gravel layers, which are widely distributed across the region.

2.3.2 Geological Structure

Tectonically, the Sulige Gas Field lies within the Yimeng Uplift, Yishan Slope, and Tianhuan Depression, predominantly positioned in the northern Yishan Slope. The

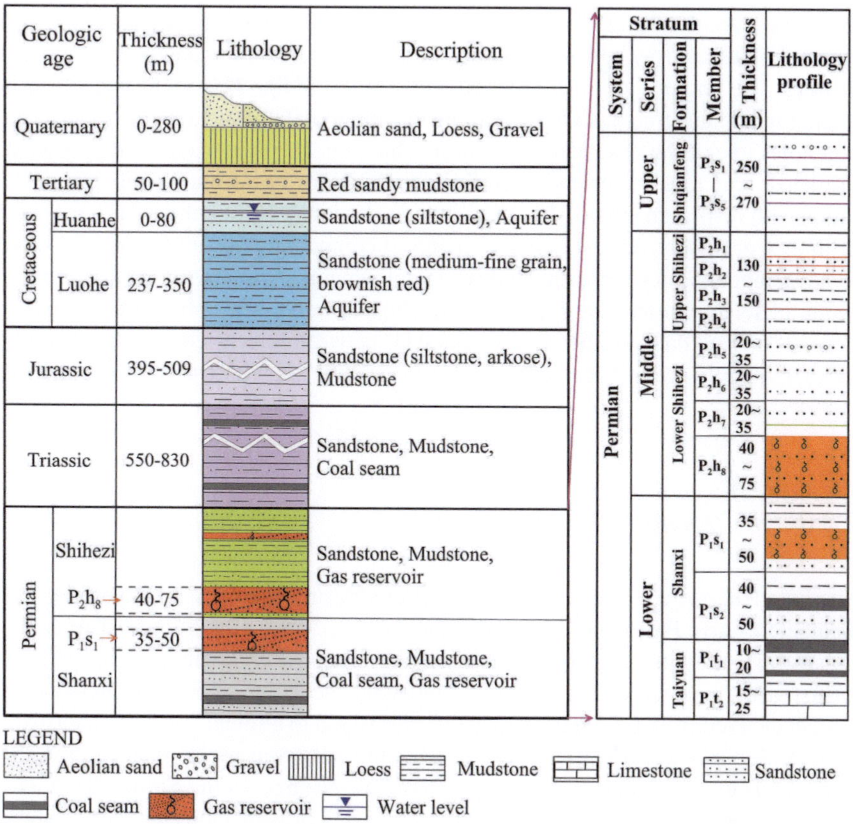

Fig. 2.4 Comprehensive stratigraphic profile of the study area

structure is generally stable, and faulting is weakly developed. On the west-dipping monocline background, the gas field develops multiple rows of nose-like structures, typically 10 m in amplitude and 2–3 km in width. The microstructural features at the top surfaces of Upper He 8, Lower He 8, and Shan 1 Members exhibit strong inheritance. Various deformation styles occur within nose-like structures, including four typical combinations and minor structural forms. The small-scale structural forms resemble those elsewhere on the Yishan Slope, although local small faults are also developed. Accurate characterization of small-scale structures is crucial for geological steering of horizontal wells. The target layers of horizontal wells need to be relatively gentle and fault-free, which facilitates well trajectory control and drilling operations. The west-dipping monocline exhibits an overall gentle structure, yet local fluctuations are more pronounced. Major structural variations along the horizontal section are mainly due to three reasons: (1) the overall structure is gentle with minor relief, but local variations are pronounced; (2) the horizontal wellbore direction is north-south, nearly perpendicular to the trend of nose structures, which intensifies structural variation; and (3) small-displacement faults are locally present, which cannot be detected by seismic methods but have significant impacts on horizontal drilling. The target layer of horizontal wells in the Sulige Gas Field is about 6 m thick. Significant structural changes along the horizontal drilling trajectory can cause boreholes to exit the reservoir top or base. Due to nearly north–south compression, local nose structures experience significant undulations, sometimes generating small faults with displacements of 10–15 m, which pose major difficulties for horizontal drilling. Conversely, relatively flat segments of compound nose-fold structures serve as advantageous sites for horizontal well deployment. Owing to the restricted resolution of seismic forecasting, the identification of micro-scale structures primarily depends on well data. Predictive discrepancies between forecasted and actual amplitudes frequently occur, which complicates horizontal well construction (Zhang et al. 2019).

2.3.3 Sedimentary Environment

The Late Paleozoic sedimentation of the Ordos Basin developed on the eroded and weathered Early Paleozoic deposits in the western North China Plate. The evolutionary history of the basin comprises three stages: epeiric sea basin (east) and rift-sag basin (west), marginal lacustrine basin, and inland lacustrine basin. The depositional system evolved from tidal flat (lagoon)-barrier island to alluvial fan-delta-lacustrine system. Sedimentation shifted from alternating carbonate rocks, coal seams, and terrigenous clastics to predominantly terrigenous clastics. In the Sulige Gas Field, the He 8 Member shows thickness variations of less than 20 m, with paleodepositional slopes of 1°–2°, reflecting a stable and gently dipping paleogeomorphology without distinct flexure zones. The paleoclimate was subtropical and humid, with strong physical weathering and sufficient sediment supply from the ancient Yishan Uplift to the north. The lacustrine basin was freshwater and open-flowing, with frequent and wide-ranging shoreline shifts. The fluvial system exhibited strong hydrodynamics, with frequent lateral channel migration. Such a broad, humid environment with ample sediment supply and strong hydrodynamics gave rise to large-scale, gentle-slope fluvial-deltaic sedimentation in the He 8 to Shan 1 Members, resulting in extensive sand body development (Xiao et al. 2019).

Compared with classical deltas, the shallow-water delta in Sulige was consistently dominated by fluvial processes. As the fluvial system transported coarse clastic material rapidly into the lacustrine basin, channel branching intensified, lateral channel extent decreased, the delta plain was widespread, while the delta front was relatively limited, reflecting a “large plain-small front” configuration.

During the deposition of the lower submember of He 8, channel belts frequently converged and diverged, forming a braided channel network. Between channel belts, paleogeomorphic highs (floodplains) existed, which were breached during floods to form crevasse channel deposits. Mid-channel bars partitioned channels within channel belts, forming a braided channel network. Rapid lateral migration of braided channels caused mid-channel bar sand bodies to laterally overlap while earlier channels were gradually abandoned and filled with fine sediments, thereby limiting sand body lateral continuity. Large-scale crevasse and floodplain deposits further segmented sand bodies vertically. Overlapping zones of compound mid-channel bars represent favorable sites for horizontal well deployment.

During the deposition of the upper submember of He 8 and the Shan 1 Member, multiple low-sinuosity meandering channels interwove. Channel sand bodies comprised channel-fill sands, point bar sands, levee sands, and crevasse-splay sands. Meandering rivers were dominated by lateral migration, forming extremely wide channel sand bodies with sheet-like distribution. Channel shifting and erosion could remove upper levees, allowing point bar deposits from different periods to stack successively. Floodplain overbank deposits isolated sand bodies and restricted their connectivity. Muddy overbank deposits were widespread and must be avoided when deploying horizontal wells (Zhang et al. 2019).

(1) Shan 1 Member and the Upper He 8 Member

The Sulige Gas Field records the deposition of meandering channel sand bodies in the Shan 1 Member, braided channel sand bodies in the Lower He 8 Member, and meandering channel sand bodies in the Upper He 8 Member, forming a base-level cycle of initial fall and later rise. The Shan 1 Member and Upper He 8 Member are dominated by meandering river delta facies, belonging to the delta plain subfacies, which can be further categorized into microfacies such as distributary channels. The cumulative vertical thickness of meandering channel sand bodies is relatively thin, appearing as banded patterns in plan view. The width of composite channels ranges from 1 to 2 km, while that of single channels is 300–500 m, showing vertical isolation and dispersion. Channel sand bodies consist of channel lag deposits, point bar sand bodies, levee sand bodies, and crevasse splay deposits.

(2) Lower He 8 Member

The Lower He 8 Member is dominated by braided river delta facies, where frequent interaction and convergence of channel belts resulted in a braided channel belt network. The Sulige Gas Field is largely situated within the braided river delta plain, which can be further divided into four microfacies and seven depositional energy units. At the base-level lowstand of the Lower He 8 Member, rapid lateral migration of braided channels led to lateral superposition of channel sand bodies, generating extensive sand bodies in planform. The composite channel bodies are 3–8 km wide, with single-channel widths typically ranging from 1 to 2 km. Multiple phases of sand body incision and stacking resulted in thick composite sand bodies, with cumulative thicknesses of several tens of meters and single-phase vertical thicknesses generally reaching 5–8 m. Thus, the sand bodies of the Lower He 8 Member are characterized by significant vertical thickness, wide lateral distribution, and large scale, typically showing sand-over-mud architecture. These features make it a prime target for horizontal drilling, with superimposed compound mid-channel bars serving as favorable zones for horizontal well placement.

2.3.4 Characteristics of Sand Body

The sedimentary facies of the Sulige Gas Field control both the developmental type and scale of its sand bodies. Braided river deposits, characterized by high channel energy and large width-to-thickness ratios (typically 80–120 m), together with the superposition of multiple sand bodies, result in the formation of composite sand bodies that are widely distributed in broad belts or continuous sheets (Hu et al. 2023). The Upper Paleozoic reservoirs of the Sulige Gas Field are mainly located in the Upper and Lower He 8 Member of the Shihezi Formation and the Shan 1 Member of the Lower Permian Shanxi Formation. These intervals can be further subdivided into seven sublayers (equivalent to sand layer sets) and fourteen sand bodies, which represent the fundamental stratigraphic units for horizontal well deployment.

The drilling encounter rate of sand bodies and the sand-to-strata ratio (the percentage of sandstone and conglomerate thickness within a stratigraphic unit) are critical for evaluating sand body characteristics and continuity. Data from completed wells in the Sulige Gas Field show that the Lower He 8 Member has the highest encounter rate, consistently above 95%, indicating large-scale sheet-like sand distribution. In comparison, the Upper He 8 and Shan 1 Members exhibit encounter rates of 65–80%, suggesting sand bodies are more commonly distributed as continuous belts or bands. Statistics reveal that the Lower He 8 Member exhibits a high ratio (around 60%), suggesting a wide, stable sand body distribution. By contrast, other sublayers have ratios of 30–50%, and given that the Upper He 8 and Shan 1 Members represent meandering river systems, their sand bodies are narrow and poorly continuous. Thus, the Lower He 8 Member constitutes the main target interval for horizontal well development. Adequate sediment supply and strong hydrodynamic conditions facilitated frequent river channel shifts, lateral sand body migration, and vertical progradation, resulting in widespread sand accumulation in the Sulige Gas Field. Composite sand bodies in the He 8 Member are typically 10–25 km wide, have vertical stacking thicknesses exceeding 20 m, and extend laterally over 150 km (Zhang et al. 2019).

Although the widespread distribution of sand bodies, the geological setting of the Sulige Gas Field is highly complex, with effective gas-bearing sand bodies being limited in scale and poorly connected. Only coarse sandstone facies, such as coarse sandstone and conglomeratic coarse sandstone, provide effective reservoirs. These sand bodies are mainly located in point bar cores and the lower portions of channel fills. Their stacking patterns fall into three categories: (1) isolated point-bar sand bodies, 300–500 m wide, with limited lateral extent; (2) point bars connected to basal coarse sandstone facies, forming relatively larger effective sand bodies, with thin coarse sandstone beds extending further and potentially linking adjacent bodies; and (3) laterally amalgamated point bars, forming locally sheet-like sand bodies with connectivity exceeding 1 km.

2.4 Reservoir

2.4.1 Porosity and Permeability

In the Sulige Gas Field, the reservoir space of Upper Paleozoic Formations is mainly provided by pores, with microfractures contributing only minimally. Reservoir samples exhibit a strong positive correlation between porosity and permeability, and physical properties derived from core analysis are consistent with well log interpretations, indicating that permeability variations are primarily controlled by pore development, a typical feature of porous reservoirs (Yang et al. 2008; Zou 2017). Secondary dissolution pores and intercrystalline pores within kaolinite dominate the pore structure, with pore sizes ranging from 5 to 400 μm and average radii between 11.98 and 107.07 μm. The range and concentration of pore size distribution vary depending on reservoir type. Owing to the deep burial and long burial history of the He 8 Member in the Sulige area, the reservoirs have undergone complex diagenetic alterations. Intense compaction-pressure solution and cementation have led to the loss of primary intergranular pores, especially in lithic sandstones rich in ductile rock fragments. However, dissolution during the middle to late diagenetic stages partially restored porosity. Dissolution of lithic and volcanic fragments generated abundant intragranular, intergranular, and moldic pores. Concurrently, the formation of authigenic kaolinite produced well-developed intercrystalline pores (Wu et al. 2021b).

Considering their frequency and contribution, pore assemblages are categorized into four types: (1) intergranular pores + micropores + dissolution pores, (2) dissolution pores + micropores + intergranular pores, (3) dissolution pores + micropores, and (4) micropores (Zhang et al. 2019). The intergranular–micropore–dissolution pore assemblage is characterized by residual intergranular pores and micropores, predominantly kaolinite intercrystalline pores, commonly occurring in quartz sandstone, with a proportion of less than 15%. The dissolution-micropore-intergranular pore assemblage represents a composite pore network, commonly observed in lithic quartz sandstone, accounting for about 30%. The main pore type is dissolution pores from soluble components, supplemented by micropores formed through alteration, particularly kaolinite intercrystalline pores. The dissolution-micropore assemblage is the most common pore type, predominantly developed in lithic sandstones, accounting for up to 40%. The micropore assemblage indicates lithologic densification and significantly deteriorated reservoir quality, contributing less than 15% of the reservoir (Fan et al. 2019; Zhang et al. 2019).

Reservoir pore structure is characterized by "large pores, small throats, few fractures, and poor pore-throat connectivity". The main challenges in reservoir development are: (1) limited hydrocarbon flow range and high threshold pressure at the initial stage, making hydraulic fracturing essential for productivity; and (2) the risk of water-blocking damage during production. In the Sulige region, the measured porosity of He 8 and Shan 1 sandstones varies from 0.3 to 21.8%, and the permeability ranges from 0.001×10^{-3} to more than 500×10^{-3} μm^2. Permeability variation is mainly controlled by pore development, showing a strong positive correlation with porosity (correlation coefficient + 0.7). The presence of fractures significantly enhances sandstone permeability, with maximum values reaching 629.5×10^{-3} μm^2. Frequency distribution plots of porosity and permeability indicate that, in the He 8 Member, samples with porosity > 8% account for 48.7%, and those with permeability > 0.5×10^{-3} μm^2 account for 32.4%. In the Shan1 Member, samples with porosity > 8% represent 46.7%, while those with permeability > 0.5×10^{-3} μm^2 constitute 23.6% (Yang et al. 2016). Overall, reservoir properties of the He 8 Member are slightly better than those of the Shan 1 Member.

2.4.2 Gas–Water Relation

Conventional gas reservoirs exhibit a clear spatial relationship between water and gas zones, usually with edge or bottom water and a well-defined gas–water boundary. Unlike conventional gas reservoirs, tight sandstone gas reservoirs are characterized by dense reservoirs and strong heterogeneity, which determine the complexity of gas accumulation and result in distinctive features in gas saturation, formation water characteristics, and gas–water distribution. Tight sandstone reservoirs are generally characterized by complex pore structures and high bound-water saturation (Yan et al. 2025). The Sulige Gas Field shares these traits but also shows unique patterns in gas–water distribution. Spatially, water content is more pronounced in the western area of the Sulige Gas Field. Vertically, gas saturation decreases upward from the Shan 1 Member to the Lower He 8 and Upper He 8 Members. During hydrocarbon generation, gas generally migrated and accumulated toward relatively higher structures, whereas formation water was more extensive in the low-lying western sector compared to the central and eastern regions. In the same sand body, water content was higher in structural lows than in highs. Meanwhile, reservoir heterogeneity strongly influences fluid distribution patterns. The central part of channel sand bodies has better physical properties and higher gas saturation, while the margins or tight sandstone zones contain more water. In settings of tight reservoirs with small-scale connected sand bodies and low structural dips, gas buoyant forces cannot effectively overcome the capillary resistance of low-porosity and low-permeability formations. Consequently, gas–water segregation is weak, and gas–water distribution is not controlled by geologic structures, with no clear edge water, bottom water, or unified gas–water contact. Formation water in the Sulige Gas Field can be categorized into three types according to its macroscopic features and vertical distribution: structurally trapped water in low positions, retained water in low-permeability zones, and capillary water. Structural low-position water is distributed at the bottom of structures, while low-permeability zone water occurs mainly at pinch-out zones of sand bodies in structural flanks (Fig. 2.5). Capillary water, influenced by heterogeneity, has uncertain distribution, occurring both in structural lows and highs, but it consistently develops in reservoir zones with poor physical properties, mostly at the margins of sand bodies. Data from completed wells and production in the Sulige Gas

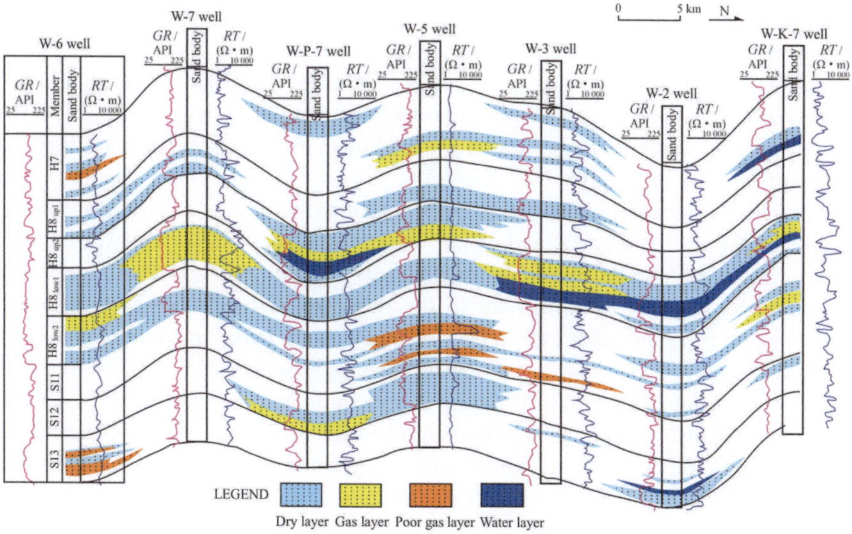

Fig. 2.5 A gas–water distribution profile of a certain well section in the Sulige Gas Field (Modified from Liu et al. 2020)

Field reveal that gas content varies significantly across zones, primarily controlled by structural evolution, hydrocarbon generation strength, accumulation timing, and the relationship with reservoir compaction timing. Gas-bearing capacity is strong in the central region, relatively weak in the east, and the western region is dominated by water-bearing reservoirs (Yang et al. 2016).

Vertically continuous sand bodies exhibit three gas–water distribution patterns: gas overlying water, water overlying gas, and mixed gas–water reservoirs. Gas overlying water reservoirs commonly occur in relatively high-permeability formations. This water is primarily residual water caused by insufficient gas charging, mainly distributed in the down-dip portions of main channels or in isolated, lens-shaped permeable sand bodies encircled by tight formations (Fig. 2.5). Water overlying gas reservoirs typically form in tighter reservoirs, where gas cannot migrate by conventional displacement. Instead, it expands and pushes formation water aside, producing a direct mechanical energy exchange between gas and water. In the absence of buoyancy effects, gas accumulates from the base of sandstone layers. The relatively tighter upper sections act as cap rocks, resulting in a water overlying gas distribution. Gas–water differentiation in tight reservoirs is controlled by pore types and the spatial arrangement of pore-throat structures. A larger pore-to-throat volume ratio raises displacement pressure, hindering the removal of capillary-held formation water during gas charging. Once formation water enters relatively large pores within complex pore-throat networks, it becomes difficult to displace. In mixed gas–water reservoirs, formation water primarily exists in the capillaries of highly heterogeneous reservoirs. Dominated by capillary forces rather than gravity, they are commonly found along channel margins or in fine-grained layers at the tops of channel cycles.

Gas preferentially accumulates in reservoirs with superior petrophysical properties. However, tight reservoirs with complex pore-throat networks have high displacement pressures, making gas-driven water displacement difficult. The buoyant force of gas (0.08–0.28 MPa) is insufficient to overcome capillary forces (0.15–2.0 MPa), leaving gas and water poorly separated and leading to mixed storage in water-bearing reservoirs (Liu et al. 2020).

Formation water critically affects both productivity and production stability of gas wells, especially in horizontal wells. Thus, well placement should avoid water-rich areas, and hydraulic fracturing should steer clear of intervals with high water saturation to improve performance. In the development of hydrocarbon-bearing basins, aquifer groundwater experiences sedimentation, burial, metamorphism, and leaching, engaging in active geochemical exchanges with hydrocarbons and host rocks. This results in the redistribution and recombination of chemical elements, forming novel hydrochemical types and unique water chemistry. Studying the chemical properties and evolution of formation water provides insights into the origin and spatial distribution of gas and water (He 2023).

2.4.3 Gas Reservoir Environment

Despite the extensive distribution of sand bodies in the Sulige Gas Field, its hydrocarbon reserves abundance remains low. The reserve abundance ranges from (0.8–1.6) $\times$ 10^8 m^3/km^2, which is significantly lower compared with similar gas fields, thus classifying it as a typical low to ultra-low abundance reservoir. During vertical well exploitation, approximately 10% of wells achieve open-flow rates above 15 $\times$ 10^4 m^3/d, whereas more than 90% fall below this threshold, most producing less than 5 $\times$ 10^4 m^3/d, which characterizes the field as a low-productivity reservoir with high development difficulty (Zhang et al. 2019). The application of horizontal wells enables communication among multiple effective sand bodies, expands the flow area, and consequently improves the degree of reserve utilization and individual well output.

Considering reservoir petrophysical properties and mercury intrusion porosimetry parameters, combined with thin-section observations and pore imaging data, the He 8 and Shan 1 reservoirs in the Sulige region are categorized into four classes. Type I gas layers: characterized by massive (partly conglomeratic) medium- to coarse-grained quartz sandstone deposits, typically exceeding 5 m in thickness. The pore system mainly consists of intergranular and dissolution pores. The capillary pressure curve is of wide-plateau or gentle-slope type, with well-sorted pore throats and displacement pressure less than 0.1 MPa. Porosity varies between 8 and 18%, commonly above 10%. Permeability is generally higher than 1.0 $\times$ 10^{-3} μm^2 and gas saturation surpasses 60%. Reservoirs of this type are the primary hosts of high-output wells (> 10 $\times$ 10^4 m^3/d). Type II gas layers: composed mainly of siliceous medium- to coarse-grained quartz sandstones, with subordinate medium- to coarse-grained lithic quartz sandstones. The dominant pore types are dissolution and intercrystalline pores, accompanied by few intergranular pores. The capillary pressure curve is of gentle-slope type, with relatively good throat sorting and displacement pressures of 0.1–0.5 MPa. Porosity is within 7–12% and permeability varies from 0.5 to 1.0×10^{-3} μm^2. Gas saturation is generally > 50% and typical production is $4.0–10.0 \times 10^4$ m^3/d. Type III gas layers: mainly medium- to coarse-grained lithic quartz sandstones, with a small proportion of lithic sandstones. The pore structure is mainly dissolution pores and micropores. Capillary pressure curves display a sloping pattern and pore-throat sorting is moderate, with displacement pressures between 0.5 and 1.0 MPa. Porosity lies between 5.0 and 9.0%; permeability $0.1–0.5 \times 10^{-3}$ μm^2; gas saturation 40–55%; production generally < 4×10^4 m^3/d. Type IV tight layers: mainly medium-grained lithic sandstones and subordinate lithic quartz sandstones, characterized by high matrix content. Micropores and micro-dissolution pores are prevalent. Capillary pressure curve shows convex-upward type and throat sorting is poor, with displacement pressures exceeding 1.0 MPa. Porosity ranges 2.0–6.0%; permeability usually below 0.1×10^{-3} μm^2; gas saturation mostly < 40%; weak gas content; daily output only a few hundred m^3 (Yang et al. 2016).

In the Sulige region, the He 8 and Shan 1 reservoirs demonstrate vertical interconnection of multiple gas layers and lateral composite continuity, with a north-south extension exceeding 100 km and an east-west width of 10–20 km. As a result, reservoir pressure exhibits considerable areal heterogeneity. For the He 8 reservoir, pressure typically varies between 22.5 and 31.5 MPa, with a pressure coefficient ranging from 0.73 to 0.93 and averaging 0.84. In the Shan 1 reservoir, pressure usually ranges from 23.2 to 35.7 MPa, with a pressure coefficient of 0.85–0.96 and an average value of 0.91. The pressure coefficient of the Shan 1 reservoir is slightly higher than that of the He 8 reservoir, yet both are overall lower than hydrostatic pressure, indicating low-pressure gas reservoirs with stable fluid properties. Reservoir temperatures are within the range of 100–110 °C. The geothermal gradient is determined as 3.06 °C/100 m.

Natural gas is mainly sourced from Carboniferous and Permian coal-bearing formations. In the He 8 reservoir, methane content varies from 86.96 to 93.72% (mean 89.82%), with relative density between 0.601 and 0.643 and CO_2 concentration around 1.56%. For the Shan 1 reservoir, methane content ranges between 85.58 and 91.79% (average 89.07%), with a methane index of 91%, relative density of 0.610–0.661, and CO_2 concentration of about 1.77%, characterizing it as a sulfur-free wet gas (Yang et al. 2016).

2.5 Hydrogeology

2.5.1 Groundwater

The Ordos Basin is a vast groundwater basin, with major aquifer systems including the Cambrian-Ordovician carbonate karst aquifer, the Carboniferous-Jurassic clastic fractured aquifer, and the Cretaceous clastic fracture-pore aquifer system. The upper basin is mostly overlain by unconsolidated porous aquifer systems, while its margins host aquifer systems of Cenozoic fault-depression basins (e.g., Guanzhong Basin, Yinchuan Plain, Hetao Plain) and bedrock mountain fracture aquifers (Fig. 2.6). In the Sulige area, groundwater predominantly occurs in the Cretaceous and Quaternary aquifers, which exhibit stable distribution, shallow burial depth, high yield, and overall favorable water quality. Geomorphologically, the aquifer systems of the study area are subdivided into the Desert Plateau groundwater system in the north and the Loess Plateau groundwater system in the south (Fig. 2.7). From the perspective of watershed, the study area primarily belongs to three river flow subsystems: the Wulanmulun River-Wuding River system, the Dusitu River-Yanchi system, and the Luo River-Yan River system (Fig. 2.6).

(1) **Desert Plateau groundwater system**

The northern Desert Plateau groundwater system is mainly composed of Quaternary porous water and Cretaceous clastic fracture-pore water. The Quaternary aquifer predominantly consists of the Upper Pleistocene Salawusu Formation, representing plateau-type discontinuous phreatic water, extending in a NE–SW direction from Baiyu Mountains into the Mu Us Desert north of the Wuding River. The aquifer lithology is dominated by fluvio-lacustrine silty fine sand, with a thickness of 30–80 m, good permeability (hydraulic conductivity 2–6 m/d), and a single-well yield of 500–3000 m^3/d. The upper Salawusu Formation consists of a thin aeolian sand layer with loose structure, favorable for precipitation infiltration. It maintains strong hydraulic connectivity with the underlying Cretaceous aquifer unit, together forming a thick, integrated aquifer body. The groundwater salinity is generally < 1 g/L, with overall good quality. Cretaceous groundwater is extensively distributed, with aquifers stratigraphically divided into the Luohe, Huanhe, and Luohandong formations from bottom to top. The Luohe Formation is the most widespread, with a thickness of 150–250 m, dominated by desert-facies fine sandstone, characterized by loose structure and good pore connectivity, providing favorable groundwater conditions. The Huanhe Formation is mainly fluvial and alluvial-fan facies, with distinct stratification and higher mud content, resulting in denser aquifer media and less favorable groundwater conditions compared to the Luohe Formation. Furthermore, the aquifer medium generally contains higher salinity and sluggish circulation, resulting in relatively poor groundwater quality. Salinity levels usually exceed those of the Luohe Formation, typically ranging 1–3 g/L. The Luohandong Formation is limited in distribution, occurring only sporadically in the

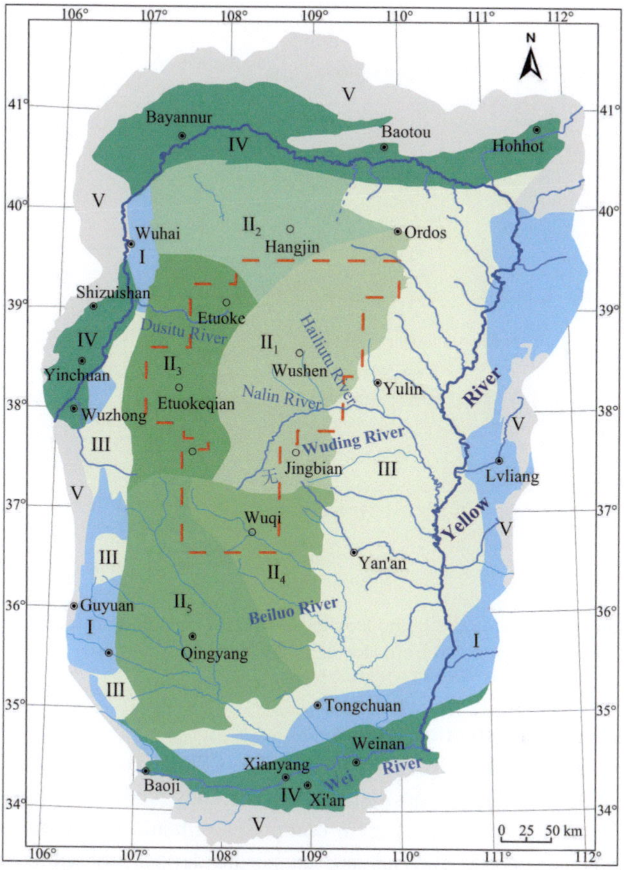

I: Cambrian–Ordovician carbonate karst aquifer system
II: Cretaceous clastic-rock fracture-pore aquifer system
(II$_1$: Wulanmulun River–Wuding River flow system;
II$_2$: Molin River–Yanhaizi flow system; II$_3$: Dusitu River–Yanchi flow system;
II$_4$: Luo River–Yan River flow system; II$_5$: Jing River–Malian River flow system)
III: Carboniferous–Jurassic clastic-rock fracture water aquifer system with pore water in the overlying unconsolidated layer
IV: Cenozoic rift-basin aquifer system
V: Fracture water aquifer system in bedrock mountainous regions.

Fig. 2.6 Zoning of the groundwater system in the study area (Modified from Hou et al. 2008)

northwestern Ordos region (Etuoke County and Etuokeqian County). Overall, in the Northern Desert Plateau, the Luohe, Huanhe, and Salawusu Formations are vertically superimposed with strong hydraulic connectivity (He 2023).

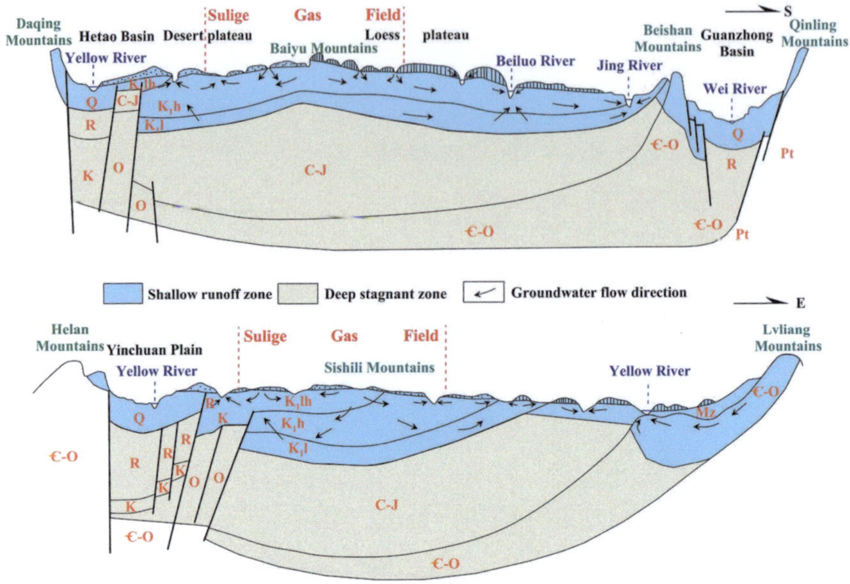

Fig. 2.7 Schematic diagrams of the hydrogeological sections in the north-south and east-west directions of the study area (Modified from Hou et al. 2008)

(2) **Loess Plateau groundwater system**

The groundwater system of the southern Loess Plateau is mainly composed of Quaternary unconsolidated aquifers and Cretaceous aquifers. The Quaternary unconsolidated groundwater system comprises phreatic water in loess layers and groundwater in the alluvial and lacustrine deposits of valleys. Phreatic water in the alluvial and lacustrine deposits is relatively rich in yield, while loess aquifers are scattered in distribution. The loess aquifers are recharged directly by atmospheric precipitation, whereas groundwater in valley alluvial-lacustrine deposits receives recharge from rainfall, surrounding loess aquifers, and subsurface baseflow. Overall, the Quaternary groundwater exhibits poor water-bearing capacity.

The Cretaceous groundwater is mainly hosted within the Huanhe and Luohe Formations. The Huanhe Formation shows multilayered structures, complex lithological compositions, and considerable facies heterogeneity, with thickness ranging from 40 to 200 m and unstable distribution. It contains relatively well-developed weathering fractures, with a fracture density of 0.25, porosity generally between 10 and 20%, and permeability coefficients typically between 0.01 and 0.1 m/d. The Huanhe Formation has poor aquifer properties, with moderate to weak water-bearing capacity and single-well yields generally less than 500 m^3/d. Groundwater in the Huanhe Formation is recharged by surface water and precipitation in exposed areas, flowing from watersheds toward rivers and ultimately discharging into the Beiluo River and its tributaries (Fig. 2.7). The Luohe Formation is generally composed of red desert-facies aeolian sandstones, with confined aquifer thicknesses of 200–450 m and phreatic zones of 0–200 m. The formation exhibits simple structures, well-developed porosity, permeability coefficients of 0.2–1.0 m/d, and favorable water-bearing capacity. The water-bearing capacity ranges from strong to weak, with single-well yields typically between 500 and 1000 m^3/d (Hou et al. 2008). Groundwater in the Luohe Formation occurs as phreatic water in outcrops and confined water in buried areas, recharged by surface runoff, precipitation, and overflow from the overlying Huanhe Formation, eventually draining into the Beiluo River.

Precipitation constitutes the dominant recharge source for groundwater in the area, accounting for over 90% of the total input. With the exception of partial evaporative discharge in northern desert endorheic regions, groundwater predominantly flows from recharge areas toward incised valleys. The Wuding River and the upper Beiluo River serve as the main discharge outlets for groundwater in the study area. Given the scarcity of surface water resources, groundwater serves as the primary source of water for agricultural, industrial, and domestic use in the study area.

2.5.2 *Water Resources Utilization*

The Sulige Gas Field covers five administrative divisions: Wuqi County (Yan'an City) and Dingbian County (Yulin City) in the southern part, along with Wushen County, Etuoke County, and Etuokeqian County (Ordos City) in the northern part. According to the *2023 Ordos Water Resources Bulletin* (Ordos City Water Conservancy Bureau 2024), the total water supply in 2023 for Etuokeqian, Etuoke, and Wushen was 1423.2, 2171.2, and 2544.9 million m^3, respectively. Of these, groundwater supplies accounted for 1261.8, 1025.4, and 1992.3 million m^3, representing 88.66%, 47.23%, and 78.30% of the total water supply, respectively (Table 2.2). In 2020, industrial water consumption in these three regions was 53.9, 655.0, and 549.2 million m^3, accounting for 4.88%, 32.12%, and 21.15% of the respective total water use. Based on the *2023 Yulin Water Resources Bulletin* (Yulin City Water Conservancy Bureau 2024), the total water resources in Dingbian County in 2023 were 272 million m^3, with a total supply of 1547.0 million m^3. Groundwater supplied 1083.6 million m^3 (70.05%), and industrial water use was 196.6 million m^3 (12.67%). The *Yan'an Water Resources Bulletin* reports that in 2023 (Yan'an City Water Conservancy Bureau 2025), Wuqi County's total water supply reached 219.7 million m^3, of which 199.8 million m^3 came from groundwater (90.94%). Industrial consumption was 103.6 million m^3, accounting for 47.16% (Table 2.3). These statistics reveal that groundwater constitutes the main water resource in the Sulige region, with industrial use accounting for a considerable share. Groundwater is essential for ensuring agricultural, industrial, and domestic water supply, making its protection particularly crucial.

Table 2.2 Water supply volume of different county within the study area (Unit: 10^4 m^3)

County	Surface water	Groundwater	Other sources	Total
Etuokeqian	1403	12,618	211	14,232
Etuoke	9703	10,254	1754	21,712
Wushen	1399	19,923	4127	25,449
Dingbian	944	10,836	3690	15,470
Wuqi	45	1998	154	2197

Table 2.3 Water consumption volume of different county within the study area (Unit: 10^4 m^3)

County	Agriculture	Industry	Resident	Ecology	Total
Etuokeqian	11,830	1518	564	320	14,232
Etuoke	11,037	8757	1037	881	21,712
Wushen	18,952	5356	765	376	25,449
Dingbian	12,291	1966	1188	24	15,469
Wuqi	366	1036	763	32	2197

2.6 Gas Field Development

The Sulige area serves as a crucial energy base in China, endowed with significant reserves of natural gas, oil, and coal, particularly characterized by abundant tight sandstone gas within the Upper Paleozoic Formations. Currently, the Sulige Gas Field has an exploration area of 6×10^4 km^2, with proven reserves reaching 4 trillion m^3 and annual output surpassing 26 billion m^3 by the end of 2020, ranking as China's largest confirmed integrated onshore natural gas field. Since its discovery in 2000, the Sulige Gas Field has undergone three development stages: preliminary evaluation, rapid production growth, and stable long-term production (Fig. 2.8). In August 2000, fracturing tests at Well Su 6 of the Upper Paleozoic strata produced an open flow of 1.2016 million m^3/day, confirming the discovery of the Sulige Gas Field and heralding the fast-paced development of China's unconventional tight sandstone gas industry (Yang et al. 2016).

From 2001 to 2005, early assessments and pilot tests accumulated valuable experience for the large-scale and cost-effective development of the gas field. In 2001, five well tests revealed poor reservoir connectivity and strong heterogeneity within the field. To improve reservoir connectivity and enhance single-well productivity, two horizontal wells were drilled and tested in 2002, but the objectives were not achieved. In 2003, Changqing Oilfield launched the Su 6 development test area, deploying 12 infill appraisal wells and conducting two-dimensional and three-dimensional reservoir seismic tests. Complementary engineering methods followed, with eight large-scale fracturing operations, eight CO_2 fracturing tests, four underbalanced drilling trials, and six slim-hole wells conducted. With further testing and evaluation, it was ultimately recognized that the Sulige Gas Field is a vast reservoir

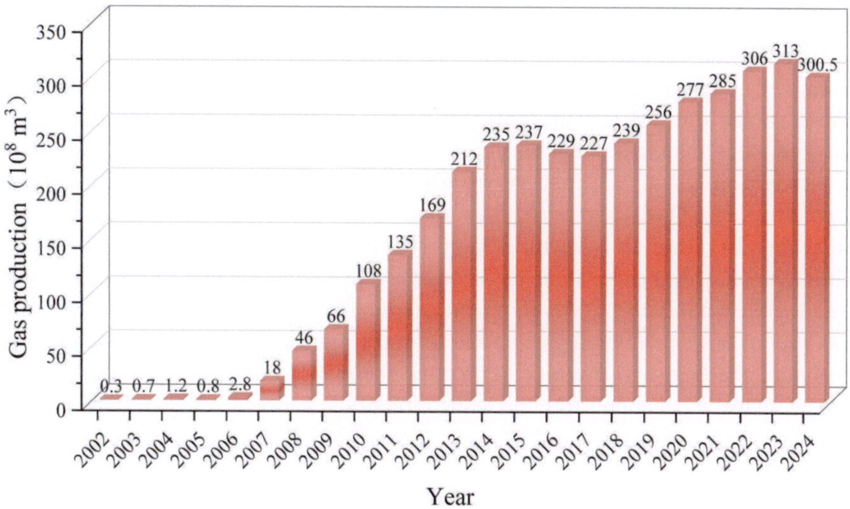

Fig. 2.8 The variation of natural gas production in Sulige Gas Field from 2002 to 2024

with extremely strong heterogeneity and wide distribution, characterized as a "three-low" field (low permeability, low pressure, and low abundance). Using conventional approaches would involve high costs and poor efficiency, making development challenging and requiring a fundamental shift in approach. Consequently, a new development strategy was proposed for Sulige: "technology-driven, mechanism innovation, simplified exploitation, and low-cost development". The development goal shifted from pursuing high single-well output to overall effectiveness, setting targets of 10,000 m^3/d per well with stable production for 3 years and a per-well cost under 8 million RMB (Yang et al. 2016; Zhang et al. 2019).

From 2006 to 2013, the field entered the phase of accelerated production. During 2006–2008, Changqing Oilfield developed twelve integrated supporting technologies, meeting two critical objectives: maintaining an 80% proportion of Type I and Type II wells, and controlling per-well costs below 8 million RMB, thus enabling efficient and economical field development. Between 2009 and 2013, efforts focused on maintaining and improving well productivity while driving ongoing technological innovation. Well types evolved from vertical and cluster wells to horizontal wells; reservoir stimulation shifted from multilayer fracturing in vertical wells to multistage fracturing in horizontal wells, with multiple fractures per stage and large-scale fracturing; production management advanced from manual patrols to digital and intelligent systems, significantly improving development efficiency and effectiveness. In 2010, annual output surpassed 10 billion m^3, reaching 13.5 billion m^3 in 2011, making it China's highest-producing onshore gas field; by 2013, annual production exceeded 20 billion m^3, marking the transition to sustained stable production.

Since 2014, the gas field has entered and maintained the phase of continuous stable output. Test areas with variable well spacing and dense well patterns were actively advanced. A three-dimensional development model was established, integrating vertical, directional, and horizontal wells through cluster deployment, large-scale well group development, differentiated design, and multi-layer exploitation. These efforts aimed to secure long-term stable production. At present, the Sulige Gas Field has maintained annual production above 20 billion m^3 for 12 consecutive years and has ranked first among Chinese gas fields for 14 consecutive years (Fig. 2.8). In 2022, daily production surpassed 100 million m^3, with annual output exceeding 30 billion m^3, reaching 30.56 billion m^3, accounting for nearly 60% of China's onshore tight sandstone gas production (He 2023; Jia et al. 2025). Over two decades of exploration and development have enabled the Sulige Gas Field to establish a set of critical technologies and to form the representative "Sulige Pattern". In the future, the field will continue to produce abundant natural gas, ensuring China's energy security and supporting the realization of the national "dual carbon" goals.

References

Fan A, Yang R, Lenhardt N et al (2019) Cementation and porosity evolution of tight sandstone reservoirs in the Permian Sulige gas field, Ordos Basin (central China). Mar Pet Geol 103:276–293. https://doi.org/10.1016/j.marpetgeo.2019.02.010

He X, Li P, Shi H et al (2022) Identifying strontium sources of flowback fluid and groundwater pollution using $^{87}Sr/^{86}Sr$ and geochemical model in Sulige gasfield, China. Chemosphere 135594. https://doi.org/10.1016/j.chemosphere.2022.135594

He X (2023) Mechanism of chemical interactions between water and rocks during hydraulic fracturing in tight sandstone reservoirs. Ph.D. thesis, Chang'an University (in Chinese with English abstract)

Hou G, Zhang M, Liu F et al (2008) Groundwater exploration in the Ordos Basin. Geological Publishing House, Beijing (in Chinese)

Hu Z, Hu M, Xiong X et al (2023) Geochemical characteristics of the braided river reservoir in block 19 of the sulige gas field. Front Energy Res 11:1219664. https://doi.org/10.3389/fenrg.2023.1219664

Jia A, Wang G, Li Y (2025) Natural gas development in China: present situation and prospect. Nat Gas Ind 45(5):31–42 (in Chinese with English abstract)

Ju Y, Wang W, Ren Z et al (2025) Multi-stage evolution of the Ordos Basin: its coupled basin-mountain systems and energy resources. Sci China Earth Sci 68(8):2426–2473. https://doi.org/10.1007/s11430-025-1627-1

Li J (2019) Study on the situation and variation characteristics of water and sediment in the Wuding River Basin. Master's thesis, Chang'an University (in Chinese with English abstract)

Liu H, Cui Y, Xia S et al (2020) Occurrence mechanism of formation water in Shihezi 8 and Shanxi 1 tight gas reservoirs and its main control factors: an example from one block in Sulige gasfield. Nat Gas Explor Dev 43(1):84–91

Ordos City Water Conservancy Bureau (2024) 2023 Ordos water resources bulletin. https://slj.ordos.gov.cn/zw/slgb/202404/t20240417_3593118.html

Shen S, Sheng Q (2008) Changes in pan evaporation and its cause in China in the last 45 years. Acta Meteorologica Sinica 66(3):452–460 (in Chinese with English abstract)

Wu X, Han S, Li H et al (2021) Identification of base flow source of the upper reaches of Beiluo River in Ordos Basin. Journal of Arid Land Resour Environ 35(10):161–168 (in Chinese with English abstract)

Wu H, Zhao J, Wu W et al (2021) Formation and diagenetic characteristics of tight sandstones in closed to semi-closed systems: typical example from the Permian Sulige gas field. J Pet Sci Eng 199:108248. https://doi.org/10.1016/j.petrol.2020.108248

Xiao H, Liu R, Zhang F et al (2019) Sedimentary model reconstruction and exploration significance of permian He 8 member in Ordos basin, NW China. Pet Explor Dev 46(2):280–292. https://doi.org/10.1016/S1876-3804(19)60008-X

Yan D, Xu H, Wang M (2025) Main controlling factors of gas–water distribution in tight gas reservoirs of Sulige Gas Field. Int J Nat Resour Environ Stud 6(2):16–25. https://doi.org/10.62051/ijnres.v6n2.02

Yan'an City Water Conservancy Bureau (2025) Yan'an water resources bulletin 2023. https://slsb.yanan.gov.cn/swyw/tzgg/1876185231388647425.html

Yang H, Fu J, Wei X et al (2008) Sulige field in the Ordos Basin: geological setting, field discovery and tight gas reservoirs. Mar Pet Geol 25:387–400. https://doi.org/10.1016/j.marpetgeo.2008.01.007

Yang H, Xi S, Wei X et al (2016) The theory of large-scale tight sandstone gas accumulation in the Ordos Basin. Science Press, Beijing (in Chinese)

Yang R, Van Loon AJ (2021) The Ordos Basin: sedimentological research for hydrocarbons exploration. Elsevier. https://doi.org/10.1016/C2020-0-01240-5

Yulin City Water Conservancy Bureau (2024) 2023 Shaanxi Yulin water resources bulletin. https://slj.yl.gov.cn/zfxxgk/fdzdgknr/zdlyxx/202412/t20241219_1938611.html

Zhang J, Yu H, Ma Z et al (2019) Quantitative characterization for tight sandstone gas reservoir in the Sulige gas field. Petroleum Industry Press, Beijing (in Chinese)

Zhao Z, Sun W, Mu X et al (2023) Temporal and spatial variation and difference of pan evaporation between hydrological station and meteorological station in the Yellow River Basin. Yellow River 45(6):24–31 (in Chinese with English abstract)

Zou C (2017) Unconventional petroleum geology, 2nd edn. Elsevier. https://www.sciencedirect.com/book/9780128122341/unconventional-petroleum-geology

Chapter 3
Characteristics of Tight Gas Reservoirs

This chapter presents a detailed study of the lithology, reservoir properties, and gas reservoir features of the tight sandstones in the lower He 8 Member of the Shihezi Formation, the main gas-bearing interval of the Sulige Gas Field. The analyses were based on core samples obtained from five tight sandstone wells and employed methods including X-ray fluorescence (XRF), X-ray diffraction (XRD), Scanning electron microscope (SEM), and casting thin section analysis. Given the inherent complexity of pore distribution in tight sandstone reservoirs, particular attention was directed toward diagenetic processes, aiming to elucidate the mechanisms and factors controlling the formation and evolution of pore structures.

3.1 Sampling

In this study, five representative wells were selected from the Sulige Gas Field, designated *R1* through *R5*, and cores were drilled from the main producing reservoir (the Lower He 8 Member of the Permian). The distribution of the sampled wells is shown in Fig. 3.1, with coring depths ranging from 2800 to 3500 m. Each core sample was prepared as a cylindrical column, 25 mm in diameter and approximately 10 cm in length, consisting primarily of grayish-white medium- to coarse-grained lithic quartz sandstones and light-gray pebble-bearing medium- to coarse-grained lithic sandstones. The cores are generally dense, containing minor gravel fragments, with some sections displaying distinct transverse fractures and interbedded sandstone-mudstone layers (Fig. 3.2).

Following numbering and low-temperature drying, the drilled cores were analyzed for elemental compositions at different locations using an X-ray fluorescence spectrometer (NITON XL2). The analyzed elements included 28 species, such as Ba, Zr, Sr, Rb, Zn, Fe, Mn, Cr, V, Ti, Ca, and K. The core samples were then crushed to 2–5 mm particle size, thoroughly homogenized, and 10 g of each core specimen was

X. He and P. Li, *Geochemical Processes in Tight Gas Hydraulic Fracturing Stimulation*, Environmental Earth Sciences, https://doi.org/10.1007/978-3-032-15180-3_3

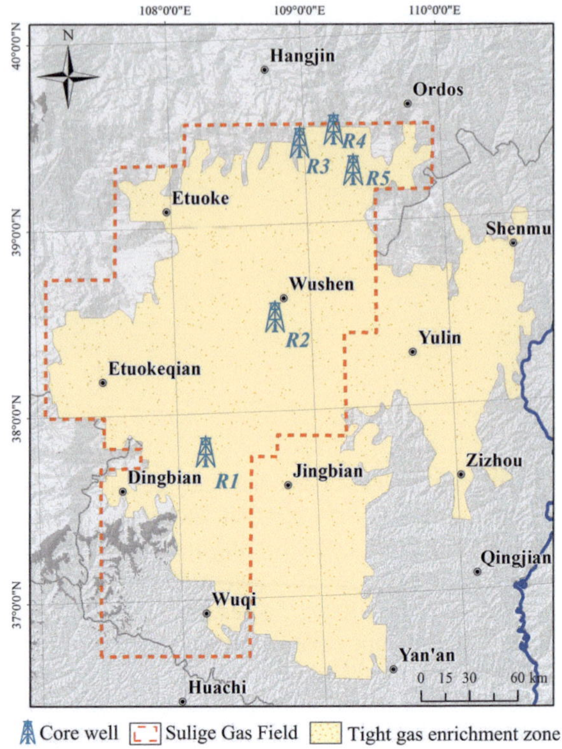

Fig. 3.1 Location of core sampling wells

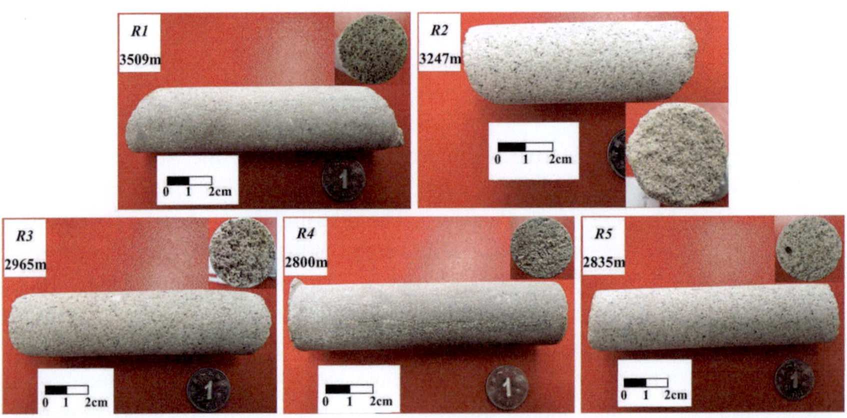

Fig. 3.2 Photos of core samples

ground and sieved through a 200-mesh screen (0.075 mm). The elemental values were measured again using XRF, and the mineral composition was determined using an X-ray diffractometer (D/max − 2500). Lastly, the microstructures of the core samples were investigated using a field-emission scanning electron microscope (TESCAN MIRA3) combined with casting thin-section analysis. SEM imaging was conducted at magnifications of 200, 500, 1000, 2000, 5000, 8000, and 10,000 × , and EDS surface scans were applied to specific micro-areas.

3.2 Lithology

3.2.1 Minerals

XRD results show that the Lower He 8 tight sandstones in the Sulige area consist primarily of quartz (45.9–81.6%) and clay minerals (16.5–47.4%), classified as lithic sandstones and lithic quartz sandstones (Table 3.1, Fig. 3.3). Clay minerals are dominated by illite (3.9–27.3%), kaolinite, and chlorite (10.9–31.8%), while smectite was not detected. In contrast to conventional sandstones, feldspar minerals are scarce in the tight sandstones of the study area. Only traces of K-feldspar (0.2%) and plagioclase (0.2%) were identified in *R2*, along with a minor amount of analcime (0.4%) in *R3*. Likewise, carbonate minerals are present in low proportions, mainly as calcite, but with notable spatial heterogeneity. Calcite content reaches up to 18.0% in *R3*, while in the remaining four wells it is less than 1.0%. Hematite also occurs widely within the tight sandstones, but its abundance is always below 1.0%. Overall, the mineral types in the tight sandstones of the study area are generally consistent with those in conventional sandstones of the Cretaceous formation (Table 3.2). However, the tight sandstones show significantly higher clay mineral contents, especially kaolinite. In contrast, conventional Cretaceous sandstones contain much higher feldspar contents and are typically categorized as feldspathic sandstones or lithic feldspathic sandstones (Fig. 3.3).

Geochemical element testing of cores indicates pronounced heterogeneity in the mineral distribution of tight sandstones in the Sulige region. Elemental contents vary

Table 3.1 The mineral composition of tight sandstones (*wt%*)

Well	Depth (m)	Quartz	Plagioclase	K-feldspar	Calcite	Illite	Kaolinite + Chlorite	Hematite	Analcite
R1	3509	74.6	/	/	0.4	4.2	19.8	1.0	/
R2	3247	65.6	0.2	0.2	0.4	9.8	22.9	0.9	/
R3	2965	45.9	/	/	18.0	3.9	31.8	/	0.4
R4	2800	51.5	/	/	0.4	27.3	20.1	0.7	/
R5	2835	81.6	/	/	1.0	5.6	10.9	0.9	/

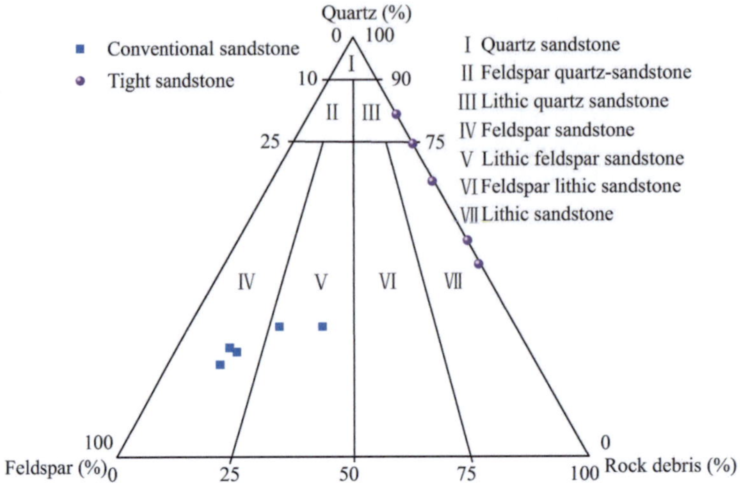

Fig. 3.3 Ternary classification diagram of conventional and tight sandstones

Table 3.2 The mineral composition of Cretaceous sandstones (*wt%*, I/S: Illite smectite interlayers, modified from Su et al. 2011)

Formation	Depth (m)	Quartz	Plagioclase	K-feldspar	Calcite	Illite	Chlorite	I/S	Analcite
Huanhe	362	24	16	18	5	14	9	/	10
Luohe	526	22	27	40	3	3	3	2	
Luohe	487	26	34	29	2	3	2	4	/
Luohe	370	31	31	19	3	5	4	7	/
Luohe	781	25	41	21	3	4	3	3	/
Luohe	685	31	13	25	21	3	/	4	3

not only between wells but also within different parts of the same well, confirming the strong heterogeneity in mineral distribution within tight sandstones. The major detected elements in reservoir cores are listed in Table 3.3. Among the five wells, the most notable differences in elemental content are observed in Ca and K. Due to the presence of calcite, Well *R3* exhibits the highest Ca concentration while having the lowest K content. The Fe content in cores is mainly governed by chlorite and hematite. Since hematite is scarce and shows little variation among the five wells, the Fe distribution is largely controlled by chlorite abundance and distribution. Owing to the relatively simple mineralogy of tight sandstones, trace elements (e.g., Zr, Sr, Rb, Mn) occur in low concentrations, while most other trace elements are below detection thresholds or are detected in only a few samples.

Table 3.3 Elemental composition of various core intervals in tight sandstones (mg/kg)

Well	Site	Zr	Sr	Rb	Zn	Fe	Mn	V	Ca	K
R1	1	54.11	19.39	8.55	14.10	12,293	118.56	30.75	662.95	2881.2
	2	132.60	147.72	35.28	17.23	11,831	94.13	/	5928.7	9572.3
R2	1	195.59	100.10	88.82	27.33	21,697	228.49	119.25	2146.9	40,274
	2	292.04	94.16	92.07	31.00	23,033	252.47	145.73	1287.1	33,821
	3	146.68	44.95	37.74	24.10	17,224	208.89	46.86	938.57	13,057
	4	84.10	48.23	34.10	19.65	12,966	163.88	43.67	938.64	15,210
	5	188.39	44.80	38.03	32.29	26,108	377.42	51.20	1358.6	15,115
	6	106.97	48.28	40.79	26.70	21,638	290.35	35.38	2296.7	18,996
	7	88.41	89.28	37.33	27.00	25,554	314.34	40.10	3412.7	16,249
R3	1	168.69	27.09	16.49	10.42	81,87.9	262.54	/	23,054	6087.3
	2	127.77	57.83	37.20	14.27	12,589	71.75	36.55	1926.3	17,221
	3	90.30	53.00	41.87	12.50	8899	71.82	29.49	1365.2	18,951
R4	1	101.81	51.86	19.56	18.62	11,970	617.89	/	41,584	7669.0
	2	169.55	34.19	15.81	19.96	15,826	534.49	/	24,591	5525.6
	3	99.38	39.91	9.83	8.84	6020.3	477.79	26.13	35,879	2902.7
R5	1	190.12	70.94	54.20	27.24	17,147	192.32	58.73	8834.4	21,045
	2	156.73	88.08	81.19	28.07	17,230	83.58	68.55	1348.6	33,769
	3	76.00	62.36	47.46	25.41	8841.3	95.50	42.68	3881.1	21,240

3.2.2 *Microstructure*

Consistent with mineralogical analyses, thin sections and SEM observations identify quartz, kaolinite, illite, and chlorite as the dominant minerals, with occasional pyrite and calcite, but no notable feldspar, montmorillonite, or illite–smectite interlayer minerals were observed. Quartz constitutes the framework of tight sandstones, whereas clay minerals are widely developed, occupying grain surfaces, intergranular, and intragranular pores (Figs. 3.4 and 3.5). XRD testing shows that chlorite and kaolinite have nearly identical characteristic peaks (d = 0.715–0.718 nm), complicating their distinction. Thin section and extensive SEM observations demonstrate that kaolinite is more prevalent than chlorite within tight sandstones. Kaolinite is extensively developed in the tight sandstone reservoirs, characterized by pseudo-hexagonal flakes and plate-like crystals of high crystallinity, with particle sizes ranging from 5–15 μm and thicknesses of 0.5–2.5 μm. Aggregates occur in vermicular, book-like, and stacked forms, displaying irregular grain sizes and disordered accumulation (Fig. 3.5b, c). Kaolinite development is one of the most prominent characteristics of the He 8 Member tight sandstones in the Sulige Gas Field.

Owing to the weakly acidic reservoir environment, kaolinite crystals are well developed, whereas illite and chlorite show poor crystallinity, often occurring as detrital matrix. Chlorite partly forms coatings on quartz grains (Fig. 3.5d), whereas

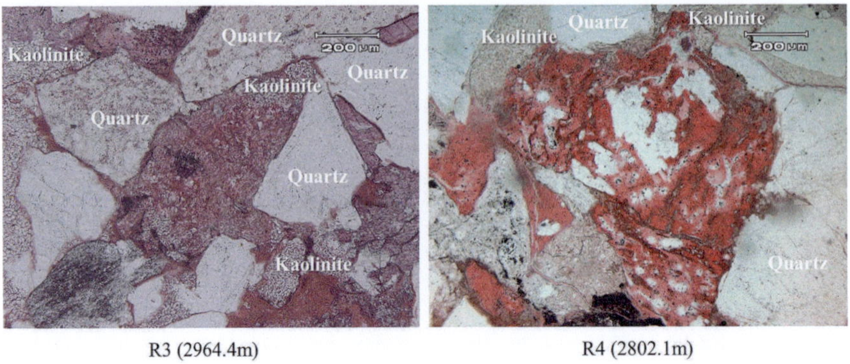

Fig. 3.4 Thin-section images of tight sandstones in the study area

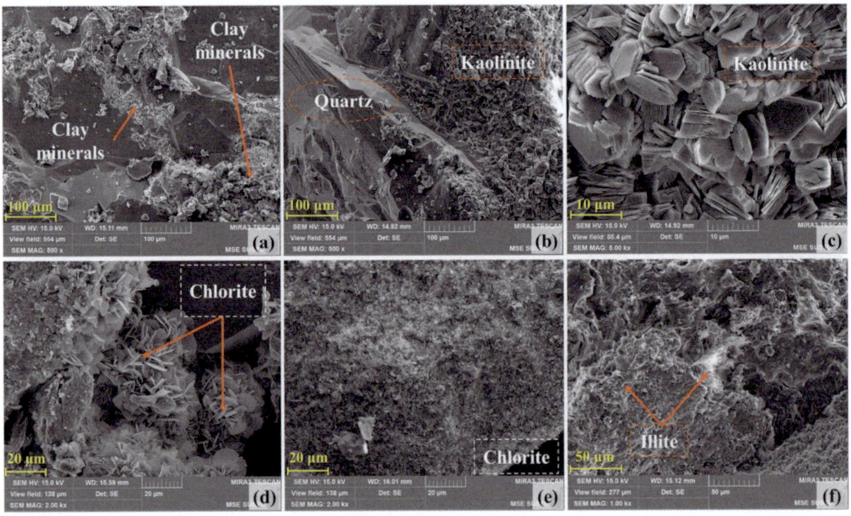

Fig. 3.5 Microscopic features of clay minerals in tight sandstones

globular chlorite is commonly developed in pore spaces (Fig. 3.5e). Illite mainly occurs as platy aggregates on grain surfaces, with minor fibrous forms filling intergranular or intragranular micropores (Fig. 3.5f). A small amount of organic matter typically occurs in tight sandstones, with TOC contents generally below 0.5% (Belyadi et al. 2019). In this study, SEM imaging identified small amounts of sheet- and band-shaped microbial-algal organic remnants embedded in clay minerals (Fig. 3.6a, b). Calcite mainly occurs as a cementing phase in intergranular pores, without distinct crystal morphology under SEM, though step-like and conchoidal fractures are occasionally observed (Fig. 3.6c). Pyrite often coexists with organic matter, being abundant in coal seams and shales, but is typically present in low amounts in tight sandstones. Influenced by reservoir heterogeneity, XRD analyses

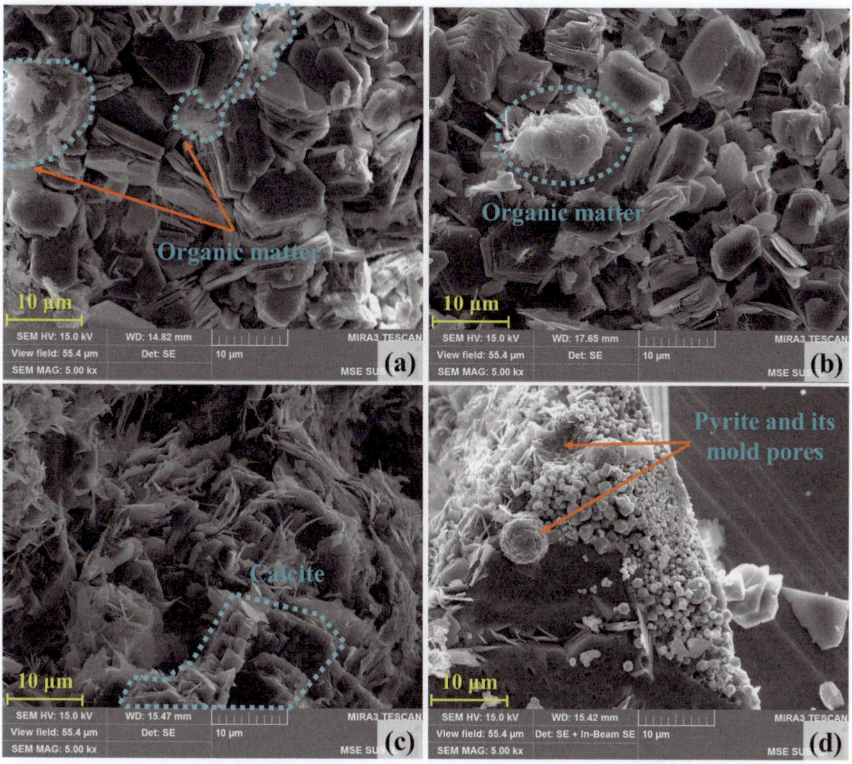

Fig. 3.6 Micro-morphologies of organic matter (**a**, **b**), calcite (**c**), and pyrite (**d**) in tight sandstones

of cores from the five wells failed to detect pyrite, but SEM imaging identified small amounts of pyrite occurring as framboidal and strawberry-shaped aggregates within pore spaces (Fig. 3.6d).

3.3 Reservoir Property

3.3.1 Pore

Pore types and structures in tight reservoirs have a direct impact on natural gas accumulation and migration, which ultimately controls reservoir quality. Thin section casting and SEM observations show that tight sandstones in the study area contain intergranular pores, intragranular dissolution pores, intergranular fractures, and minor pyrite moldic pores (Figs. 3.4, 3.5, 3.6 and 3.7). Quartz grains in the tight

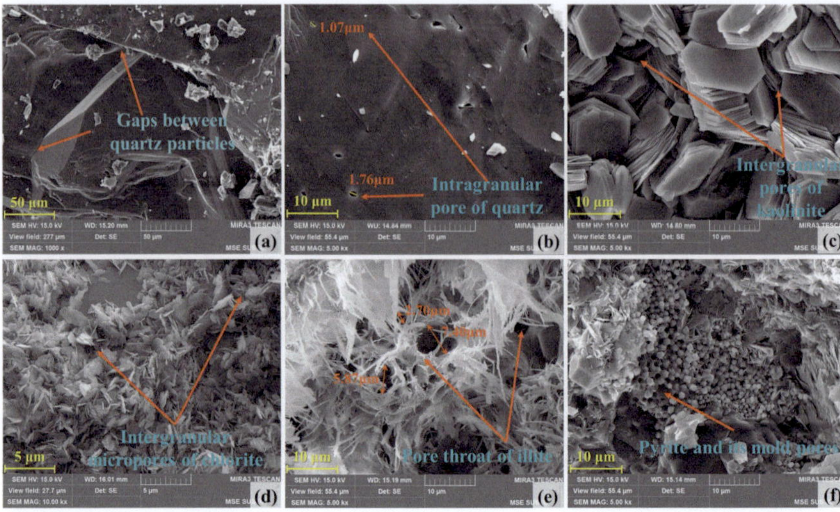

Fig. 3.7 Pores of tight sandstones

sandstone reservoirs mostly exhibit linear contacts, which give rise to intergranular microcracks (Fig. 3.7a). Moreover, quartz crystals contain limited intragranular vesicles (Fig. 3.7b) and dissolution pores (Fig. 3.4). These vesicles vary from nanometer-sized pores in linear arrays to micrometer-sized pores in clustered distributions. For clay minerals, kaolinite and chlorite generate abundant intercrystalline micropores (Fig. 3.7c, d), whereas fibrous illite commonly produces bundle-shaped pore throats with pore sizes varying from 2.7 to 7.4 μm (Fig. 3.7e). These intercrystalline pores formed by clay minerals constitute crucial reservoir space for gas accumulation. Nodular pyrite crystals form moldic pores with diameters of 1–1.5 μm, while strawberry-shaped aggregates display pore sizes mainly between 3 and 7 μm (Figs. 3.6d, 3.7f). Overall, the pore system of the Lower He 8 Member reservoir is primarily composed of secondary dissolution pores and intercrystalline pores dominated by kaolinite.

3.3.2 *Porosity and Permeability*

The porosity of tight sandstone reservoirs is typically less than 10% (Zou 2017). In the present study, the average porosity of tight sandstone cores from five gas wells falls within the range of 7.7–12.6%. Reservoir tight sandstones are primarily composed of lithic quartz sandstone and lithic sandstone, where quartz typically accounts for more than 50%. The existence of rigid quartz particles helps preserve part of the primary pores in tight reservoirs, thereby producing a notable correlation between porosity and permeability in reservoir cores (Fig. 3.8). Across the wells, the average

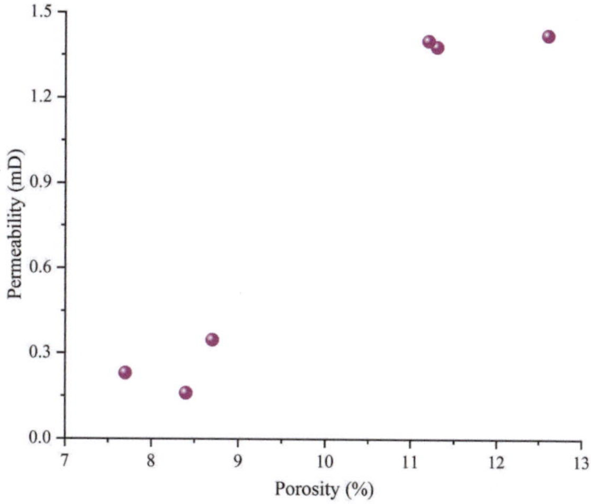

Fig. 3.8 The relationship between porosity and permeability of tight sandstones

permeability of tight sandstone reservoirs varies between 0.16 and 1.42 mD. Further investigations reveal that cores with elevated clay mineral content typically possess low porosity and permeability in the study area.

Taking Well *R1* as an example, porosity and permeability remain positively correlated along the vertical section (Fig. 3.9). As shown in Fig. 3.9, significant hydrocarbon content occurs only when permeability is greater than 0.1 mD. Therefore, compared with porosity, permeability serves as a more direct determinant of reservoir quality. Across the Sulige region, reservoir storage space is mainly provided by macropores with throat radii exceeding 0.1 μm and micropores with throat radii below 0.1 μm. Approximately 61% of tight sandstone reservoirs have porosity less than 8%, and about 83% exhibit permeability below 0.5 mD, indicating that the reservoirs are generally characterized by low porosity and low permeability.

3.4 Tight Gas

In the Sulige Gas Field, tight sandstone gas is characterized by high methane (CH_4) content, typically exceeding 93%, followed by ethane (3.5%), propane (0.6%), and higher carbon-number alkanes, with the total hydrocarbon content generally averaging over 98% (Table 3.2). The natural gas methane coefficient (CH_4/total hydrocarbons × 100%) exceeds 95% on average, showing characteristics of predominantly dry gas with subordinate wet gas. The non-hydrocarbon gases are primarily CO_2 and N_2, with CO_2 usually below 3% and N_2 below 1%. Trace amounts of H_2 and He are also detected among the non-hydrocarbon gases, though their contents are extremely low, usually under 0.1%. H_2S concentrations are typically low in the reservoirs, classifying them as slightly or low-sulfur gas reservoirs. The average density of tight sandstone gas is approximately 0.715 g/L, with a relative density of 0.55.

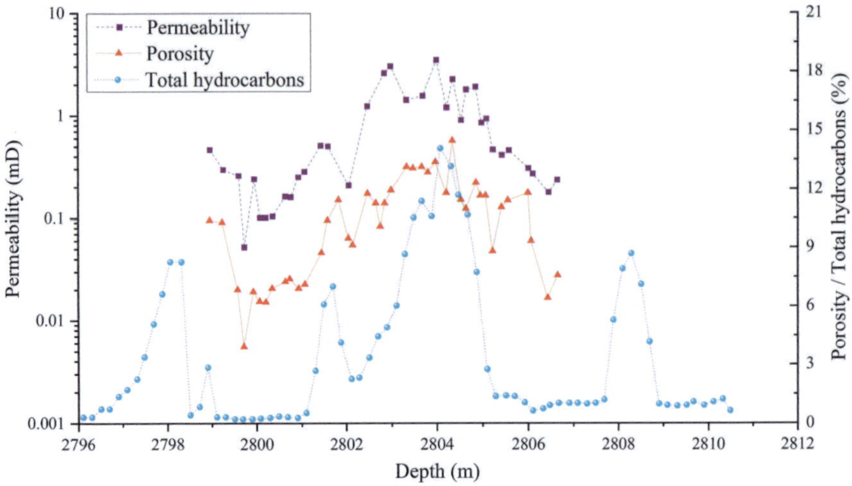

Fig. 3.9 Physical property characteristics of the tight sandstones in Well *R1*

Table 3.4 presents the gas composition and density features of tight sandstone gas from selected producing wells.

Table 3.4 The composition and density features of tight sandstone gas from selected producing wells

No.	CH_4 (%)	C_2H_6 (%)	C_3H_8 (%)	C_{4+} (%)	H_2 (%)	He (%)	N_2 (%)	CO_2 (%)	Density (g/L)
1	93.720	3.474	0.620	0.421	0.017	0.033	0.586	1.017	/
2	94.636	2.200	0.310	0.121	0.008	0.030	0.201	2.464	/
3	94.525	3.375	0.493	0.141	0.047	0.049	0.779	0.470	/
4	94.480	3.191	0.535	0.300	0.027	0.024	0.759	0.680	0.703
5	92.810	4.569	0.854	0.420	0.016	0.020	0.664	0.652	0.717
6	93.050	4.273	0.720	0.358	0.000	0.037	0.718	0.842	0.715
7	92.802	4.711	0.832	0.336	0.016	0.061	0.547	0.695	0.717
8	93.323	4.159	0.645	0.377	0.032	0.000	0.603	0.862	0.714
9	93.552	2.073	0.417	0.169	0.017	0.051	0.572	3.149	0.722
10	88.316	6.965	1.878	0.978	0.014	0.033	0.599	1.217	0.767

3.5 Diagenesis

Following burial, sandstone deposits experience diagenetic processes driven by variations in temperature, pressure, and fluid properties. These involve water–rock interactions and mineral transformations, gradually densifying the sandstone reservoirs and forming gas-rich tight sandstones. Understanding the diagenetic processes involved in tight sandstone reservoir formation and their influence on petrophysical properties provides insights into pore evolution and the potential impacts on hydraulic fracturing (Lai et al. 2018; Li et al. 2023). Therefore, it is necessary to investigate the diagenetic processes of tight sandstones. Diagenesis comprises multiple water–rock interaction processes, including dissolution and cementation, which alter pore structure and distribution, ultimately affecting reservoir permeability.

3.5.1 Compaction

Compaction is a key process in the densification of gas-bearing sandstones in the Sulige Gas Field, mainly manifested as mechanical compaction under geological forces. The compaction features are characterized by grains forming tight point and linear contacts, closer packing, and blurred grain boundaries. During compaction, quartz and feldspar particles tend to fracture, whereas clay minerals exhibit bending deformation (Luo et al. 2019). In the study area, quartz grains in tight sandstones predominantly display linear contacts, forming numerous intergranular fractures. Some grains show point contacts and local fracturing, suggesting strong compaction during diagenesis (Fig. 3.10a, b). SEM images further reveal that kaolinite aggregates have been partly compacted, leading to the destruction of intercrystalline pores (Fig. 3.10c). Similarly, some illite occurs in compact platy aggregates, while chlorite minerals undergo marked bending deformation after compaction (Fig. 3.10d). Compaction serves as the dominant cause of reservoir densification in the study area, leading to a significant reduction in primary intergranular porosity. Hence, compaction hinders the development of pores in tight sandstone reservoirs.

3.5.2 Cementation

Cementation refers to the process by which minerals (cement) precipitate from pore fluids and consolidate the surrounding loose sediments (Lai et al. 2018; Luo et al. 2019). Petrographic thin sections and SEM images reveal that clay mineral cementation is common in the lower part of He 8 Member reservoirs in the Sulige Gas Field, while carbonate (mainly calcite) and authigenic quartz cementation are observed in some samples (Fig. 3.11). Clay mineral cementation in the study area primarily includes kaolinite, illite, and chlorite cementation. Kaolinite cementation is

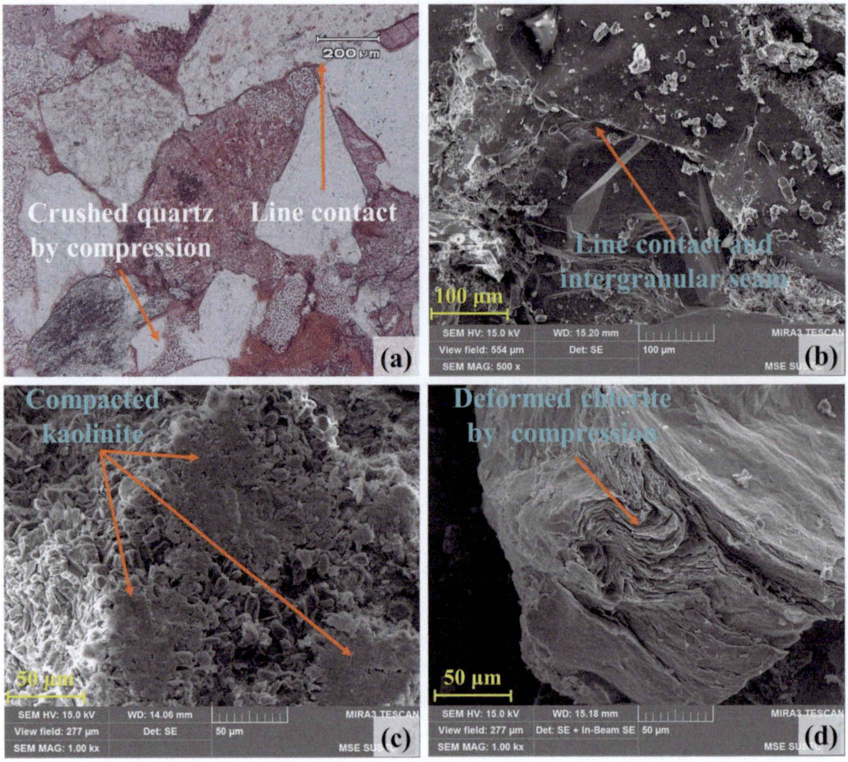

Fig. 3.10 Types and features of compaction in tight sandstone

widespread, characterized by book-like kaolinite crystals precipitating in intergranular pores of quartz grains. These kaolinite cements, formed by pore fluid precipitation (Fig. 3.11), exhibit well-developed crystallinity, loose packing, and significant micropore preservation. Another type of kaolinite cement is less abundant, filling secondary dissolution pores of feldspars, formed by in-situ precipitation of kaolinite after feldspar dissolution. Because of the restricted crystal growth space, kaolinite crystals accumulated in situ, with weaker microporosity development. According to XRD results, calcite is the principal carbonate mineral in the tight sandstone, with carbonate cementation mainly represented by calcite infilling intergranular pores (Fig. 3.11a). Siliceous cement is mainly composed of authigenic quartz, with SEM images showing quartz crystallites and columnar single quartz crystals (Fig. 3.11b, c). Feldspar dissolution, clay mineral transformation, and pressure solution of quartz grains release SiO_2, providing abundant material for siliceous cementation. Cementation generally reduces primary porosity in reservoirs, but some cements enhance the rock framework's compressive strength, mitigating compaction-induced damage to intergranular pores. Therefore, cementation is generally considered to play a role in partially preserving porosity.

Fig. 3.11 Types and features of cementation in tight sandstone

3.5.3 Dissolution

Dissolution denotes the process whereby minerals are leached and dissolved during diagenesis due to changes in the diagenetic environment (Yang et al. 2017). Based on mineral composition, thin section, and SEM data, the tight sandstones underwent significant feldspar dissolution and clay mineral-induced replacement of quartz during diagenesis. SEM images show widespread quartz crystal dissolution by kaolinite and chlorite (Fig. 3.12a, b). Feldspar content is very low in the Lower He 8 Member tight sandstones, whereas kaolinite is the most abundant clay mineral. During thermal evolution, organic matter produced water-soluble organic acids before hydrocarbon generation, which aggressively dissolved and corroded diverse minerals (Lai et al. 2018). Feldspar minerals readily undergo dissolution in acidic fluids, generating kaolinite, SiO_2, and K^+ (Eq. 3.1). In-situ replacement of feldspars by kaolinite results in abundant kaolinite but scarce feldspar in the tight sandstones. The released SiO_2 crystallizes into authigenic quartz or attaches to quartz grains as overgrowths, whereas K^+ is carried away by fluids or participates in the kaolinite-to-illite transformation (Eq. 3.2) to form illite. SEM images show kaolinite crystals with fibrous terminations and decreasing thickness, suggesting their transition from kaolinite to illite (Fig. 3.12c, d).

$$\begin{aligned} &2\,KAlSi_3O_8(K\text{—}feldspar) + 2H^+ + H_2O \\ &= Al_2Si_2O_5(OH)_4(Kaolinite) + 4\,SiO_2 + 2K^+ \end{aligned} \tag{3.1}$$

$$\begin{aligned} &3\,Al_2Si_2O_5(OH)_4(Kaolinate) + 2K^+ \\ &= 2KAl_3Si_3O_{10}(OH)_2(Illite) + 2H^+ + 3H_2O \end{aligned} \tag{3.2}$$

In addition to feldspar dissolution, carbonate minerals and certain lithic cements are likewise susceptible to dissolution when exposed to acidic fluids. Dolomite is nearly absent in the Lower He 8 Member, and although the calcite content exceeds that of feldspar minerals, it remains significantly lower than in typical sandstones. Additionally, SEM images reveal limited step-shaped fracture surfaces in calcite,

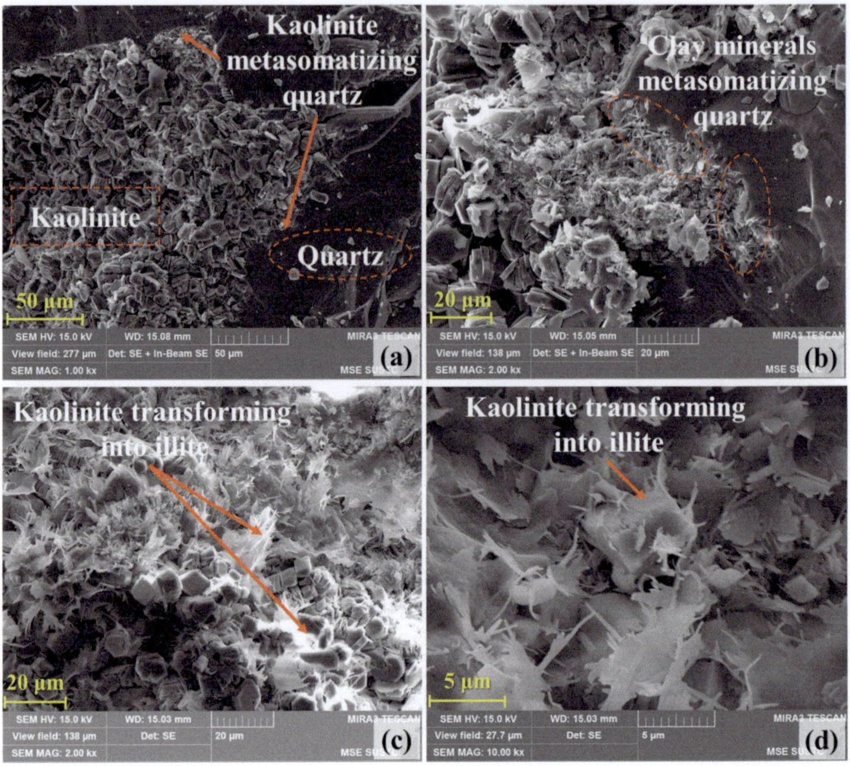

Fig. 3.12 Types and features of dissolution in tight sandstone

suggesting carbonate minerals underwent partial dissolution in the course of diagenesis. Furthermore, dissolution features in some quartz crystals led to the development of intragranular pores within quartz grains (Fig. 3.4).

Dissolution plays the dominant role in secondary porosity development of clastic rocks. Feldspar dissolution creates dissolution pores and simultaneously generates numerous kaolinite intercrystalline pores during kaolinite formation (Fig. 3.4). Consequently, dissolution tends to enhance pore development in tight reservoirs throughout diagenesis. Large-scale dissolution usually occurs during the stage when organic matter undergoes thermal evolution to generate organic acid fluids, creating abundant secondary pores while hydrocarbons charge into them. Conversely, when acidic formation water coexisted with CO_2, the relatively weak acidity of the reservoir limited the scope and strength of dissolution (Luo et al. 2019).

Overall, the Lower He 8 Member tight sandstones in the Sulige area are highly diagenetically altered. Compaction results in grain-supported fabrics with predominantly linear contacts, leading to severe destruction of primary intergranular pores. Cementation is dominated by clay mineral cements, while carbonate and siliceous

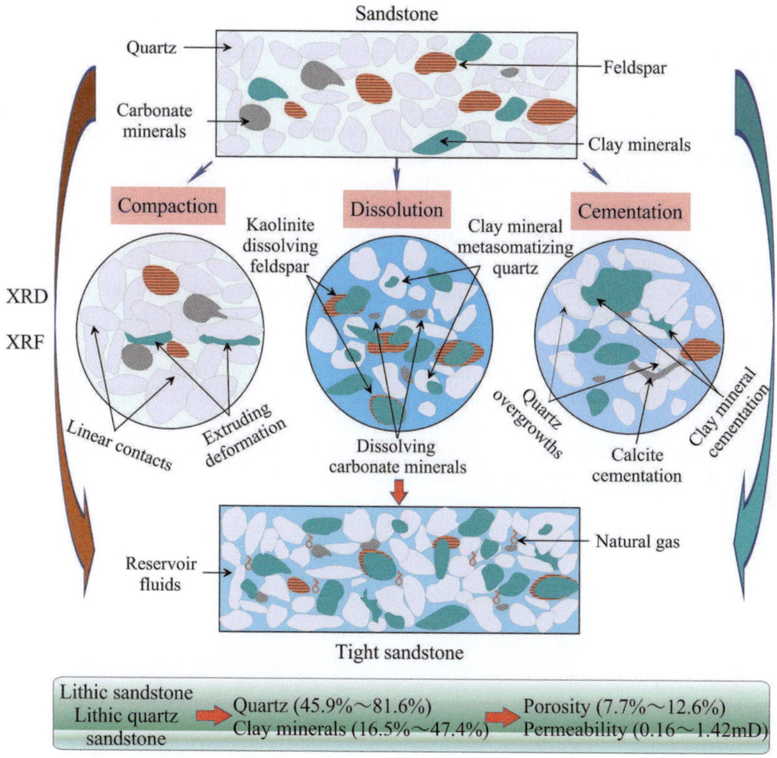

Fig. 3.13 Schematic diagram of tight sandstone diagenetic processes

cements are comparatively less developed. Compaction and cementation are unfavorable for pore development in tight sandstones, with clay mineral cements commonly filling intergranular pores and impairing reservoir pore structure. With the modification of compaction and cementation, large primary intergranular pores are filled and destroyed, further densifying the reservoir and producing numerous clay mineral intercrystalline pores. Dissolution tends to increase porosity, as diagenetic dissolution and replacement almost completely remove feldspar, leaving kaolinite formed through in-situ replacement. Later dissolution events contributed to porosity recovery, leading to the formation of numerous intragranular and intergranular pores (Fig. 3.13).

References

Belyadi H, Fathi E, Belyad F (2019) Hydraulic fracturing in unconventional reservoirs: theories, operations, and economic analysis. Elsevier, Cambridge. https://doi.org/10.1016/C2018-0-01643-6

Lai J, Wang G, Wang S et al (2018) Review of diagenetic facies in tight sandstones: diagenesis, diagenetic minerals, and prediction via well logs. Earth Sci Rev 185:234–258. https://doi.org/10.1016/j.earscirev.2018.06.009

Li P, He X, Zhou C et al (2023) Mineral compositions and microstructural characteristics of the tight sandstone reservoir in the Sulige area and their potential influence on hydraulic fracturing. Hydrogeol Eng Geol 50(3):1–11 (In Chinese with English abstract)

Luo L, Meng W, Gluyas J et al (2019) Diagenetic characteristics, evolution, controlling factors of diagenetic system and their impacts on reservoir quality in tight deltaic sandstones: typical example from the Xujiahe formation in Western Sichuan Foreland Basin, SW China. Mar Pet Geol 103:231–254. https://doi.org/10.1016/j.marpetgeo.2019.02.012

Su X, Wu C, Dong W, Hou G (2011) Strontium isotope evolution mechanism of the cretaceous groundwater in Ordos desert plateau. J Chengdu Univ Technol (Sci Technol Ed) 38(3):348–358 (in Chinese with English abstract)

Yang L, Xu T, Liu K et al (2017) Fluid-rock interactions during continuous diagenesis of sandstone reservoirs and their effects on reservoir porosity. Sedimentology 64(5):1303–1321. https://doi.org/10.1111/sed.12354

Zou C (2017) Unconventional petroleum geology, 2nd ed. Elsevier. https://www.sciencedirect.com/book/9780128122341/unconventional-petroleum-geology

Chapter 4
Hydraulic Fracturing Experiment and Hydrochemistry of Flowback Water

Hydraulic fracturing flowback fluids are generated in considerable quantities with poor quality and complex constituents, carrying high environmental pollution risks. These fluids complicate water treatment and can induce soil and groundwater pollution, among other issues. What is the formation mechanism of flowback fluids in hydraulic fracturing, and how does their hydrochemistry evolve? Analyzing the hydrochemical and water quality features of flowback fluids not only uncovers geochemical processes during fracturing but also clarifies solute origins and water quality evolution, offering critical guidance for treatment strategies. In this chapter, flowback fluid samples were obtained from six tight sandstone wells after hydraulic fracturing with different well types (horizontal and directional) and fracturing fluid systems (guar, variable-viscosity slickwater, slickwater, and biopolymer) to identify hydrochemical and water quality characteristics and to clarify the sources and transformation patterns of solutes. Meanwhile, a comparative analysis was conducted on the hydrochemistry and water quality of flowback fluids from shale gas and coalbed methane wells.

4.1 Field Hydraulic Fracturing Design

In this study, six tight sandstone gas wells (Well *S1*–Well *S6*) were fractured in the Sulige Gas Field. The fracturing fluids used included four systems: guar for Well *S1* and Well *S2*, variable-viscosity slickwater for Well *S3* and Well *S4*, slickwater for Well *S5*, and biopolymer for Well *S6*. Among the six wells, Well *S3* was a directional (vertical) well, while the other five were horizontal wells. The spatial distribution of these six wells is shown in Fig. 4.1.

In the field trials, both pre-fracturing make-up freshwater and post-fracturing flowback fluid samples from various time intervals were collected and subjected to hydrochemical and isotopic analyses. At the onset of flowback, intensive sampling

X. He and P. Li, *Geochemical Processes in Tight Gas Hydraulic Fracturing Stimulation*, Environmental Earth Sciences, https://doi.org/10.1007/978-3-032-15180-3_4

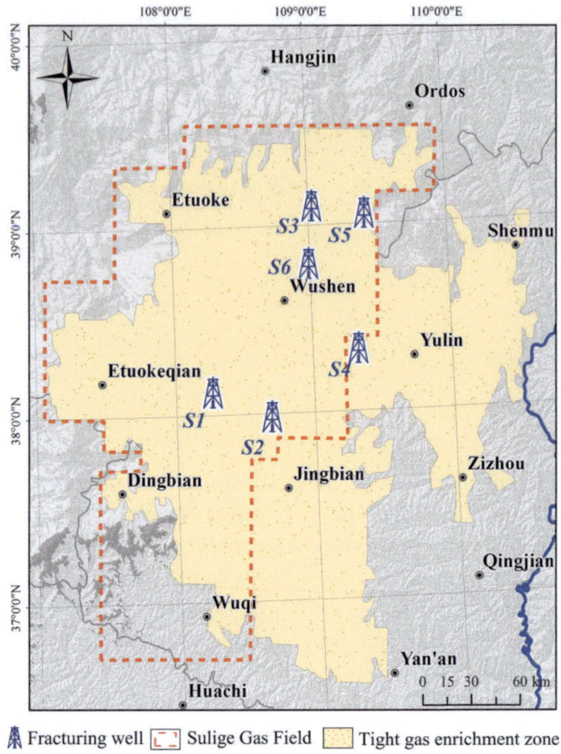

Fig. 4.1 Location of hydraulic fracturing wells

was carried out, collecting one sample every 2 h. With increasing flowback duration, sampling intervals were progressively lengthened. Well *S1* served as a production well with a 33-day sampling period, while the other five wells were test wells with a 7-day sampling period. Ultimately, 65, 16, 22, 34, 31, and 14 flowback samples were collected from Well *S1* through Well *S6*, respectively. Furthermore, a fracturing fluid make-up water sample was collected from each well. The make-up freshwater for fracturing fluids was sourced from shallow groundwater near the gas wells. In total, 182 flowback fluid samples and 6 fracturing fluid make-up water samples were collected from the field trials (Fig. 4.2).

4.1.1 Hydraulic Fracturing Wells

The basic information of the six fractured wells is presented in Table 4.1. The *S1* horizontal well reaches a depth of 3455 m, with a horizontal interval of 700 m. Logging curves (gamma ray) indicate that the sandstone penetration rate of the horizontal well is 100%, with 599 m of gas-bearing sandstone, corresponding to an effective reservoir encounter rate of 85.6%. The gas-bearing interval of Well *S1* lies in the Lower

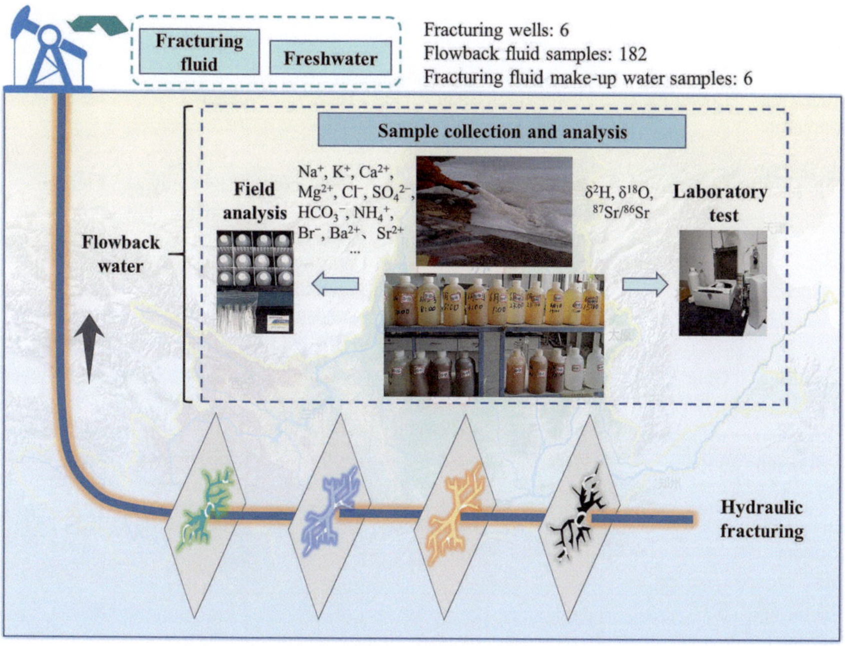

Fig. 4.2 Schematic diagram of field hydraulic fracturing design

He 8 Member, consisting primarily of light gray fine sandstone, grayish-white fine sandstone, and argillaceous sandstone, with the gas-bearing sections dominated by light gray and grayish-white fine sandstone. In the Well *S1* block, hydrocarbons account for 98.1%, with methane at 93.6% and ethane at 3.52%. CO_2 accounts for 1.05% of the gas, while H_2S ranges from 0.8 to 10.69 mg/m^3, with an average of 7.01 mg/m^3. The original formation pressure of Well *S1* is approximately 30.7 MPa, with formation temperatures between 104 and 109 °C. On March 8, 2021, the *S1* horizontal well was fractured in six stages, injecting a total of 2501 m^3 of guar gum fracturing fluid, with 1790 m^3 recovered, yielding a flowback rate of 71.57%.

The *S2* horizontal well reaches 3210 m in depth, with a horizontal interval of 1050 m. Logging results indicate a total sandstone thickness of 916 m, including 813 m of gas-bearing sandstone, corresponding to an effective reservoir penetration rate of 77.4%. The gas-bearing intervals in Well *S2* are located in the lower He 8 and Shan 1 formations, dominated by gray-white fine sandstone. Being in close proximity to Well *S1*, Well *S2* has a comparable reservoir gas composition, with methane averaging 94%, CO_2 at about 1%, and H_2S concentrations ranging between 0.8 and 19.53 mg/m^3, averaging 6.33 mg/m^3. Based on nearby wells and regional data, the original reservoir pressure of Well *S2* is estimated at around 29.5 MPa, with formation temperatures between 95 and 100 °C. On March 14, 2021, Well *S2* was fractured in eight stages using guar-based fracturing fluid, totaling 4755.5 m^3 injected and 1175 m^3 flowback volumes.

Table 4.1 Information of six hydraulic fracturing wells

Well	*S1*	*S2*	*S3*	*S4*	*S5*	*S6*
Fracturing formation	Lower He 8	Lower He 8, Shan 1	Shan 1, He 8, He 6, He 4	He 8	Shan 2	He 8
Lithology	Light gray, grayish-white fine sandstone	Gray-white fine sandstone	Light-gray fine sandstone	Light-gray fine sandstones, gray-white medium sandstone	Light gray, grayish-white fine sandstone	Light-gray fine sandstone, gray-white fine-medium sandstone
Fracturing fluid	Guar	Guar	Variable-viscosity slickwater	Variable-viscosity slickwater	Slickwater	Biopolymer
Well depth (m)	3455	3210	3067	2909	2901	3096
horizontal segment (m)	700	1050	Directional well	1500	1000	1334
Fracturing sections	6	8	5	11	7	8
Reservoir temperature (°C)	104–109	95–100	88.80–90.36	82	88–90	90–95
Reservoir pressure (MPa)	30.7	29.5	26.10–26.90	23.9–25.1	26	27.7
Fracturing fluid volume (m^3)	2501	4755.5	1257.8	4841	3547.9	3647.7
Flowback fluid volume (m^3)	1790	1175	637.3	/	985	345

Well *S3*, situated in the eastern Sulige Gas Field, is a directional well with a total drilling depth of 3067 m. Gas-bearing strata are extensively developed in the Shihezi and Shanxi formations, where the reservoirs are primarily composed of light-gray fine sandstones. In this block, H_2S concentrations range from 0 to 4.25 mg/m^3, while CO_2 accounts for 0.08–4.67%. The pressure coefficient is 0.94, with formation pressure around 29.0 MPa. The He 8 Member has a formation temperature near 89 °C, while the Shanxi Formation reaches about 90 °C. This well underwent five-stage fracturing on May 27, 2021, injecting 1257.8 m^3 of variable-viscosity slickwater, yielding a cumulative flowback of 637.3 m^3.

Well *S4* has a depth of 2909 m and a horizontal section length of 1500 m. Logging indicates a sandstone interval of 1318 m, with a sandstone encounter rate of 87.9%. Of these, 824 m are gas-bearing sandstones, giving an effective reservoir encounter rate of 54.9%. The gas-bearing interval belongs to the He 8 Member, with lithology dominated by light gray fine sandstones and gray-white medium sandstones. The block exhibits methane contents of 91.35–94.10%, CO_2 contents of 0.64–2.8%, and H_2S levels of 0.33–17.76 mg/m^3. Based on neighboring well pressure data and the average pressure coefficient of the block, the formation pressure of this well is estimated at 23.9–25.1 MPa, with a formation temperature of approximately 82 °C. This well was fractured on September 14, 2020, with 11 horizontal stages and a total injection of 4841 m^3 of variable-viscosity slickwater.

Well *S5*, located in the eastern Sulige Gas Field, has a depth of 2901 m, a horizontal section length of 1000 m, and penetrates 830 m of effective reservoir. The gas-bearing interval occurs in the Shan 2 Member of the Shanxi Formation, with lithology dominated by light-gray and gray-white fine sandstones. In this block, methane averages 92.79%, H_2S ranges from 0 to 4.0 mg/m^3, and CO_2 averages 0.66%. The formation pressure is about 26.0 MPa, with temperatures ranging from 88 to 90 °C. It was fractured on October 23, 2020, in seven stages, using slickwater. The total injected volume was 3547.9 m^3, with a cumulative flowback of 985 m^3, giving a flowback rate of 27.76%.

Well *S6*, with a depth of 3096 m and a horizontal length of 1334 m, encountered 412 m of effective reservoir, giving a gas-bearing encounter rate of 30.8%. The fractured interval is located in the He 8 Member, with lithology consisting of light-gray fine sandstones and gray-white medium- to fine-grained sandstones. Formation pressure is approximately 27.7 MPa, and temperature ranges between 90 and 95 °C. On September 27, 2020, Well *S6* underwent eight-stage fracturing using a biopolymer-based fracturing fluid, with a total injection of 3647.7 m^3 and a cumulative flowback of 345 m^3, resulting in a flowback rate of 9.46%.

4.1.2 Sampling and Analysis

After the hydraulic fracturing tests, the collected samples were sealed, stored at low temperature, and immediately sent to the laboratory for physicochemical analysis. The main analytical parameters of flowback and make-up water included pH, K^+, Na^+, Mg^{2+}, Ca^{2+}, Cl^-, SO_4^{2-}, HCO_3^-, CO_3^{2-}, NO_3^-, NH_4^+, Br^-, Ba^{2+}, Sr^{2+}, Fe, Al, boron (B), Total dissolved solids (TDS), Permanganate index (COD_{Mn}), Total hardness (TH), δ^2H, $\delta^{18}O$, and $^{87}Sr/^{86}Sr$. Among them, only eight major ions (K^+, Na^+, Mg^{2+}, Ca^{2+}, Cl^-, SO_4^{2-}, HCO_3^-, CO_3^{2-}) were analyzed for *S4*, *S5*, and *S6* wells, while *S1* and *S2* wells additionally measured NO_2^-, and Well *S3* was tested for Cr^{6+}. In addition, two make-up freshwater samples (from Well *S1* and Well *S3*) and 38 flowback water samples (16 from Well *S1* and 22 from Well *S3*) were analyzed for $^{87}Sr/^{86}Sr$. Major ion analyses were conducted at the Shaanxi Institute of Engineering

Prospecting Environmental Testing Co., Ltd. for analysis, based on Chinese national water quality standards (Jia et al. 2021; Shi et al. 2025).

Stable isotopes δ^2H and $\delta^{18}O$ were measured using thermal conversion elemental analysis-isotope ratio mass spectrometry (TC/EA-IRMS) on a MAT 253 instrument, with precisions of ± 1.0‰ and ± 0.2‰, respectively. The δ^2H and $\delta^{18}O$ values are expressed in per mil deviations relative to Vienna Standard Mean Ocean Water (VSMOW):

$$\delta(‰) = \frac{R_{\text{sample}} - R_{\text{standard}}}{R_{\text{sample}}} \times 1000 \tag{4.1}$$

$^{87}Sr/^{86}Sr$ ratios were determined using a multi-collector inductively coupled plasma mass spectrometer (Neptune Plus MC-ICP-MS). NBS 987 served as the strontium isotope standard, yielding a repeatability of 0.710249 ± 0.000013 (2σ, n = 13). For convenience, the $^{87}Sr/^{86}Sr$ ratios can be normalized to the global seawater value of 0.709169:

$$\varepsilon_{\text{Sr}} = 10^4\left(\left(^{87}\text{Sr}/^{86}\text{Sr}\right)_{\text{sample}}/\left(^{87}\text{Sr}/^{86}\text{Sr}\right)_{\text{seawater}} - 1\right) \tag{4.2}$$

4.2 Hydrochemistry of Different Fluids

Produced water and flowback fluid are two key fluid types in oil and gas development. Flowback fluid refers to the liquid returned to the surface after hydraulic fracturing, primarily consisting of injected fracturing fluids mixed with some formation water. Produced water typically denotes formation water that is continuously brought to the surface during hydrocarbon production, sometimes still carrying traces of fracturing fluid. Thus, a clear temporal distinction between flowback fluid and produced water is lacking, "flowback water" can be used as a general term.

4.2.1 Hydraulic Fracturing Fluids

Fracturing fluid is a general term for the fluids employed in fracturing stimulation of low-permeability reservoirs. Fracturing fluids vary widely in system design and additive composition. In practical operations, based on injection order and function, fracturing fluids are further divided into pad fluids, carrying fluids, and displacement fluids. Consequently, fracturing fluids represent complex mixtures whose exact chemical makeup is challenging to ascertain directly at well sites. After decades of development, a complete fracturing fluid system has been formed, including water-based, oil-based, foam, emulsion, alcohol-based, and acid-based fracturing fluids

(Barati and Liang 2014; Abdelaal et al. 2021; He et al. 2026). Among various fluid systems, water-based fracturing fluids are most widely used, consisting mainly of water and additives, with water typically accounting for more than 95% (Steliga et al. 2015). Variations in source water and additives lead to differences in fracturing fluid chemistry. In this study, a guar-based fracturing fluid was prepared in the laboratory using deionized water, and hydrochemical composition data of fracturing fluids used in low-permeability reservoir stimulation were also collected (Table 4.2). Overall, water-based fracturing fluids prepared with freshwater exhibit chemical compositions largely determined by the make-up water, with additives having little effect on major ion contents (e.g., K^+, Na^+, Mg^{2+}, Ca^{2+}, Cl^-, SO_4^{2-}, HCO_3^-), resulting in low ionic concentrations. Conversely, when produced water is reused as make-up water, the resulting fracturing fluids tend to have elevated concentrations of Cl^-, Fe, and similar ions. The primary variation among fracturing fluids lies in pH. Crosslinked systems (e.g., guar) are usually alkaline, slickwater fluids are near neutral (pH 6–8), while certain reservoirs with high carbonatite employ acidic stimulation fluids with pH values typically less than 3 (Table 4.2).

When the specific composition of the fracturing fluid is unknown, the make-up water used for fracturing fluid preparation can be used to characterize its hydrochemical properties. In this study, make-up freshwater samples from six hydraulic fracturing wells were collected, and their physicochemical parameters were analyzed. The results show that ion concentrations in the make-up freshwater were generally low. Among cations, K^+ ranged from 1.03 to 8.73 mg/L, Na^+ from 8.02 to 66.8 mg/L, Ca^{2+} from 34.1 to 68.1 mg/L, and Mg^{2+} from 9.72 to 23.1 mg/L (Fig. 4.3a). For anions, Cl^- concentrations ranged from 10.0 to 40.0 mg/L, SO_4^{2-} from 24.0 to 62.4 mg/L, and HCO_3^- from 183 to 354 mg/L, while CO_3^{2-} was not detected (Fig. 4.3b). Ca^{2+} was the predominant cation, and HCO_3^- was the predominant anion, categorized as $HCO_3 - Ca$ type water. Since the make-up freshwater was sourced from local groundwater, it reflects the low-ion concentration characteristics of the regional groundwater.

Among other ions, NO_3^- concentrations were 3.19–17.86 mg/L, while NH_4^+ and NO_2^- were not detected. Fe and B were detected only in the make-up freshwater from Well *S1*, at 1.58 mg/L and 0.51 mg/L, respectively. Br^- concentrations ranged from 0.14 to 0.42 mg/L, Sr^{2+} from 0.32 to 0.51 mg/L, Ba^{2+} from 0.32 to 0.51 mg/L, Al ranged from < 0.02–3.08 mg/L, while Mn was not detected. pH values ranged from 7.76 to 8.19, with TDS ranging from 202.5 to 501.0 mg/L. In terms of isotopes, δ^2H values of the make-up freshwater ranged from − 63.24 to − 56.82‰, and $\delta^{18}O$ values from − 9.79 to − 4.78‰. The $^{87}Sr/^{86}Sr$ ratios ranged from 0.712055 to 0.713274, close to the corresponding isotopic values of the local groundwater (He et al. 2022).

The basic chemical compositions of five typical additives in guar gum fracturing fluid were tested (Table 4.3). Results show that the pH regulator is alkaline, causing the fracturing fluid to be alkaline. The pH control primarily contains CO_3^{2-} and Na^+, suggesting sodium carbonate as its main constituent. Similarly, the bactericide and crosslinker displayed elevated pH values, yielding an alkaline solution. The bactericide and surfactant showed low concentrations of inorganic ions, suggesting that their major constituents are organic substances. The clay stabilizer contained

Table 4.2 Chemical compositions of different fracturing fluids (mg/L)

Parameter	Crosslinked fracturing fluid[a]	Slickwater[a]	Guar-based fracturing fluid[b]	Gelled fracturing fluid[c]	Unknown fracturing fluid using in the shale reservoir[d]	Unknown fracturing fluid using in the tight sandstone reservoir[e]	Unknown fracturing fluid using in the shale reservoir[f]	Unknown fracturing fluid using in the tight oil reservoir[g]	Guar-based fracturing fluid (this study)
pH	8.7	6.3	–	–	1.2	2.47	–	–	10.07
Na^+	824	34	1430	145.7	66.67	561	34	34	716
K^+	792	57	64	26.3	2.35	46.4	4	4	5.1
Ca^{2+}	51.3	24	82	18.5	40.48	20.7	26	26	1.0
Mg^{2+}	10.2	8.2	–	3.7	40.60	112	10	10	0.33
Cl^-	914	145	470	360	3263.88	729	51	51	368
SO_4^{2-}	38	21	604	8.96	475.59	0	132.1	132.1	1.0
CO_3^{2-}	24	1.2	–	–	–	–	–	–	458
HCO_3^-	195	243	0	600	215.00	1460	–	–	61
Fe	0.21	1.58	–	–	99.75	0.053	0.3	0.3	< 0.05
Al	0.2	1.5	–	–	0.03	0.096	0.267	0.267	< 0.02
Br^-	–	–	< 10	5.6	0.02	–	< 0.1	< 0.1	< 0.10
Sr^{2+}	0.012	0.011	–	0.06	0.61	–	0.38	0.38	0.08
Ba^{2+}	< 0.01	0.012	–	0.02	0.27	–	0.12	0.12	3.04
B	–	–	–	–	1.23	–	–	–	82.75
Mn	0.028	0.251	–	–	–	–	–	–	< 0.05
NO_3^-	1.4	0.21	–	–	–	–	–	–	–
NH_4^+	5.6	0.56	–	–	–	–	–	–	–

Note: **a**: Steliga (2015); **b**: Birkle (2016); **c**: Rosenblum et al. (2017); **d**: Cui et al. (2020); **e**: Herz-Thyhsen et al. (2020); **f**: Phan et al. (2020); **g**: Tian et al. (2023)

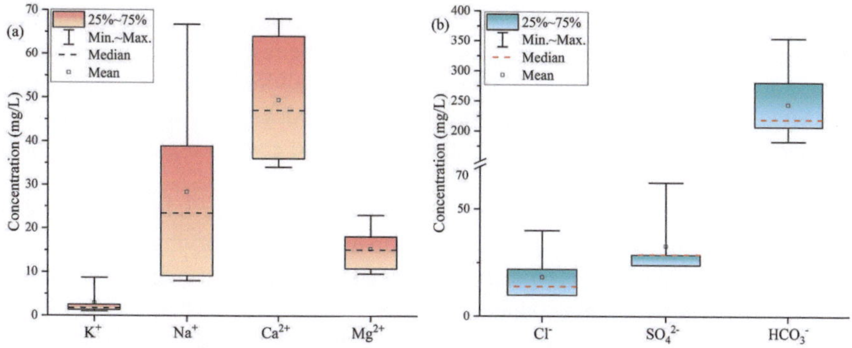

Fig. 4.3 Boxplots of ions in make-up freshwater (**a**: cations; **b**: anions)

Table 4.3 Basic inorganic chemical compositions of fracturing fluid and their additives (mg/L)

Fluids	pH	CO_3^{2-}	HCO_3^-	F^-	Cl^-	SO_4^{2-}	Na^+	K^+	Ca^{2+}	Mg^{2+}
Fracturing fluid	10.07	457.9	0	40.19	367.7	1.00	716	5.1	1.00	0.33
Bactericide	9.72	80.02	0	0.15	2.93	0	1.4	0	0	0.22
Clay stabilizer	6.77	0	16.11	2.56	381.2	1.11	30.2	0	0.10	0.37
Surfactant	7.75	0	19.33	0.24	4.79	0	5.6	0	0.05	0.29
pH control	11.09	243.2	0	0.68	2.82	1.04	272	1.4	0.90	0.43
Crosslinker	9.68	270.2	0	38.57	0.70	0.39	462	3.7	0.55	0.16

relatively high Cl^- concentrations compared with other additives, suggesting the use of quaternary ammonium salt-type stabilizers. Unlike the other four additives, the crosslinker exhibited more complex chemical compositions, with abundant F^- in solution, resulting in elevated F^- concentrations in the fracturing fluid. Overall, inorganic ion contents in fracturing fluids were generally low. Among additives, the crosslinker and pH control had the greatest influence on the fluid's inorganic ion composition and solution environment.

4.2.2 *Formation Water*

During sandstone deposition and diagenetic compaction, the water stored in reservoir pores also undergoes changes. In unconventional hydrocarbon reservoirs, formation water is often associated with oil and gas. The hydrochemical signature of formation water often records evidence of hydrocarbon migration, accumulation, and reservoir formation (Li et al. 2005). However, formation water and its hydrochemistry influence not only the physical properties of reservoirs but also the effectiveness of hydraulic fracturing in tight sandstones. In reservoirs containing large volumes of formation water, gas–water flow during fracturing imposes strong resistance to gas migration, leading to liquid buildup in the wellbore and ultimately reducing gas well output. The Sulige tight sandstone reservoirs exhibit varying degrees of water production, with especially evident occurrence in the western and eastern regions of the field (Yang et al. 2022).

Influenced by sedimentary processes, the primary gas-bearing intervals (He 8 and Shan 1) in the Sulige area are enriched with formation water to varying extents. Based on salinity, formation water is typically classified into three types: conventional brine, diluted formation water, and condensate water. In the Sulige area, formation water generally exhibits high salinity, mostly belonging to conventional brine. The Shanxi Formation has an average salinity of 47.27 g/L, slightly higher than that of the Shihezi Formation with a mean of 43.13 g/L (Zou 2017). In this study, formation water data from several gas wells in the Sulige Gas Field (provided by Changqing Oilfield) were collected, with basic hydrochemical data summarized in Table 4.4. Formation water cations are mainly Na^+ and Ca^{2+}, while Mg^{2+} is comparatively low. Anions are dominated by Cl^-, and $SO_4{}^{2-}$ concentrations are generally low, though some wells exhibit higher values likely associated with elevated H_2S levels in the reservoir. The Na/Cl coefficient (γNa/γCl) serves as an indicator of the extent of formation water evolution and the conditions for hydrocarbon preservation (Li et al. 2005). Generally, formation waters that have undergone cation exchange and intensive water–rock interactions exhibit γNa/γCl values below 0.87 (Yang et al. 2022). In this study, γNa/γCl values of formation water samples ranged from 0.09 to 0.64, suggesting intense alteration, good reservoir sealing, and favorable hydrocarbon preservation. Formation water in the study area is typically light yellow to brick red, while oil-bearing formation water is dark brown. Hydrochemical characteristics of formation water vary significantly across regions and stratigraphic intervals. Additionally, TDS in formation water tends to increase with burial depth (Fig. 4.4), with the Shanxi Formation (deeper strata) showing higher TDS values than the overlying Shihezi Formation.

4.2.3 *Fracturing Flowback Fluid*

Flowback fluid provides an excellent medium for investigating water–rock interactions during hydraulic fracturing. Testing and analyzing physicochemical properties and isotopic characteristics of flowback fluids enables identification of the types, processes, and mechanisms of water–rock interactions during fracturing. For the six fractured wells, flowback fluids contained major cations ranked by average concentration as $Na^+ > Ca^{2+} > K^+ > Mg^{2+}$ (Fig. 4.5). Na^+ varied between 90 and 10,514 mg/L, averaging 2876 mg/L (median 2737 mg/L), with the maximum value observed in Well *S4*. Ca^{2+} ranged from 60.1 to 6212 mg/L, averaging 1268 mg/L (median 1202 mg/L), with the highest value found in Well *S3*. K^+ concentrations ranged from 1.08 to 1489 mg/L, with an average of 308 mg/L (median 158 mg/L). The highest

Table 4.4 The basic hydrochemistry of formation water in the Sulige Gas Field (mg/L)

No	Formation	$Na^+ + K^+$	Ca^{2+}	Mg^{2+}	Cl^-	SO_4^{2-}	HCO_3^-	pH	Color
1	Shihezi	3651	2385	207	9377	1225	290	6.0	Yellow
2	Shihezi	3261	2028	205	7846	1621	292	6.0	Pale yellow
3	Shihezi	762.36	6709.95	419.31	12,824.6	564.44	2246.18	6.06	Brick-red
4	Shihezi	402.11	1307.26	91.93	2281.01	148.17	1800.34	5.84	Tawny
5	Shihezi	319.51	909.4	80.44	1827.68	135.82	1023.8	6.37	Brick-red
6	Shanxi	6696	4309	923	18,634	2430	355	6.0	Pale yellow
7	Shanxi	4848	4732	615	15,900	2025	418	6.0	Pale yellow
8	Shanxi	3278	4225	308	12,253	1215	472	6.0	Brick-red

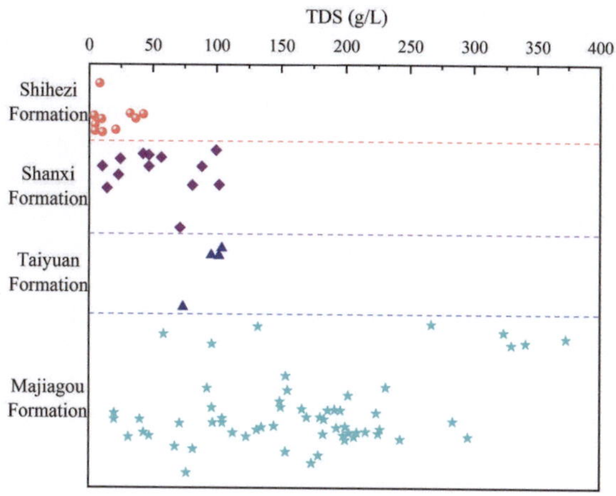

Fig. 4.4 TDS values of different formations in the study area (Modified from Li et al. 2005)

concentration occurred in Well *S1* in the southern study area. Mg^{2+} concentrations spanned 6.08–395 mg/L, with an average of 86.4 mg/L (median 60.8 mg/L). The values in Well *S5* were consistently higher than in other wells.

Cl^- is the dominant anion, and the major anions, in descending order of mean concentration, are $Cl^- > HCO_3^- > SO_4^{2-} > CO_3^{2-}$ (Fig. 4.5). Cl^- varied between 22.0 and 24,000 mg/L, averaging 7130 mg/L (median 6650 mg/L), with the highest level recorded in Well *S3*. HCO_3^- concentrations range from 0 to 3966 mg/L, averaging 984 mg/L (median 894 mg/L), with the maximum in Well *S2*. SO_4^{2-} concentrations are generally low, ranging 24.0–2690 mg/L, with an average of 204 mg/L (median 120 mg/L). A maximum SO_4^{2-} value of 2690 mg/L was found in an early (6 h) flowback sample from Well *S1*, while other samples remained < 1000 mg/L. CO_3^{2-} was absent from all flowback samples, attributed to prevailing pH conditions.

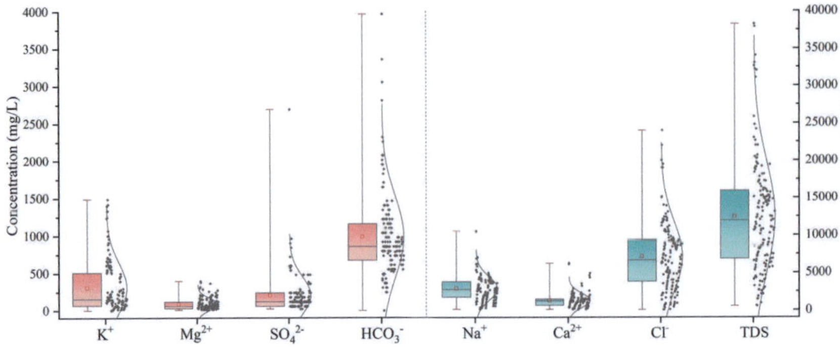

Fig. 4.5 Boxplots of ions in flowback fluid

Flowback fluids are characterized by high salinity, with TDS ranging 0.53–38.3 g/L (average 12.4 g/L, median 11.9 g/L), the maximum in Well *S3*. With high Ca^{2+} levels, TH in flowback fluids ranged from 200 to 16,030 mg/L, with a mean of 3524 mg/L. These results indicate substantial variation in ion concentrations among different wells and fracturing fluid systems. Although different fracturing fluid systems were applied, post-fracturing flowback fluids were weakly acidic overall (pH 2.17–8.91, mean 6.35, median 6.33). A Piper diagram of six wells (Fig. 4.6) revealed most samples plotted in regions D, G, and 7, signifying dominance of Cl^- and Na^+. Flowback fluids shifted from pre-fracturing HCO_3–Ca water to Cl–Na type.

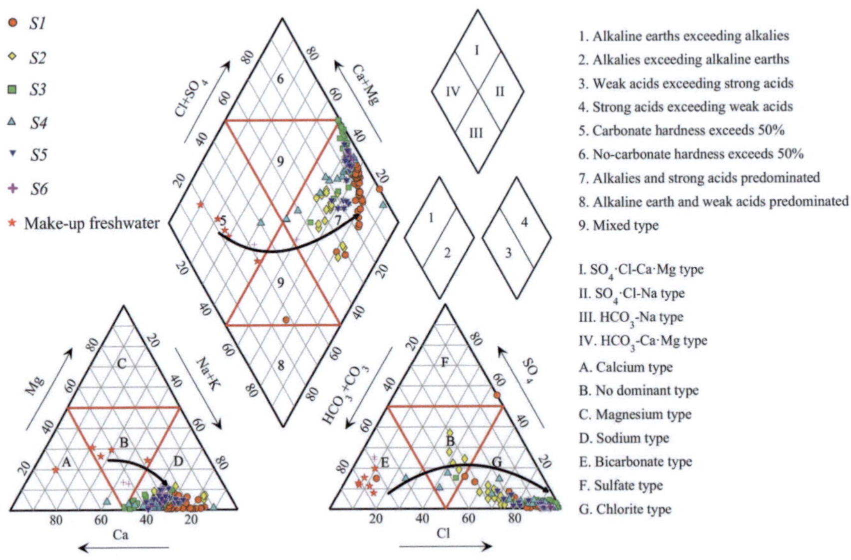

Fig. 4.6 Piper plot illustrating hydrochemical type of different fluids

Comprehensive analyses of flowback fluids from *S1*, *S2*, and *S3* wells (Table 4.5) showed high NH_4^+ levels (2.57–634 mg/L), with the maximum in Well *S1*, but consistently higher averages (174 mg/L) in Well *S3* using variable-viscosity slickwater fracturing fluid. By contrast, NO_3^- and NO_2^- were minor; NO_3^- ranged from < 2.50 to 24.01 mg/L, averaging 5.46 mg/L (median 4.86 mg/L). NO_2^- was detected only once in Well *S1* (17.0 mg/L, with NO_3^- of 24.01 mg/L), and all other samples were < 0.003 mg/L. Br^- concentrations spanned < 0.10–205 mg/L (mean 68.2 mg/L), with markedly higher values in Well *S1* compared to other wells. Flowback fluids from *S1* and *S2* wells (guar-based fracturing fluids) showed high B concentrations (0.52–66.5 and 43.2–57.3 mg/L). In contrast, Well *S3* (variable-viscosity slickwater) showed distinctly lower B (0.74–3.21 mg/L).

Total Fe concentrations were 0.39–204.24 mg/L, with Well *S3* yielding higher averages (81.14 mg/L). Al concentrations ranged 0.10–23.90 mg/L (average 5.62 mg/L, median 5.12 mg/L). Similar to B, Al was higher in guar-based fluid wells than in variable-viscosity slickwater well. Elevated Mn occurred in *S1* and *S2* wells, while Mn (<0.05 mg/L) and Cr^{6+} (<0.005 mg/L) were absent in Well *S3*. Mn concentrations were < 0.05–6.18 mg/L in Well *S1* (mean 2.17 mg/L, median 2.83 mg/L), and < 0.05–27.60 mg/L in Well *S2* (mean 9.60 mg/L, median 6.09 mg/L). Sr^{2+} and Ba^{2+} were also elevated. Sr^{2+} ranged 0.42–363 mg/L (mean 86.6 mg/L, median 89.7 mg/L). Ba^{2+} ranged 2.25–934 mg/L (average 133 mg/L, median 112 mg/L). Well *S3* had higher Sr^{2+} and Ba^{2+} concentrations than *S1* and *S2* wells (Table 4.5).

COD_{Mn}, indicating organic values, was 120–5200 mg/L, with guar-based fluid wells showing much higher values than variable-viscosity slickwater well. COD_{Mn} in Well *S1* varied 560–2880 mg/L (mean 1169 mg/L), and in Well *S2* ranged 520–5200 mg/L (mean 2220 mg/L). COD_{Mn} in Well *S3* was relatively low, between 120 and 304 mg/L. Overall, flowback fluids exhibit high pollution potential.

Stable hydrogen and oxygen isotope values of flowback fluids in *S1* and *S2* wells are close. In Well *S1*, δ^2H varies between − 62.09 and − 21.16‰ (mean − 37.53‰, median − 35.27‰), and $\delta^{18}O$ ranges from − 9.33 to − 3.42‰ (mean − 5.35‰, median − 5.13‰). For Well *S2*, δ^2H is between − 65.26 and − 25.70‰ (mean − 41.87‰, median − 40.02‰), and $\delta^{18}O$ ranges from − 8.46 to − 2.84‰ (mean − 5.29‰, median − 5.18‰). Well *S3*, located in the north, shows δ^2H between − 58.05 and − 25.87‰, and $\delta^{18}O$ from − 2.65 to 6.87‰, values consistently higher than in *S1* and *S2* wells located in the south of the study area. Most flowback samples plot to the lower right of the Global Meteoric Water Line (GMWL), and are generally enriched compared to local groundwater and precipitation (Fig. 4.7). *S1* and *S2* wells exhibit similar isotope features, aligning linearly below and to the right of the GMWL. Conversely, isotopes in Well *S3* deviate markedly from the GMWL (Fig. 4.7). Since short-term water–rock interaction does not markedly change δ^2H and $\delta^{18}O$, the contrasting isotopic features likely result from differing geological settings between north and south.

Table 4.5 Statistics results of flowback fluid hydrochemical parameters (*S1*, *S2*, *S3* wells)

Parameter	Unit	*S1* (Guar-based fluid)				*S2* (Guar-based fluid)				*S3* (Variable-viscosity slickwater)			
		Min	Max	Mean	Median	Min	Max	Mean	Median	Min	Max	Mean	Median
pH		2.17	7.98	6.26	6.14	4.25	8.91	6.01	5.94	5.69	7.92	6.9	6.86
K^+	mg/L	16.9	1489	632	592	19.4	299	146	132	19.6	172	86.0	75.8
Na^+	mg/L	266	6879	3405	3630	543	2855	1831	1759	515	7054	3256	2572
Ca^{2+}	mg/L	60.1	3507	1225.4	1353	70.1	1428	742.71	676	240	6212	2228.64	1478
Mg^{2+}	mg/L	6.08	365	75.6	60.8	18.2	243	95.9	83.6	18.2	273	101	66.9
Cl^-	mg/L	74.0	19,000	8184	8800	400	5600	2975	2750	875	24,000	9821.59	7150
SO_4^{2-}	mg/L	24.0	2690	176	120	120	901	520	540	120	961	335	240
HCO_3^-	mg/L	0	2014	1087	1037	610	3966	2075	2075	244	1220	538	458
Br^-	mg/L	< 0.10	205	93.7	107	0.45	74.5	30.0	24.8	0.63	39.0	20.6	21.1
B	mg/L	0.52	66.5	31.8	31.4	43.2	57.3	48.4	48.4	0.74	3.21	1.66	1.53
NH_4^+	mg/L	3.22	634	18.41	7.88	2.57	19.3	7.95	5.79	138	195	173.45	173.5
NO_3^-	mg/L	< 2.50	24.01	6.24	5.83	3.15	11.41	5.62	4.88	< 2.50	5.31	2.78	2.81
Fe	mg/L	1.48	79.33	20.17	13.53	2.95	204.24	63.1	36.53	0.39	178.00	81.14	89.38
Al	mg/L	0.36	14	5.89	5.29	4.79	23.9	11.64	10.20	0.10	3.45	0.46	0.21
Mn	mg/L	< 0.05	6.18	2.17	2.83	< 0.05	27.60	9.60	6.09	< 0.05	< 0.05	/	/
Sr^{2+}	mg/L	0.42	163	78.0	92	1.35	61.7	30.9	26.3	11.9	363	143	110
Ba^{2+}	mg/L	2.25	180	94.3	113	3.12	42.6	27.6	30.1	11.3	934	325	155
TDS	g/L	0.93	33.04	14.18	14.82	21.04	12.48	7.37	6.78	3.01	38.27	16.89	12.57
TH	mg/L	200	9509	3372	3628	250	4504	2250	2001.5	701	16,014	5983	4004
COD_{Mn}	mg/L	560	2880	1169	1040	520	5200	2220	1840	120	304	185	156
δ^2H	‰	− 62.09	− 21.16	− 37.53	− 35.27	− 65.26	− 25.70	− 41.87	− 40.22	− 58.05	− 25.87	− 36.46	− 37.00

(continued)

Table 4.5 (continued)

Parameter	Unit	*S1* (Guar-based fluid)				*S2* (Guar-based fluid)				*S3* (Variable-viscosity slickwater)			
		Min	Max	Mean	Median	Min	Max	Mean	Median	Min	Max	Mean	Median
$\delta^{18}O$	‰	− 9.33	− 3.42	− 5.35	− 5.13	− 8.46	− 2.84	− 5.29	− 5.18	− 2.65	6.87	1.56	− 0.01
$^{87}Sr/^{86}Sr$		0.713226	0.715002	0.714651	0.714807	–	–	–	–	0.714375	0.714599	0.714532	0.714567

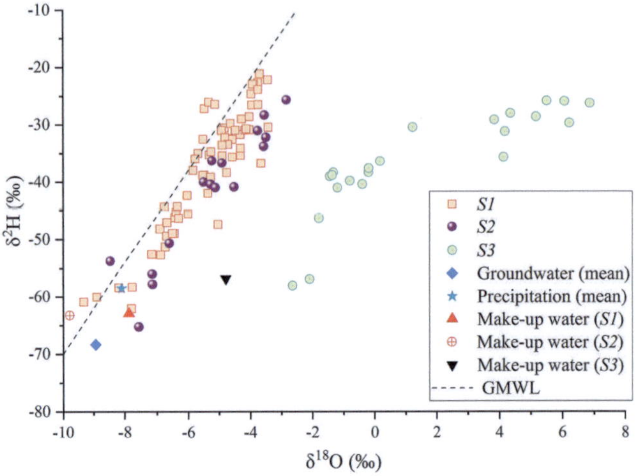

Fig. 4.7 Relationships between δ^2H and $\delta^{18}O$ in different fluids

4.3 Components Variation of the Flowback Fluids

Figure 4.8 illustrates the variation of major ion concentrations in six wells during flowback. In general, major ion concentrations increased with flowback time across the six wells and stabilized after 70 h. Na^+, Ca^{2+}, K^+, and Cl^- show consistent patterns, rising with time then stabilizing, except K^+ in Well *S1*, which declined at the late stage. Within the same well, these four ions often display similar fluctuation patterns. Mg^{2+} also tends to increase with time but fluctuates greatly. Especially, it first increased then decreased in Well *S5*. Among anions, SO_4^{2-} generally decreases with time. SO_4^{2-} also exhibits large fluctuations, similar to Mg^{2+}, because of its low levels. HCO_3^- increased then stabilized in guar-based fluid wells (Well *S1* and Well *S2*); decreased then stabilized in variable-viscosity slickwater wells (Well *S3* and Well *S4*); and showed no clear trend in slickwater (Well *S5*) and biopolymer (Well *S6*) wells. Notably, ion concentrations sometimes increase or decrease abruptly. Taking Cl^- as an example, in Well *S1* (70–238 h) it fluctuated dramatically, from 10,400 mg/L (138 h) to 19,000 mg/L 4 h later. In Well *S3*, Cl^- was 19,500 mg/L at 103 h, but fell sharply to 8800 mg/L 4 h later. Similarly, in Well *S4* over 108 h, Cl^- rose gradually but spiked at 64 h from 5600 to 20,000 mg/L, then dropped to 6400 mg/L after 4 h.

The variation pattern of Br^- is similar to that of Cl^-, but it exhibits much greater fluctuations. Both Sr^{2+} and Ba^{2+} exhibit almost the same trend as Cl^-, characterized by a rise in concentration with flowback time, followed by stabilization (Fig. 4.9). The concentrations of Fe, Mn, and Al are relatively low, with considerable fluctuations during the entire flowback process, showing no clear increasing or decreasing trends. NH_4^+ and NO_3^- generally exhibit a decreasing-then-increasing trend, though their concentrations fluctuate considerably throughout flowback. B displays distinct variation patterns in wells with different fracturing fluid systems. In the guar-based

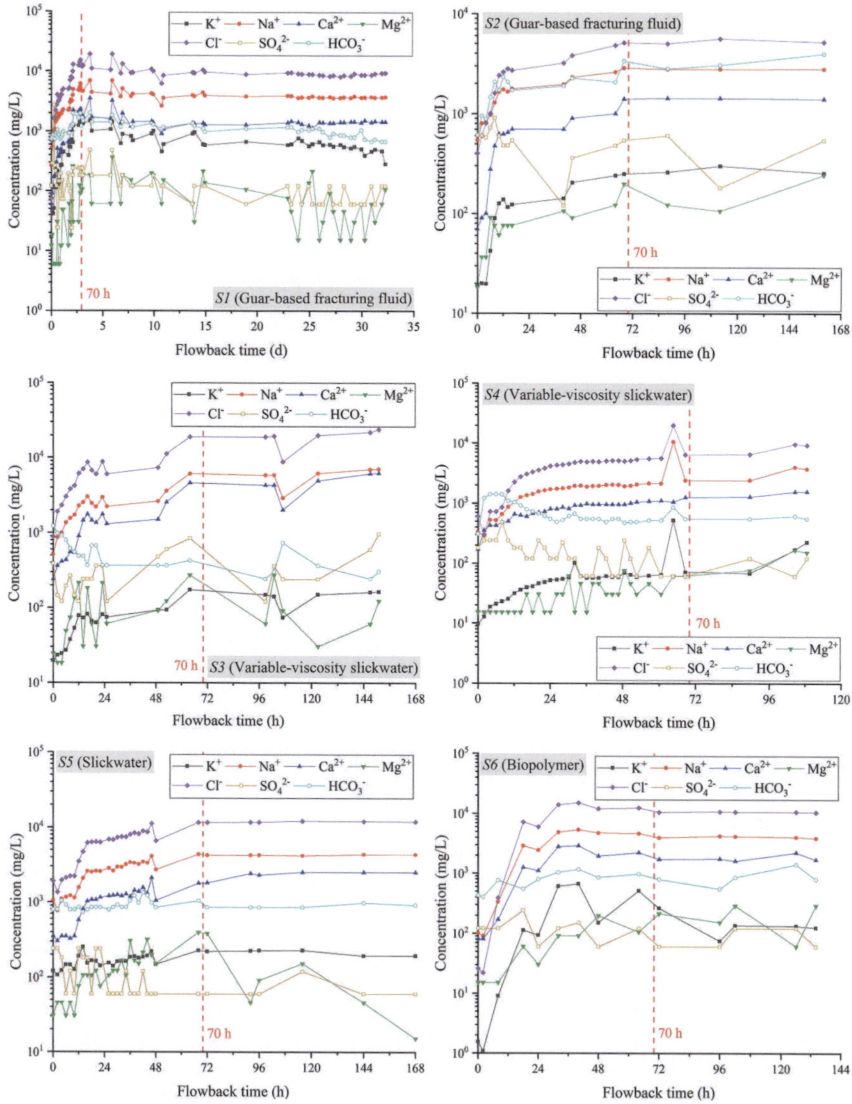

Fig. 4.8 Evolution patterns of major ion concentrations in flowback fluids across six fracturing wells

fracturing fluid system (*S1* and *S2* wells), the use of boron-containing crosslinkers leads to high B concentrations in flowback fluids, which gradually decrease over time (Fig. 4.10). In contrast, in the variable-viscosity slickwater system, where no boron additives are used, B concentrations are much lower but increase with flowback time.

COD_{Mn} follows a similar trend to B. In the early flowback of *S1* and *S2* wells, incompletely degraded guar gum colloids contribute to elevated COD_{Mn} levels, which

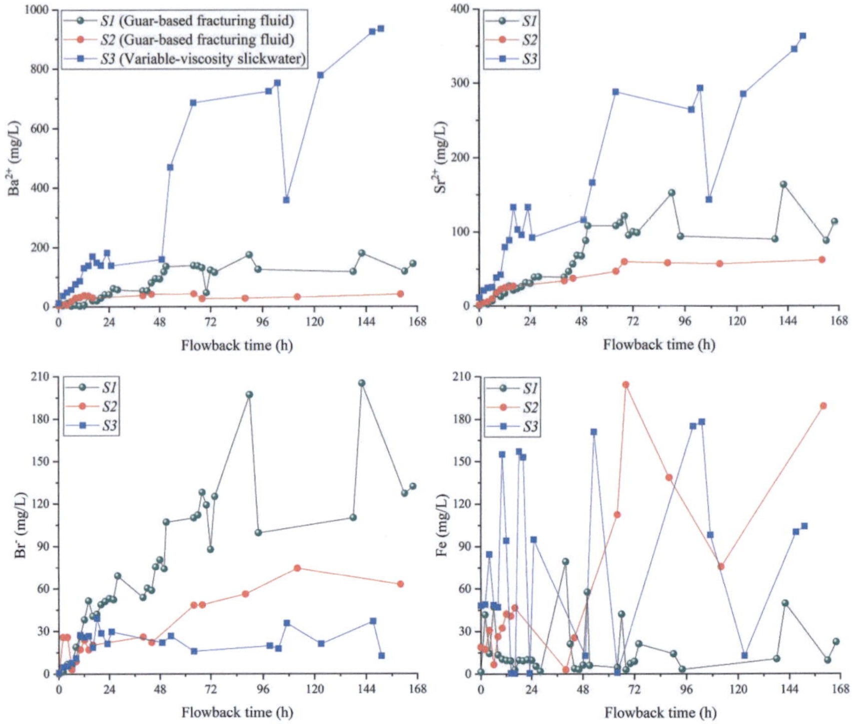

Fig. 4.9 Concentration evolution of Br^-, Sr^{2+}, Ba^{2+}, and Fe

decrease as fracturing fluid components diminish (Fig. 4.11). This trend suggests that in the initial stage, flowback fluid composition is dominated by injected fracturing fluid. In the variable-viscosity slickwater system, COD_{Mn} levels are significantly lower than in the guar system, showing a moderate increasing trend during flowback. pH exhibits distinct variation trends across flowback fluids from different fracturing fluid systems. In *S1* and *S2* wells (guar system), pH first increases and then decreases, whereas in Well *S3* (variable-viscosity slickwater system), pH shows a fluctuating decreasing trend (Fig. 4.11).

Both δ^2H and $\delta^{18}O$ increase over time (Fig. 4.12), but δ^2H exhibits more pronounced fluctuations than $\delta^{18}O$ during flowback. In the late flowback stage of Well *S1*, $\delta^{18}O$ values continue to rise rather than stabilizing like other conservative indicators. This may be attributed to reservoir temperature recovery and strong evaporation effects, causing partial isotopic enrichment of $\delta^{18}O$. Similarly, $^{87}Sr/^{86}Sr$ values grow with time, at a faster rate than δ^2H and $\delta^{18}O$ (Fig. 4.13). In *S1* and *S3* wells, $^{87}Sr/^{86}Sr$ values increase rapidly after flowback, then level off, experience short-term declines, and subsequently rise again until stabilization. Taking Well *S1* as an example, the initial $^{87}Sr/^{86}Sr$ value is 0.713226, close to the make-up water

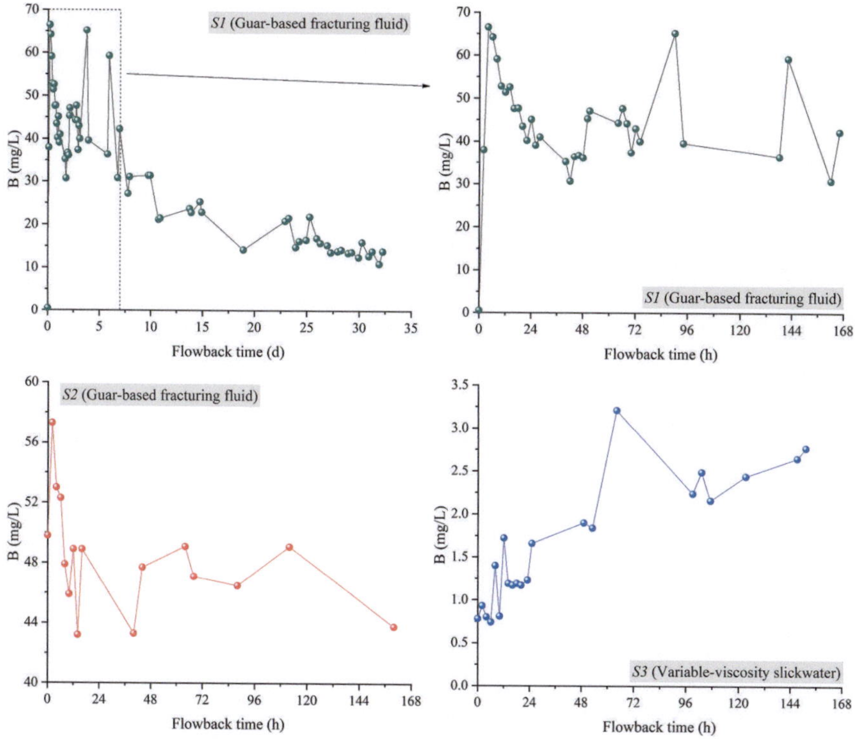

Fig. 4.10 Concentration evolution of boron

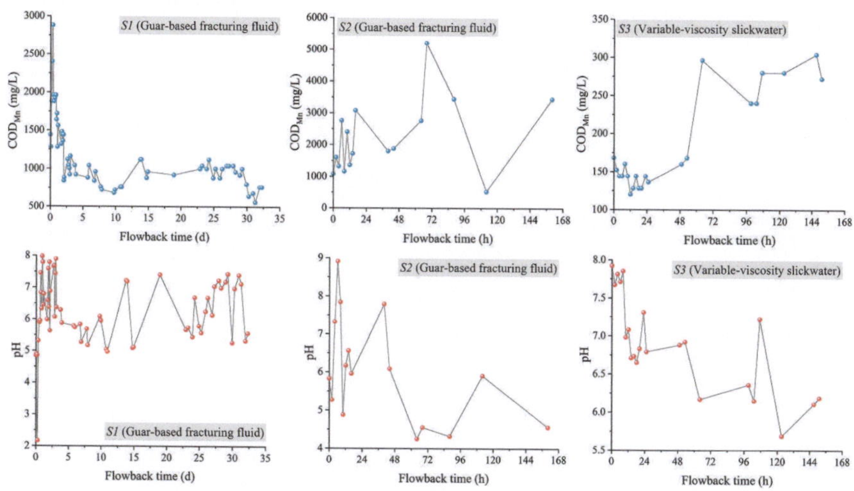

Fig. 4.11 Concentration evolution of COD_{Mn} and pH

ratio (0.712055). Within 4 h, it rises sharply to 0.714804, fluctuates during 22–190 h, yet remains consistently above 0.714. Ultimately, stability is achieved after approximately 190 h.

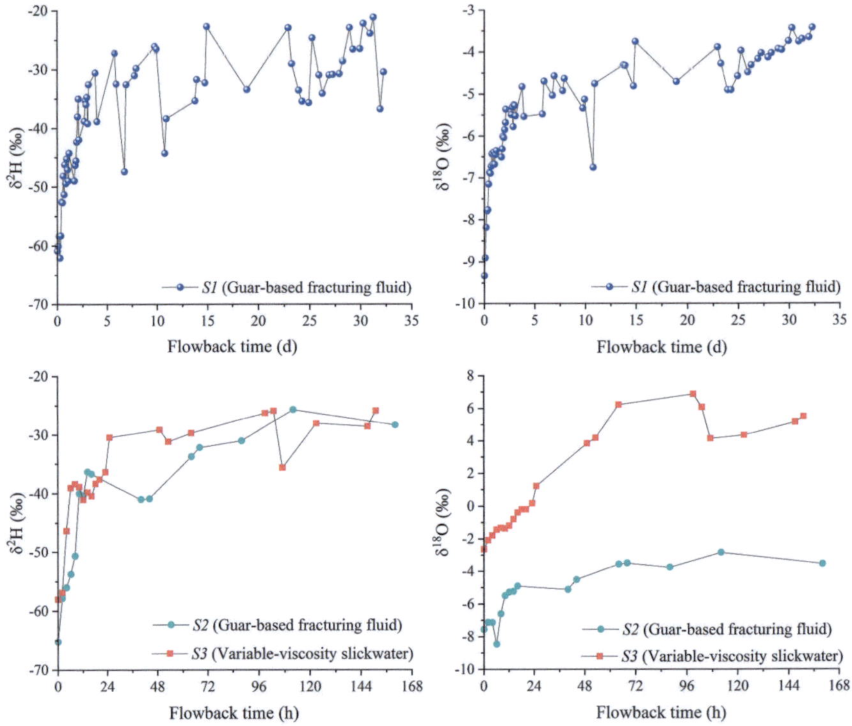

Fig. 4.12 Evolution of δ^2H and $\delta^{18}O$ values

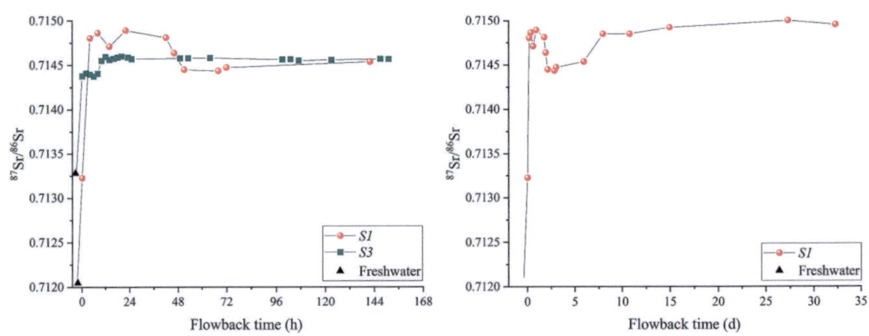

Fig. 4.13 Evolution of $^{87}Sr/^{86}Sr$ values

4.4 Components Sources of the Flowback Fluids

The chemical and isotopic compositions of fracturing fluids resemble low-ion freshwater, whereas flowback fluids exhibit characteristics of high salinity. Thus, the origin of flowback fluid constituents requires further examination. Based on the hydrochemical characteristics and compositional evolution of flowback fluids, the mixing of fracturing fluids with formation waters is likely the key factor influencing the observed changes. In natural waters, Cl^- primarily originates from halite dissolution, and the effects of precipitation or dissolution can generally be neglected, making Cl^- widely regarded as a conservative tracer (Llewellyn et al. 2015; Zhang et al. 2022). In fracturing fluids, Cl^- concentrations are usually below 500 mg/L, whereas in flowback fluids they can reach up to 24 g/L within a week (Table 4.5). However, no significant halite minerals are detected in the tight sandstone reservoirs (Table 3.1), indicating that Cl^- in flowback fluids is not primarily derived from halite dissolution within the sandstone. Like Cl^-, Br^- is also a conservative ion, mainly occurring in geothermal waters, seawater, and brines through evaporation and concentration processes, while being typically scarce in uncontaminated freshwater (Cui et al. 2020). Elevated Br^- in flowback fluids indicates a strong signature of formation brines. Previous work has demonstrated that condensate waters are isotopically lighter in δ^2H and $\delta^{18}O$, while formation brines are isotopically heavier (Rowan et al. 2015). In the present study, δ^2H and $\delta^{18}O$ in flowback fluids increase markedly over time, reflecting the mixing of fracturing fluids with formation waters. A correlation analysis demonstrates that Cl^-, Na^+, Ca^{2+}, Sr^{2+}, and Ba^{2+} in flowback fluids are well correlated (Fig. 4.14), implying that mixing with high-salinity formation brines governs their concentrations and evolution.

Sr and its isotopes are further employed to trace the sources of flowback fluid components. Silicates and carbonates are the most common Sr-bearing minerals, and their dissolution is the main geochemical process controlling the sources and evolution of Sr (Xie et al. 2013; Gao et al. 2020). $^{87}Sr/^{86}Sr$ ratios are typically low in carbonates but relatively high in silicate rocks. Sr derived from silicate dissolution usually has $^{87}Sr/^{86}Sr$ ratios of 0.716–0.720, while Sr from calcite and dolomite dissolution shows ratios of 0.7075–0.7080 and 0.7080–0.7100, respectively (Roy et al. 1999; Zheng et al. 2021). Previous studies indicate that calcite in the Sulige area has $^{87}Sr/^{86}Sr$ values of 0.711058–0.711350 (Rao et al. 2009), thus 0.711 is used as the boundary between silicate- and carbonate-derived Sr. The results show that flowback fluid samples all plot within the silicate dissolution field (Fig. 4.15a), indicating both that flowback fluids originate from formation water and that Sr in formation water is mainly derived from silicate dissolution in tight sandstones.

The relationship between $^{87}Sr/^{86}Sr$ and Mg/Ca can further constrain the source of Sr. Sr derived from carbonate rocks (calcite and dolomite) generally exhibits lower $^{87}Sr/^{86}Sr$ values, with Mg/Ca ratios being lower in calcite aquifers (0.01–0.26) and higher in dolomite aquifers (>0.85). Sr originating from silicate rocks is characterized by higher $^{87}Sr/^{86}Sr$ values and intermediate Mg/Ca ratios (Pu et al. 2012; Sun et al. 2021). Flowback fluid samples fall between the silicate and calcite domains, close

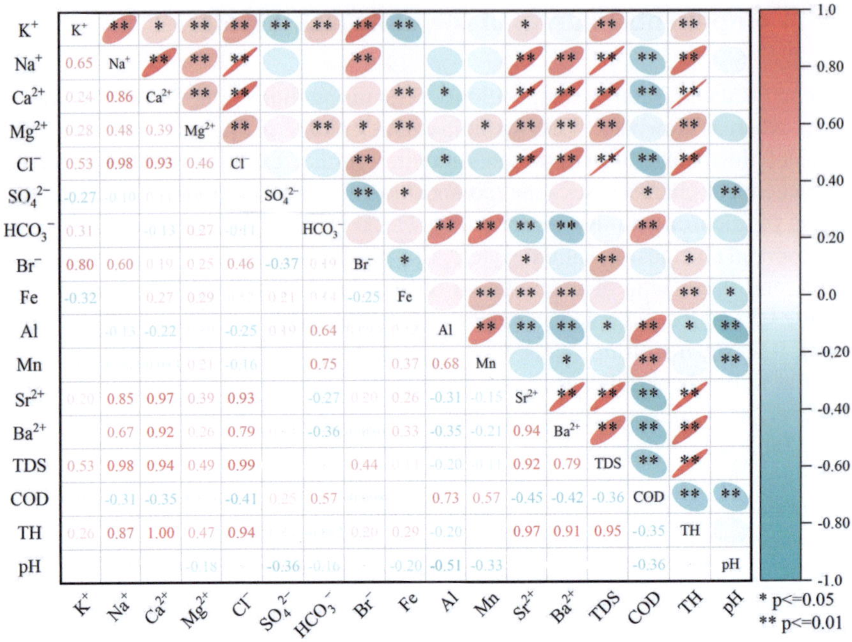

Fig. 4.14 Pearson correlation matrix of chemical components in flowback fluids

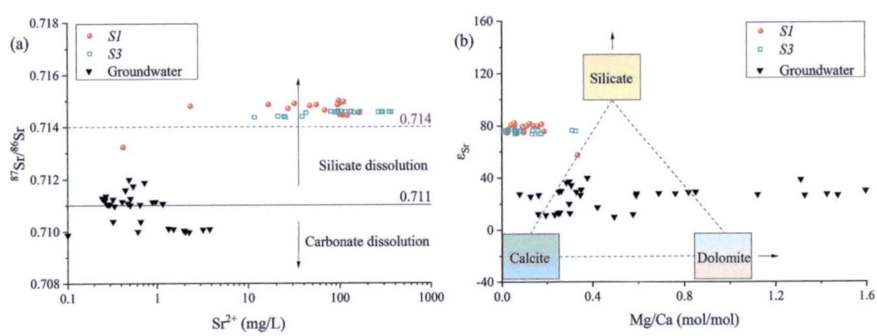

Fig. 4.15 Source analysis of strontium in flowback fluids

to the silicate end-member, supporting the inference that Sr in reservoir formation water mainly originates from silicate dissolution (Fig. 4.15b). Unlike flowback fluid samples, groundwater from the overlying Cretaceous strata plots in the carbonate–silicate transition zone, where weathering of both carbonates and silicates likely governs Sr input. Some samples are strongly influenced by dolomite dissolution, providing additional support for this inference. Unlike the closed system of gas reservoirs, groundwater is significantly recharged by precipitation. The $^{87}Sr/^{86}Sr$ ratio of local rainfall is about 0.710760 (Rao et al. 2015), and recharge from this low-ratio precipitation contributes significantly to the lower values observed in groundwater.

Based on the above hydrochemical and isotopic analyses, it can be concluded that flowback fluids are primarily the product of mixing between fracturing fluids and formation water. As flowback progresses, the proportion of fracturing fluid decreases, while the proportion of formation water increases. However, if Sr^{2+} in flowback fluids were solely derived from the mixing of fracturing fluids and formation water, its concentration should fall within the range defined by the two end-members. Nevertheless, deviations of Sr^{2+} concentrations and $^{87}Sr/^{86}Sr$ ratios from the expected mixing range indicate that additional Sr^{2+} in flowback fluids is contributed by interactions between fracturing fluids and Sr-bearing minerals. In *S1* and *S2* wells using guar-based fracturing fluids, Al, Mn, and other metal concentrations in flowback fluids are markedly higher than those in Well *S3* fractured with slickwater. The presence of elevated Ba^{2+} and depleted $SO_4{}^{2-}$ further confirms water–rock interactions such as mineral dissolution and precipitation during fracturing. Hence, products of water–rock interactions represent an additional key control on the chemistry of flowback fluids. In summary, the mixing between fracturing fluids and formation waters governs the chemical composition of flowback fluids, while water–rock interactions during fracturing exert significant additional effects.

4.5 Water Quality

Hydrochemical and source analyses reveal that hydraulic fracturing of Sulige tight sandstones produces flowback fluids with elevated ionic contents and degraded water quality. Comparison with the "*Groundwater Quality Standard GB/T 14,848 – 2017*" shows that major exceedances include COD_{Mn}, NH_4-N, Cl^-, Na^+, TDS, Ba^{2+}, Fe, TH, B, Al, Mn and pH (Table 4.6). COD_{Mn} and NH_4-N classify all samples as Class V water, while more than 90% of the samples also fall into Class V for Cl^-, Na^+, TDS, Ba^{2+}, Fe, and TH. The high salinity and metal content of flowback fluids pose a serious threat of groundwater contamination, making direct discharge unsuitable.

When compared to the "*Surface Water Environmental Quality Standard GB3838 – 2002*" and "*Irrigation Water Quality Standard GB5084 – 2021*", flowback fluids also greatly exceed permissible limits, with B, NH_4-N, Fe, Ba^{2+}, and COD_{Mn} exceeding Class III values in all tested samples. If discharged into surface waters, flowback fluids would inevitably degrade environmental water quality. High salinity, elevated B and Cl^-, together with slightly acidic pH, also make flowback fluids inappropriate for direct use in irrigation (Table 4.7). Specialized treatment of flowback fluids is thus essential. A comparison across fracturing fluid systems shows that flowback fluids from guar-based fluids are of notably poorer quality than those from slickwater. Flowback fluids from guar systems generally contain higher concentrations of metal ions and COD, resulting in stronger pollution potential.

Table 4.6 Water quality parameters of flowback fluids compared with groundwater standard (unit: mg/L)

Parameter	Number of samples	Class III standard for surface water	Number of exceeding standard	Class IV standard for surface water	Number of exceeding standard	Class V standard for surface water	Number of exceeding standard	Proportion of Class V (%)
pH	103	6.5 ~ 8.5	46	5.5 ~ 6.5 8.5 ~ 9.0	37	< 5.5 > 9.0	20	19.42
Na^+	182	≤ 200	2	≤ 400	4	> 400	176	96.70
Cl^-	182	≤ 250	3	≤ 350	1	> 350	178	97.80
SO_4^{2-}	182	≤ 250	151	≤ 350	4	> 350	27	14.84
B	103	≤ 0.5	0	≤ 2.0	16	> 2.0	87	84.47
NH_4-N	103	≤ 0.5	0	≤ 1.5	0	> 1.5	103	100
NO_3-N	103	≤ 20	103	≤ 30	0	> 30	0	0
NO_2-N	81	≤ 1.0	80	≤ 4.8	0	> 4.8	1	1.23
Fe	103	≤ 0.3	0	≤ 2.0	6	> 2.0	97	94.17
Al	103	≤ 0.2	11	≤ 0.5	10	> 0.5	82	79.61
Mn	103	≤ 0.1	45	≤ 1.5	8	> 1.5	50	48.54
Ba^{2+}	103	≤ 0.7	0	≤ 4.0	4	> 4.0	99	96.12
Cr^{6+}	22	≤ 0.05	22	≤ 0.1	0	> 0.1	0	0
TDS	182	≤ 1000	3	≤ 2000	3	> 2000	176	96.70
TH	103	≤ 450	7	≤ 650	2	> 650	94	91.26
COD_{Mn}	103	≤ 3.0	0	≤ 10	0	> 10	103	100

Table 4.7 Water quality parameters of flowback fluids compared with standards (unit: mg/L)

Parameter	Number of samples	Class III standard for surface water	Number of exceeding standard (proportion)	Threshold values for irrigation (dryland)	Number of exceeding standard (proportion)
pH	103	6 ~ 9	40 (38.83%)	5.5 ~ 8.5	21 (20.39%)
Cl^-	182	≤ 250	179 (98.35%)	≤ 350	178 (97.80%)
SO_4^{2-}	182	≤ 250	31 (24.22%)	/	/
B	103	≤ 0.5	103 (100%)	≤ 1.0	97 (94.17%)
NH_4-N	103	≤ 1.0	103 (100%)	/	/
NO_3-N	103	≤ 10	0	/	/
Fe	103	≤ 0.3	103 (100%)	/	/
Mn	103	≤ 0.1	58 (56.31%)	/	/
Ba^{2+}	103	≤ 0.7	103 (100%)	/	/
Cr^{6+}	22	≤ 0.05	0	≤ 0.1	0
TDS	182	/	/	≤ 1000	179 (98.35%)
COD_{Mn}	103	≤ 6	103 (100%)	/	/

4.6 Flowback Waters from Tight Gas, Shale Gas, and Coalbed Methane

Based on the U.S. Geological Survey National Produced Waters Geochemical Database (Blondes et al. 2023), additional published literature data were compiled to construct a global geochemical dataset encompassing tight gas, shale gas, and coalbed methane systems. Statistical analyses reveal that shale gas and tight gas reservoirs are typically host high-salinity formation brines due to greater buried depths, leading to generally higher salinity and ion concentrations in their flowback waters (Table 4.8). The flowback fluids exhibit elevated concentrations of K^+, Na^+, Ca^{2+}, Cl^-, Br^-, Mn, Ba^{2+}, Sr^{2+}, B, and Li^+, alongside low levels of Mg^{2+}, HCO_3^-, and F^- (Table 4.8), suggesting significant potential for extracting resources such as Li and Br. The maximum TDS of shale gas flowback water can reach 394,600 mg/L, with average and median values of 46,131.03 and 17,100, respectively. Tight gas flowback water exhibits slightly lower salinity, with a maximum TDS of 300,038 mg/L, and mean and median values of 10,558.24 and 8590 mg/L, respectively. Coalbed methane reservoirs are generally shallow and closely connected with groundwater aquifers and atmospheric precipitation. Consequently, coalbed methane flowback water aligns with the GMWL, whereas shale gas and tight gas deviate significantly from it (Fig. 4.16). Shale gas samples mostly plot above the meteoric line, likely reflecting isotopic exchange with H_2S gas, while tight gas samples are concentrated below the line, showing strong water–rock interactions and evaporation effects (Fig. 4.16, Pang et al. 2017). Coalbed methane flowback waters display relatively low salinity and ion contents, with a maximum TDS of 17,700 mg/L, and mean and median values of 3125.78 and 1920 mg/L, respectively, and are relatively enriched in Mg^{2+}, HCO_3^-, and F^- (Table 4.8). Isotopic analyses further indicate that shale and tight gas flowback waters are enriched in heavy isotopes due to reservoir environmental effects. For instance, the $^{87}Sr/^{86}Sr$ ratios in shale gas flowback waters range up to 0.721840 (mean 0.713255, median 0.713814), while those in tight gas reach 0.729700 (mean 0.714178, median 0.711347). Overall, although shale and tight gas flowback waters are geochemically complex and challenging to treat, they exhibit substantial potential for resource recovery and utilization.

Table 4.8 Statistics results of chemical compositions in different fracturing fluids (mg/L)

Parameter	Tight gas				Shale gas				Coalbed methane			
	Max	Mean	Median	Number of samples	Max	Mean	Median	Number of samples	Max	Mean	Median	Number of samples
pH	11.7	7.18	7.2	7077	11.8	7.54	7.6	3062	10.42	8.01	8.1	4062
K^+	15,000	131.39	36.6	5729	4080	372.00	266.9	482	991	16.68	3.91	3903
Na^+	82,722	3384.84	2840	6891	117,000	15561.46	6593	1891	51,700	1177.72	764.42	4143
Ca^{2+}	93,848	412.41	50	7125	162,324	4239.53	152	3349	13,900	1242.28	1027.22	4140
Mg^{2+}	4756	35.49	9	6536	5560	393.72	24	1659	20,100	523.75	87.6	4009
Cl^-	184,520	5627.37	4100	7301	196,000	18,159.33	7927	3264	110,000	420.13	14.4	4163
$SO_4{}^{2-}$	47,000	109.46	22	6425	3580	188.49	150	2755	2407	136.98	8	1210
$HCO_3{}^-$	55,266	1385.23	952	6979	13,880	1133.21	1184	2,810	43,310	1593.03	1150	3886
Sr^{2+}	2206	204.74	59.7	261	15,400	1399.21	388	597	565	6.62	1.882	639
Mn	35	4.72	3.25	107	51.3	4.49	2	411	6	0.12	0.05772	2141
Fe	3491	34.79	13.4	5446	432	37.81	23	1583	4180	11.33	2.6	3047
Br^-	1609.27	134.35	53	259	3340	506.67	211	695	346	5.33	1.8	1192
Ba^{2+}	2196	34.32	8.29	1893	22,400	750.74	21.4	1139	5750	14.94	0.90	1058
B	334	43.19	40	180	155	19.39	11.514	399	9.6	0.45	0.26	1844
F^-	3.2	3.15	3.15	2	29	3.05	1.1	88	20	2.85	1.82	442
Li^+	204	12.91	5.4	219	633.93	50.71	34.95	562	36	0.48	0.2	565
TDS	300,038	10,558.24	8590	7215	394,600	46,131.03	17,100	1581	177,000	3125.78	1920	3822
δ^2H	−15	−45.54	−40.94	208	−15	−37.67	−38	284	−9.91	−75.13	−76.28	273
$\delta^{18}O$	15	−2.52	−3.96	216	2.5	−3.47	−2.59	290	0.9	−17.10	−10.9	301
$^{87}Sr/^{86}Sr$	0.721840	0.713255	0.713814	149	0.729700	0.714178	0.711347	344	0.709179	0.709179	0.709179	1

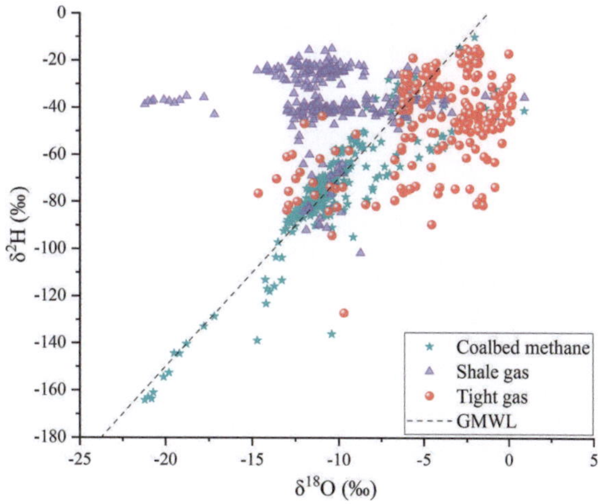

Fig. 4.16 Relationships between δ^2H and $\delta^{18}O$ in different flowback waters

References

Abdelaal A, Aljawad MS, Alyousef Z, Almajid MM (2021) A review of foam-based fracturing fluids applications: from lab studies to field implementations. J Nat Gas Sci Eng 95:104236. https://doi.org/10.1016/j.jngse.2021.104236

Barati R, Liang J (2014) A review of fracturing fluid systems used for hydraulic fracturing of oil and gas wells. J Appl Polym Sci 131(16):1–11. https://doi.org/10.1002/app.40735

Birkle P (2016) Geochemical fingerprinting of hydraulic fracturing fluids from Qusaiba Hot Shale and formation water from Paleozoic petroleum systems. Saudi Arabia Geofluids 16(3):565–584. https://doi.org/10.1111/gfl.12176

Blondes MS, Knierim KJ, Croke MR et al (2023) U.S. geological survey national produced waters geochemical database (Ver. 3.0, December 2023). U.S. Geological survey data release. https://doi.org/10.5066/P9DSRCZJ

Cui X, Zheng Z, Zhang H et al (2020) Impact of water-rock interactions on indicators of hydraulic fracturing flowback fluids produced from the Jurassic shale of Qaidam Basin, NW China. J Hydrol 590:125541. https://doi.org/10.1016/j.jhydrol.2020.125541

Gao J, Zou C, Li W et al (2020) Hydrochemistry of flowback water from changning shale gas field and associated shallow groundwater in Southern Sichuan Basin, China: Implications for the possible impact of shale gas development on groundwater quality. Sci Total Environ 713:136591. https://doi.org/10.1016/j.scitotenv.2020.136591

He X, Li P, Qian H et al (2026) Biogeochemistry during hydraulic fracturing: a critical review of reservoirs, fluids, processes, and implications. Geoenergy Sci Eng 256:214143. https://doi.org/10.1016/j.geoen.2025.214143

He X, Li P, Shi H et al (2022) Identifying strontium sources of flowback fluid and groundwater pollution using $^{87}Sr/^{86}Sr$ and geochemical model in Sulige gasfield, China. Chemosphere 135594. https://doi.org/10.1016/j.chemosphere.2022.135594

Herz-Thyhsen RJ, Kaszuba JP, Dewey JC (2020) Mineral dissolution and precipitation induced by hydraulic fracturing of a mudstone and a tight sandstone in the Powder River Basin, Wyoming, USA. Appl Geochem 119:104636. https://doi.org/10.1016/j.apgeochem.2020.104636

Jia H, Qu W, Ren W, Qian H (2021) Impacts of chemical weathering and human perturbations on dissolved loads of the Wei River, the Yellow River catchment. J Hydrol 603:26950. https://doi.org/10.1016/j.jhydrol.2021.126950

Li X, Hou D, Hu G et al (2005) Formation fluid characteristics and natural gas accumulation of gas fields in the central Ordos Basin. Geological Publishing House, Beijing (in Chinese)
Llewellyn GT, Dormanb F, Westlandb JL et al (2015) Evaluating a groundwater supply contamination incident attributed to Marcellus Shale gas development. PNAS 112(20):6325–6330. https://doi.org/10.1073/pnas.1420279112
Pang Z, Kong Y, Li J, Tian J (2017) An isotopic geoindicator in the hydrological cycle. Procedia Earth Planet Sci 17:534–537. https://doi.org/10.1016/j.proeps.2016.12.135
Phan TT, Hakala JA, Sharma S (2020) Application of isotopic and geochemical signals in unconventional oil and gas reservoir produced waters toward characterizing in situ geochemical fluid-shale reactions. Sci Total Environ 714:136867. https://doi.org/10.1016/j.scitotenv.2020.136867
Pu J, Yuan D, Zhang C et al (2012) Identifying the sources of solutes in karst groundwater in Chongqing, China: a combined sulfate and strontium isotope approach. Acta Geol Sin 86(4):980–992. https://doi.org/10.1111/j.1755-6724.2012.00722.x
Rao W, Chen J, Yang J et al (2009) Sr isotopic and elemental characteristics of calcites in the Chinese deserts: implications for eolian Sr transport and seawater Sr evolution. Geochim Cosmochim Acta 73:5600–5618. https://doi.org/10.1016/j.gca.2009.06.028
Rao W, Jin K, Jiang S et al (2015) Chemical and strontium isotopic characteristics of shallow groundwater in the Ordos Desert Plateau, North China: implications for the dissolved Sr source and water-rock interactions. Geochemistry 75:365–374. https://doi.org/10.1016/j.chemer.2015.07.003
Rosenblum J, Nelson AW, Ruyle B (2017) Temporal characterization of flowback and produced water quality from a hydraulically fractured oil and gas well. Sci Total Environ 596:369–377. https://doi.org/10.1016/j.scitotenv.2017.03.294
Rowan EL, Engle MA, Kraemer TF et al (2015) Geochemical and isotopic evolution of water produced from Middle Devonian Marcellus shale gas wells, Appalachian basin. Pennsylvania AAPG Bull 99(2):181–206. https://doi.org/10.1306/07071413146
Roy S, Gaillardet J, Allègre CJ (1999) Geochemistry of dissolved and suspended loads of the Seine river, France: Anthropogenic impact, carbonate and silicate weathering. Geochim Cosmochim Acta 63(9):1277–1292. https://doi.org/10.1016/S0016-7037(99)00099-X
Shi X, Gao Y, Qian H et al (2025) Elucidating the hydrochemistry and REE evolution of surface water and groundwater affected by acid mine drainage. Environ Pollut 366:125495. https://doi.org/10.1016/j.envpol.2024.125495
Steliga T, Kluk D, Jakubowicz P (2015) Analysis of chemical and toxicological of fluids for shale hydraulic and flowback water. Pol J Environ Stud 24(5):2185–2196. https://doi.org/10.15244/pjoes/43501
Sun J, Takahashi Y, Strosnider WHJ et al (2021) Identification and quantification of contributions to karst groundwater using a triple stable isotope labeling and mass balance model. Chemosphere 263:127946. https://doi.org/10.1016/j.chemosphere.2020.127946
Tian J, Wang L, Ni L et al (2023) Temporal variations in geochemistry of hydraulic fracturing fluid and flowback water in a tight oil reservoir. Pet Sci 20:3013–3021. https://doi.org/10.1016/j.petsci.2023.05.004
Xie X, Wang Y, Ellis A et al (2013) Delineation of groundwater flow paths using hydrochemical and strontium isotope composition: a case study in high arsenic aquifer systems of the Datong basin, northern China. J Hydrol 476:87–96. https://doi.org/10.1016/j.jhydrol.2012.10.016
Yang R, Dong L, Zhang J et al (2022) Origin, distribution and controlling factors of stratigraphic water Inthe western Sulige Gas Field. Acta Sedimentol Sin 40(1):267–280 (In Chinese with English abstract)
Zhang L, Li P, He X (2022) Interactions between surface water and groundwater in selected tributaries of the Wei River (China) revealed by hydrochemistry and stable isotopes. Hum Ecol Risk Assess 28(1):79–99. https://doi.org/10.1080/10807039.2021.2016054
Zheng Z, Cui X, Zhu P et al (2021) Sensitivity assessment of strontium isotope as indicator of polluted groundwater for hydraulic fracturing flowback fluids produced in the Dameigou Shale

of Qaidam Basin. J Groundw Sci Eng 9(2):93–101. https://doi.org/10.19637/j.cnki.2305-7068.2021.02.001

Zou C (2017) Unconventional petroleum geology, second ed. Elsevier. https://www.sciencedirect.com/book/9780128122341/unconventional-petroleum-geology

Chapter 5
Laboratory Experiment and Fracturing Fluid-Tight Sandstone Interactions

Post-fracturing flowback fluids constitute a diagnostic tracer for investigating geochemical processes during hydraulic fracturing. However, analyses of flowback fluids from field trials are limited in their ability to characterize post-fracturing mineralogical changes and microstructures of tight sandstones, as many geochemical processes are masked by the mixing of fracturing fluids with formation water. Laboratory water–rock experiments thus compensate for these shortcomings and serve as a valuable supplement to field-scale investigations. Therefore, water–rock interaction experiments between fracturing fluids and tight sandstones were conducted using drilled core samples in this chapter. To verify the influence of individual fracturing fluid additives on water–rock interactions, experiments were also conducted between additives and reservoir tight sandstones, with deionized water used as a control. Additionally, by adjusting pH, temperature, and solid-to-liquid ratios, the study examined how environmental parameters influence water–rock interactions between fracturing fluids and tight sandstones.

5.1 Laboratory Experiments Design

5.1.1 Experiment Processes

The experimental workflow is presented in Fig. 5.1. The experimental solutions included deionized water, fracturing fluid, and fracturing fluid additive solutions. A guar-based fracturing fluid system, widely applied in tight sandstone reservoirs, was used, and the fluid and additives were provided by Changqing Oilfield. Deionized water was prepared using a purification system (resistivity 18.25 MΩ·cm), and rock cores were tight sandstone samples obtained from five wells described in Chap. 3. In the Sulige Gas Field, guar-based fracturing fluids account for more than 50% of

X. He and P. Li, *Geochemical Processes in Tight Gas Hydraulic Fracturing Stimulation*, Environmental Earth Sciences, https://doi.org/10.1007/978-3-032-15180-3_5

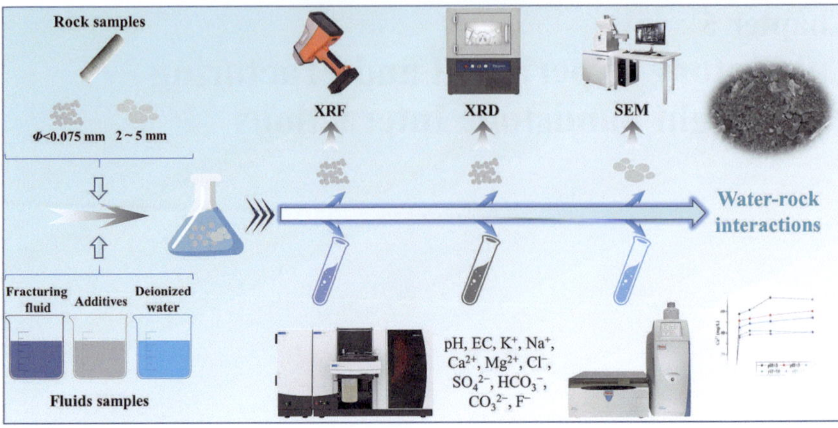

Fig. 5.1 Flow chart of the water–rock interaction experiment

usage. In addition to hydroxypropyl guar gum, guar fracturing fluids generally incorporate five types of additives: pH control agents, bactericides, crosslinkers, surfactants, and clay stabilizers. Crosslinkers (borates, aluminum, titanium, and zirconium compounds) are key additives that react with guar gum to form a network structure, thereby enhancing viscosity and stability. pH control agents (e.g., acetic acid, Na/KOH) maintain the effectiveness of cross-linkers. Biocides (e.g., sodium hypochlorite, glutaraldehyde) inhibit microbial growth, preventing degradation of fracturing fluids during storage and use. Clay stabilizers, such as tetramethylammonium chloride and potassium chloride, prevent clay swelling and migration in reservoirs. Surfactants, including ethoxylated alcohols and isopropanol, optimize viscosity control and aid in flowback after fracturing (He et al. 2026). Laboratory solutions were prepared with deionized water according to their proportions in guar-based fracturing fluid, with concentrations varying between 0.1 and 0.3 wt%.

In the laboratory, water–rock interaction experiments were designed with three categories of solutions: deionized water, guar-based fracturing fluid, and fracturing fluid additives including biocides, clay stabilizers, pH regulators, flowback aids, and crosslinkers. Taking into account the number of core samples, analytical requirements, and established experimental designs for shale reservoirs worldwide (Zolfaghari et al. 2016; Huang et al. 2020), the laboratory water–rock ratio was fixed at 4:1 (mL: g). Accordingly, each experiment involved 60 mL of solution with 15 g of tight sandstone samples. Experimental core samples comprised a mixture of 75% granules (2–5 mm) and 25% powders (< 0.075 mm) (Zolfaghari et al. 2016; Herz-Thyhsen et al. 2019). Since temperature and pressure fluctuate significantly during hydraulic fracturing and vary spatially, exact reservoir conditions are difficult to obtain. Thus, laboratory experiments were performed under room temperature and atmospheric pressure. Experiments were carried out at ambient temperature (20 ± 2 °C) and atmospheric pressure (0.1013 MPa). Solution and rock samples were taken after 1, 3, 6, 10, and 20 days of reaction for testing relevant parameters. Since

Table 5.1 Controlled experiment of fluid-rock interactions

Tight sandstone sample	Experimental group	Time (day)	Water/rock
R1–R5	Fracturing fluid	1, 3, 6, 10, 20, (30)	(60 mL)/(15 g)
R1-R5	Deionized water	1, 3, 6, 10, 20, (30)	(60 mL)/(15 g)
R2	Bactericide clay stabilizer	1, 3, 6, 10, 20, 30	(60 mL)/(15 g)
R3	Surfactant	1, 3, 6, 10, 20	(60 mL)/(15 g)
R4	pH control	1, 3, 6, 10, 20	(60 mL)/(15 g)
R5	Crosslinker	1, 3, 6, 10, 20	(60 mL)/(15 g)

Well *R2* contained a larger number of cores, the experiment duration was extended to 30 days (Table 5.1).

To explore how environmental variations influence results, supplementary experiments were performed under different pH conditions (3, 5, 7, 10, 12), temperatures (20, 50, 90 °C), and solid-to-liquid ratios (1/10, 1/8, 1/5 g/mL). Acidic conditions were adjusted using acetic acid (analytical grade), alkaline conditions with NaOH (analytical grade), and high-temperature environments (50–90 °C) were maintained via water bath heating. Considering that environmental changes may significantly accelerate water–rock interactions, solution and core samples were collected at 4, 12, 24, and 48 h for analysis.

5.1.2 Samples Analysis

Hydrochemical parameters (e.g., pH, Electrical conductivity (EC), K^+, Na^+, Ca^{2+}, Mg^{2+}, Cl^-, SO_4^{2-}, HCO_3^-, CO_3^{2-}, F^-) were tested for the initial fracturing fluids, additive solutions, and post-reaction solutions. pH and EC were measured with a portable tester. HCO_3^- and CO_3^{2-} were determined by a dual-indicator neutralization titration. Cations (Na^+, K^+, Ca^{2+}, Mg^{2+}) were analyzed using an atomic absorption spectrometer (PinAAcle™ 900H), and anions (Cl^-, SO_4^{2-}, F^-) were analyzed using ion chromatography (ICS − 600). Furthermore, X-ray fluorescence (XRF), X-ray diffraction (XRD), and field-emission scanning electron microscopy (SEM) were used to examine elemental composition, mineralogy, and microstructures of core samples before and after experiments, consistent with the procedures in Sect. 3.1 of Chap. 3 (Fig. 5.1).

5.2 Fracturing Fluid-Tight Sandstone Interactions

Following the water–rock reactions between the fracturing fluid and tight sandstone, the concentrations of various ions in the fracturing fluid exhibited a rising tendency. Meanwhile, the elemental contents in the core decreased compared to the original rock. This indicates that fracturing fluid with low ionic concentrations exhibits strong dissolution capacity toward tight sandstone. Due to the neutralization effect of rock minerals, the pH of the fracturing fluid displayed an overall decreasing trend with increasing reaction time. In all five wells, pH followed a consistent trend, decreasing markedly during the initial stage (1–6 days) from 10.07 in the fresh fracturing fluid to about 9.30. By the 10th day, pH rose slightly to about 9.70 and subsequently fluctuated slightly, influenced by carbonate equilibrium in the solution. Given the similar mineralogy of the five wells, post-reaction ionic concentrations were close to each other, and their temporal evolution followed the same trends (Fig. 5.2). Similar to the pH trend, the F^- concentration in the fracturing fluid decreased from 40.19 mg/L in the initial fluid to about 32 mg/L after reaction. The decrease in F^- concentration may be attributed to adsorption by the solid phase and fluorite precipitation.

The elevated concentrations of CO_3^{2-} and F^- in the initial fluid restricted the overall Ca^{2+} levels in the post-reaction solution. Post-reaction Ca^{2+} concentrations

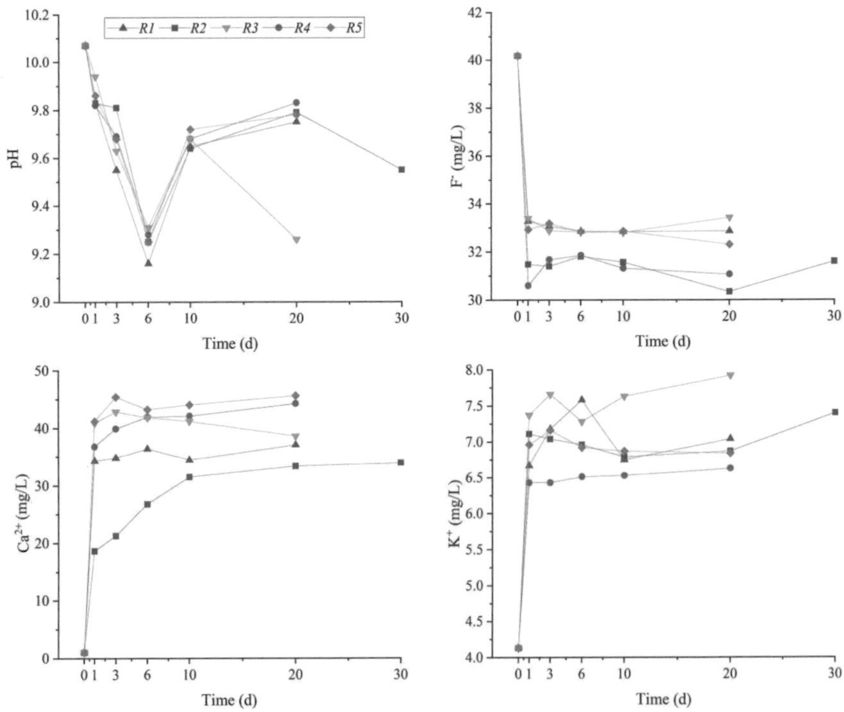

Fig. 5.2 The changes in typical parameters after fracturing fluid-rock reactions

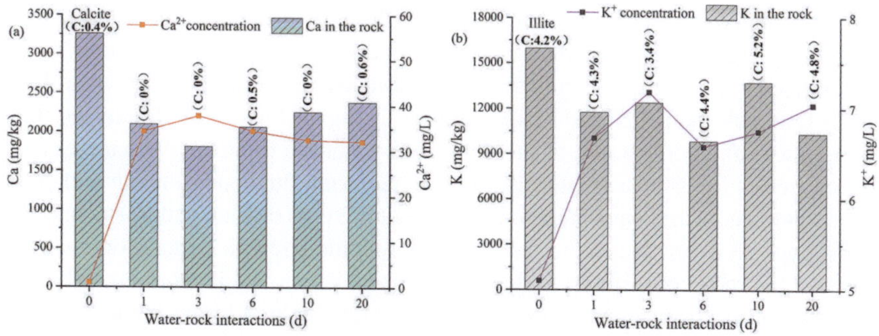

Fig. 5.3 Ca (**a**) and K (**b**) evolution after fracturing fluid-rock reactions

were mainly within 30–50 mg/L, accompanied by a significant reduction of Ca in the cores, averaging 885–17,919 mg/kg. In wells with abundant calcite, the solution contained higher Ca^{2+} levels after reaction, while the cores experienced more pronounced Ca depletion than those with lower calcite content. As calcite dissolved initially, Ca^{2+} concentrations rose to a threshold, beyond which CO_3^{2-} and F^- likely precipitated with Ca^{2+}, leading to declines in CO_3^{2-} and F^- while limiting further Ca^{2+} enrichment. For example, in Well *R1*, Ca^{2+} concentrations increased while Ca in the core decreased after reaction, with no calcite detected. By Day 10, a small amount of calcite was detected, coinciding with reduced Ca^{2+} in the solution and increased Ca in the rock, suggesting secondary calcite precipitation at the late stage (Fig. 5.3).

Tight sandstone generally lacks K-feldspar, and illite is thus the dominant K-bearing mineral. After reaction, K^+ concentrations slightly increased (by about 2 mg/L). In Well *R4*, with the highest illite content (27.3%), the increase in K^+ concentration and K depletion in the core were minimal. In contrast, Well *R3*, with the lowest illite content (3.9%), showed the greatest K^+ increase and K depletion. These results suggest that illite remained largely stable and did not experience noticeable dissolution in the fracturing fluid-tight sandstone interactions. Taking Well *R1* as an example, the reaction led to a modest rise in K^+ (around 2 mg/L) and an average K reduction of 4343 mg/kg in the core. Illite content showed little change, with a slight overall increase (Fig. 5.3).

5.3 Impacts of Additives on Water–Rock Interaction

5.3.1 Bactericide

In the bactericide environment, the pH of the solution decreases after water–rock interaction. Since both the bactericide and fracturing fluid environments exhibit relatively high initial pH values, the pH of the two solutions shows a similar trend under the buffering effect of sandstone minerals, declining with increasing reaction time (Fig. 5.4). After water–rock interaction, the increase in Na^+ concentration in the bactericide environment is significantly greater than in the deionized water and fracturing fluid environments. Ca^{2+} concentration is higher than in the fracturing fluid environment and comparable to the deionized water environment. The increase in Cl^- concentration is also greater than in the fracturing fluid environment and close to the deionized water environment. The remaining ions display patterns consistent with deionized water. This indicates that the low-ionic-strength bactericide exerts an effect on the rock core similar to that of deionized water, and is more favorable for the leaching of chemical elements from the rock compared to the fracturing fluid environment.

Table 5.2 summarizes the variations in residual element contents within the rock core before and after water–rock reactions under the bactericide condition. All elements in the rock core exhibit leaching behavior, with continuous release during the first three days, followed by stabilization with slight fluctuations. For example, after three days of water–rock reaction, the residual Ca in the core decreased from 2316 mg/kg in the original core to 1360 mg/kg, and then remained between 1400 and 1500 mg/kg. The variation patterns of Fe and Mn are similar, as are those of Ca and

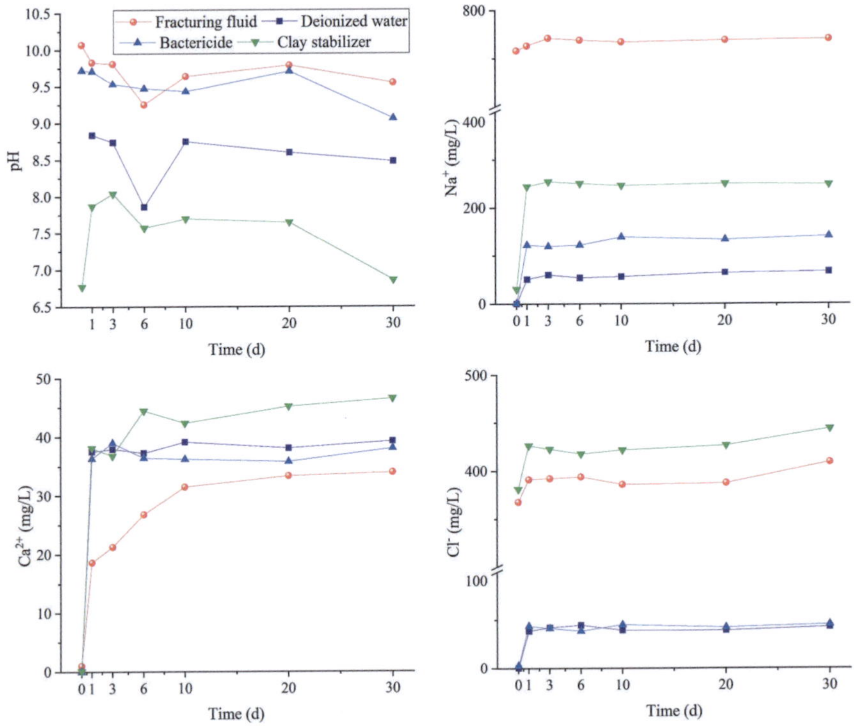

Fig. 5.4 The changes in typical parameters after bactericide/clay stabilizer-rock reactions

Sr, indicating that Fe and Mn, as well as Ca and Sr, are paragenetically associated in tight sandstones. Compared with the fracturing fluid and deionized water environments, the elemental leaching from the rock core in the bactericide environment is close to that in deionized water and greater than in the fracturing fluid environment. Mineralogical analysis indicates that minor amounts of calcite (0.4%) and K-feldspar (0.2%) dissolved completely during the bactericide-rock interaction, with calcite dissolving rapidly and disappearing entirely post-reaction. K-feldspar dissolved more slowly, disappearing after three days of reaction. Plagioclase content exhibited no substantial variation. Quartz exhibited a slight decline (-0.18%) after the reaction, whereas in the deionized water and fracturing fluid environments, quartz content increased by + 1.47% and + 0.68%, respectively (Fig. 5.5). For clay minerals, illite increased after reaction (+4.35%), whereas chlorite + kaolinite decreased (−3.55%). Except for illite, the biocide condition exhibited noticeable dissolution effects on sandstone core minerals.

5.3.2 *Clay Stabilizer*

In the clay stabilizer environment, influenced by the dissolution of calcite minerals, the pH of the solution initially (1–3 days) increased from 6.77 to 8.04 after the water–rock interaction, followed by a subsequent decline (Fig. 5.4). Compared with deionized water and fracturing fluid environments, the clay stabilizer exhibited stronger leaching capacity for core elements, with higher increases in Na^+, Ca^{2+}, K^+, and Cl^- concentrations after reaction. The enhanced dissolution is likely due to trace inorganic ion components in the clay stabilizer, which promote the dissolution of readily soluble minerals within the sandstone.

Owing to its strong dissolution capacity, the clay stabilizer environment yielded generally lower residual element contents in the core compared with deionized water and fracturing fluid environments. The decreases in Ca and Sr were most significant (Table 5.2), dropping from 2316 mg/kg and 112.3 mg/kg in the original core to 1021 mg/kg and 69.39 mg/kg, respectively. In the clay stabilizer environment, both calcite and K-feldspar dissolved completely and were not detected after the reaction. The change in quartz content was similar to that in the fracturing fluid environment, with an increase of 0.58%. Illite variation was close to that in the biocide environment, increasing by 4.38%. Chlorite + kaolinite decreased by 4.17%, resembling the deionized water environment but slightly higher than in the biocide environment (Fig. 5.5).

Table 5.2 Elemental variation of the rock in the fracturing fluid, deionized water, bactericide, and clay stabilizer (mg/kg)

Environment	Reactive time (d)	Zr	Sr	Rb	Zn	Fe	Mn	V	Ca	K
Natural rock	/	426.0	112.3	97.93	42.34	31,273	350.0	114.6	2316	32,037
Bactericide	1	409.2	103.0	90.05	52.71	26,491	320.0	81.11	1527	20,663
	3	373.2	94.02	78.45	47.63	24,326	291.5	74.73	1360	20,024
	6	390.2	91.28	80.79	40.24	22,617	225.1	80.74	1455	19,999
	10	374.8	86.95	71.78	46.33	23,333	318.1	65.09	1539	16,229
	20	530.1	99.05	85.20	49.40	26,556	308.0	80.01	1409	19,231
	30	366.2	94.06	82.36	47.83	26,895	283.6	71.76	1509	22,635
Clay stabilizer	1	393.5	73.03	77.18	49.77	23,435	289.9	74.65	1414	20,594
	3	464.1	79.03	93.27	50.44	29,289	347.5	74.18	1524	23,014
	6	361.8	69.39	73.45	44.11	21,778	294.6	62.15	1021	16,976
	10	395.5	72.11	73.91	35.77	23,285	296.7	62.53	1202	16,643
	20	356.1	71.01	74.75	48.73	24,736	283.7	54.48	1208	19,868
	30	399.1	71.08	80.63	42.19	26,754	312.9	80.74	1208	21,060
Fracturing fluid	1	399.1	92.24	92.28	36.60	27,461	344.8	105.6	1916	28,331
	3	578.7	81.89	86.24	40.64	24,630	277.6	93.47	1275	21,771
	6	406.8	83.36	80.45	47.71	26,431	297.5	81.60	1547	23,061
	10	419.7	78.10	81.26	44.91	25,624	273.5	71.09	1441	21,695
	20	403.3	82.01	79.50	53.2	26,815	309.4	67.29	1195	18,540
	30	402.6	77.47	78.79	47.23	24,568	274.8	88.78	1214	20,925
Deionized water	1	387.5	113.6	103.1	56.83	31,045	335.6	79.82	1718	24,273
	3	321.9	77.33	66.46	41.50	21,537	220.3	47.20	1493	18,812
	6	483.9	98.65	82.12	51.52	25,032	317.9	65.48	1604	16,846
	10	417.7	91.70	78.27	45.21	22,987	261.9	82.01	1264	18,079
	20	355.6	88.06	76.82	51.34	24,244	238.0	77.41	1685	21,636
	30	386.0	83.03	72.6	47.08	23,159	286.3	65.39	1465	20,261

5.3.3 *Surfactant*

In the surfactant environment, also influenced by carbonate dissolution, the pH increased on the first day of reaction and then gradually decreased. For cations, K^+ remained at low levels and was only trace-detected in the late stage of the reaction. Na^+, Ca^{2+}, and Mg^{2+} concentrations stabilized after 3–6 days of reaction. For anions, Cl^- concentration generally increased during the first 3–6 days and then stabilized. SO_4^{2-} concentration increased with reaction time, suggesting that the flowback aid environment promotes the dissolution of sulfate minerals. In the deionized water environment, rapid dissolution of carbonate minerals caused a high initial pH (8.43),

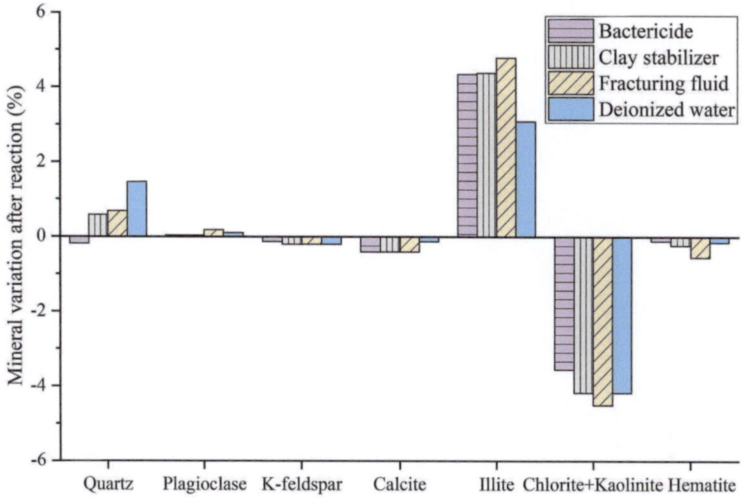

Fig. 5.5 Mineral variation after bactericide/clay stabilizer-rock reactions

followed by a decrease, similar to the fracturing fluid environment. In general, the low ionic strength of surfactant confers strong dissolution capability for tight sandstone, exceeding that of deionized water (Fig. 5.6).

Elemental leaching in the surfactant solution was sustained for the first 1–3 days. Although shorter than in fracturing fluid and deionized water environments, the overall variation pattern remained similar to deionized water. Rock cores retained higher residual elemental contents compared with deionized water and fracturing fluid conditions (Table 5.3). For example, the average reduction in Ca under the surfactant environment was 2772 mg/kg, compared with 2306 mg/kg in deionized water and 2078 mg/kg in the fracturing fluid environment. As for mineralogical variations, calcite was not fully dissolved in the flowback aid condition, and minor amounts were still detected at all stages. Compared to bactericide and clay stabilizer conditions, calcite dissolution was relatively limited.

5.3.4 pH Control

In the pH control environment, pH continuously decreased during the first 1–6 days of immersion and stabilized thereafter, reflecting the neutralization effect of the core on the alkaline solution. As the pH of fracturing fluid is primarily governed by the pH control, the pH variation trends of the two solutions during reactions with sandstone were essentially identical. EC decreased continuously during the first 6 days of reaction and stabilized thereafter. Among cations, the concentrations of K^+, Na^+, and Mg^{2+} increased during the first 3 days, then slightly decreased and stabilized. In contrast, Ca^{2+} concentration markedly increased after 1 day of

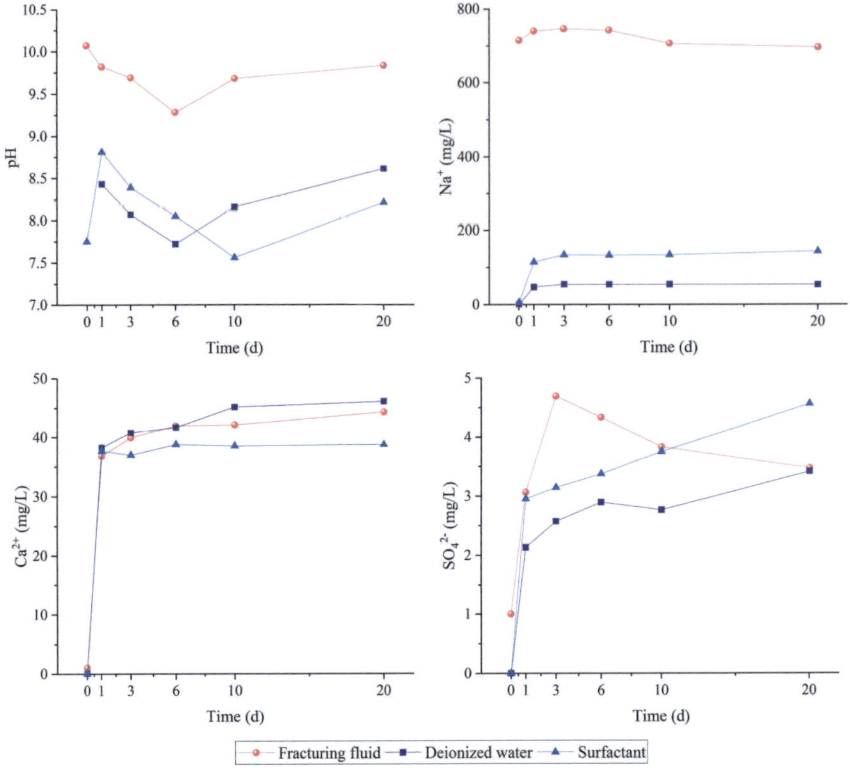

Fig. 5.6 The changes in typical parameters after surfactant-rock reactions

reaction but subsequently decreased, accompanied by a gradual reduction in CO_3^{2-} concentration (Fig. 5.7), suggesting significant calcite precipitation during the later stage of the water–rock reaction. For anions, the concentrations of Cl^-, F^-, and SO_4^{2-} continuously increased throughout the reaction.

The overall leaching of elements in the pH control environment was lower than in the guar-based fracturing fluid and deionized water environments. The alkaline environment induced by the pH control inhibited calcite dissolution in sandstone and also suppressed the release of other elements. Taking Sr as an example, its content in the original rock was 82.16 mg/kg, which dropped to 53.07 mg/kg after 3 days of reaction in the pH control environment and later rose to 70.39 mg/kg after 20 days (Table 5.4). Conversely, in the fracturing fluid and deionized water conditions, the residual Sr in the core exhibited a continuous dissolution trend. Since the core used in the pH control experiment contained a relatively high calcite content (18%), mineralogical analysis after the reaction revealed calcite dissolution with a decrease of 8.38%, similar to that observed in the deionized water and fracturing fluid settings. The elevated dissolution ratio may be attributed to reservoir core heterogeneity and

Table 5.3 Elemental variation of the rock in the fracturing fluid, deionized water, and surfactant (mg/kg)

Environment	Reactive time (d)	Zr	Sr	Rb	Zn	Fe	Mn	V	Ca	K
Natural rock	/	279.8	128.4	103.0	71.20	29,066	272.1	104.7	5527	39,314
Surfactant	1	301.7	114.0	93.89	53.19	24,847	189.3	56.52	3296	28,489
	3	277.1	110.2	91.19	48.81	24,554	210.2	55.27	3953	25,848
	6	277.8	99.33	82.82	48.36	21,265	178.1	64.13	3908	26,735
	10	283.2	93.15	75.91	47.57	19,512	175.1	61.39	2764	21,759
	20	260.7	98.00	78.72	61.57	21,531	169.1	47.39	3323	20,892
Fracturing fluid	1	418.2	91.16	92.86	63.07	24,223	156.9	54.40	3354	26,087
	3	249.4	73.59	73.80	41.72	18,363	153.9	45.11	2676	23,826
	6	291.5	73.06	76.42	41.64	18,057	159.9	56.28	2573	22,844
	10	234.4	79.01	79.31	53.60	20,654	164.1	53.08	3192	23,306
	20	328.9	75.02	76.46	41.04	18,322	127.7	39.12	1978	20,867
Deionized water	1	311.8	97.79	86.29	52.39	22,154	167.6	63.71	3595	27,054
	3	238.4	90.59	75.29	52.44	19,925	179.8	63.32	2996	23,735
	6	316.7	99.46	82.21	48.86	21,295	160.8	66.57	3518	24,230
	10	284.6	99.34	81.42	51.00	21,138	158.0	47.15	3155	21,222
	20	322.9	100.1	84.99	50.44	21,163	176.2	45.88	2838	20,411

reductions in other mineral phases, resulting in a calcite decrease markedly higher than in other well cores.

5.3.5 *Crosslinker*

In the crosslinker environment, the solution pH decreased during the first 1–3 days of reaction and then stabilized at approximately 9.4, showing a variation pattern similar to that in the fracturing fluid environment. Owing to the elevated Na^+ concentration (467 mg/L) in the original crosslinker solution, Na^+ levels in the solution remained nearly unchanged during the entire reaction. The concentrations of K^+, Ca^{2+}, and Mg^{2+} increased continuously during the first 3 days, then slightly decreased after 6 days and stabilized. The Ca^{2+} concentration trend in the crosslinker environment was similar to that in the fracturing fluid environment but slightly lower than in the deionized water environment. For anions, Cl^- concentration increased continuously during the first 3 days, decreased on day 6, and then stabilized overall. A fluctuating downward trend in F^- concentration was observed post-reaction, attributable to adsorption and mineral precipitation processes. $SO_4{}^{2-}$ concentration rose with prolonged immersion, showing more significant increases in the later stages of the

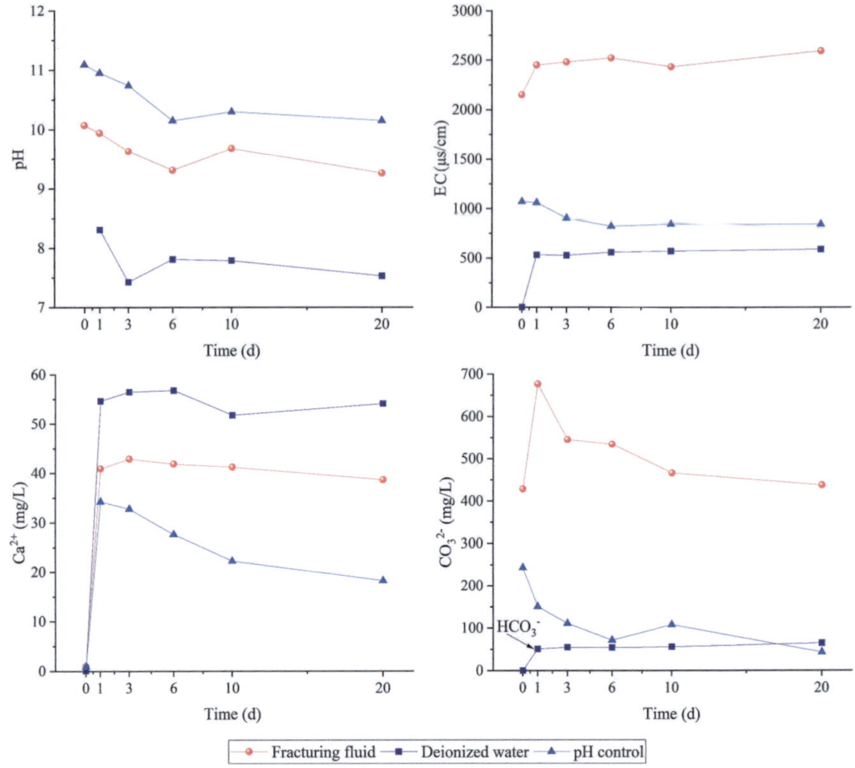

Fig. 5.7 The changes in typical parameters after pH control-rock reactions

reaction. A comparison of the fracturing fluid, crosslinker, and deionized water environments showed that pH in all three solutions first declined and then rose slightly. The guar-based fracturing fluid exhibited a higher initial pH and a longer duration of decline (6 days), whereas in the crosslinker and deionized water environments, pH decreased during the first 3 days and then slightly increased. In the deionized water system, carbonate mineral dissolution led to the largest pH increase (Fig. 5.8).

In the crosslinker environment, both major and trace elements in the residual core exhibited similar trends: continuous leaching during the first 3 days, partial recovery after 6 days, and subsequent fluctuations (Table 5.5). Changes in the remaining Ca content in the core corresponded well to the Ca^{2+} concentration trends across the environments. The mineralogical changes indicated a sharp reduction in kaolinite content, which disappeared completely during some post-reaction periods, especially in the fracturing fluid environment. In alkaline environments, kaolinite is prone to dissolution, which caused further reduction or even disappearance of kaolinite in the *R5* core where its initial content was relatively low (Fig. 5.9).

Overall, in the five additives, surfactant, bactericide, and clay stabilizer had low ion concentrations, and their water–rock interaction effects resembled those of the

Table 5.4 Elemental variation of the rock in the fracturing fluid, deionized water, and pH control (mg/kg)

Environment	Reactive time (d)	Zr	Sr	Rb	Zn	Fe	Mn	Ca	K
Natural rock	/	131.8	82.16	29.92	39.58	16,930	1429	50,787	9202
pH control	1	168.6	54.82	21.30	32.55	10,413	1062	32,467	6207
	3	151.9	53.07	19.34	25.82	9182	945.4	31,986	5430
	6	162.6	62.59	22.97	32.54	11,701	1128	32,210	6023
	10	159.4	65.05	23.08	36.18	12,634	1204	38,253	6496
	20	155.7	70.39	23.24	37.98	12,275	1180	31,936	5727
Fracturing fluid	1	175.3	54.90	23.49	42.62	14,554	1235	38,933	6530
	3	160.9	52.81	25.29	34.37	13,650	1122	35,553	7167
	6	154.3	51.75	21.65	34.92	11,151	1035	33,110	5952
	10	150.6	44.31	24.25	25.49	9852	930.4	29,488	5986
	20	184.5	50.95	22.54	26.30	9770	973.7	27,260	5025
Deionized water	1	154.3	52.54	27.32	32.52	10,983	1162	36,192	6099
	3	140.5	45.04	17.76	28.38	8518	869.0	29,680	5454
	6	156.6	45.90	21.90	36.39	9786	950.3	30,186	5986
	10	128.9	54.54	24.47	31.07	10,474	1012	36,250	6147
	20	153.9	60.90	25.09	40.93	12,904	1179	32,196	5822

deionized water environment, primarily promoting mineral dissolution in tight sandstone. Conversely, the crosslinker and pH control exerted stronger influences on the chemical composition and pH of guar-based fracturing fluid, making their effects after water–rock reactions closer to those observed in the fracturing fluid environment. These two additives exerted more significant effects on fracturing fluid-tight sandstone interactions, suggesting that their selection should be given particular attention when formulating fracturing fluids.

5.4 Impacts of Environmental Factors on Water–Rock Interaction

Influenced by construction conditions, reservoir environments, and fracturing fluid types, the fracturing fluid-reservoir rock interactions typically occur in a dynamically changing environment. To investigate how changes in fracturing conditions may affect water–rock reactions, this study conducted experiments on fracturing fluid-tight sandstone interactions under different temperatures, pH, and solid–liquid ratios, aiming to explore how environmental factors affect these interactions.

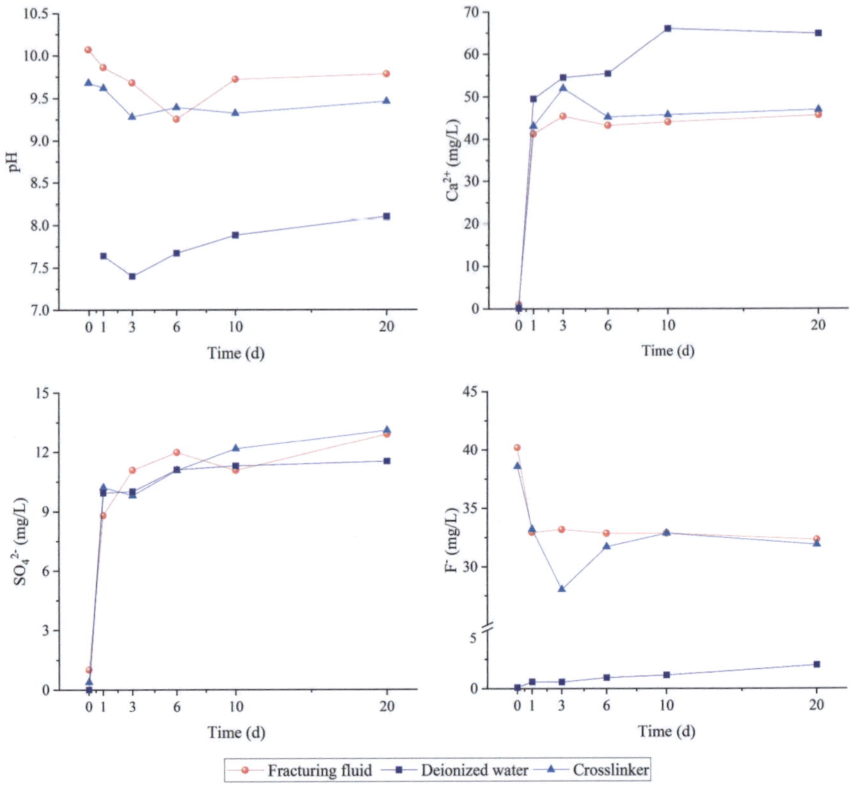

Fig. 5.8 The changes in typical parameters after crosslinker-rock reactions

5.4.1 pH

The pH level is a key parameter controlling water–rock interactions, and variations in pH conditions may lead to different interaction pathways. The experiments showed that, owing to the neutralizing effect of tight sandstone, the solution pH increased under acidic conditions after reaction, while under alkaline conditions it decreased with time (Fig. 5.10). At pH = 3, representing acidic conditions, the concentrations of most ions increased markedly, except for F^-. F^- concentration was highly sensitive to pH, decreasing substantially in acidic environments. F^- readily combines with H^+ to form the weak acid HF, lowering free F^- concentration; meanwhile, the rock matrix exhibits strong adsorption of F^- under acidic conditions. This effect was most pronounced at pH = 3, where F^- concentration decreased from 40.19 mg/L in alkaline conditions to 7 mg/L. When pH was 5, less weak acid formed, and the F^- concentration was approximately 30 mg/L. Compared with pH = 3, Ca^{2+} concentration was significantly reduced under weakly acidic conditions (pH = 5). The decrease in H^+ reduced the dissolution of Ca-bearing minerals such as calcite.

Table 5.5 Elemental variation of the rock in the fracturing fluid, deionized water, and crosslinker (mg/kg)

Environment	Reactive time (d)	Zr	Sr	Rb	Zn	Fe	Mn	Ca	K
Natural rock	/	218.9	84.50	58.42	33.52	19,717	397.2	14,964	21,756
Crosslinker	1	202.7	73.98	56.57	31.91	18,183	312.1	8752	16,143
	3	213.0	59.17	44.56	27.14	11,384	225.1	5819	11,483
	6	284.8	72.93	55.81	30.10	18,544	394.2	13,043	16,858
	10	244.4	69.30	48.83	32.42	16,459	322.0	12,235	14,361
	20	236.9	65.74	50.27	47.77	15,508	299.9	9584	13,136
Fracturing fluid	1	196.9	56.65	56.68	31.12	15,023	229.9	7662	15,737
	3	227.6	53.62	52.68	31.64	18,074	309.3	8861	18,569
	6	245.4	54.83	48.09	35.38	17,020	316.4	11,354	16,495
	10	181.9	42.12	40.08	31.10	12,626	263.6	9192	12,351
	20	212.6	51.01	46.31	26.13	14,266	291.1	8840	12,707
Deionized water	1	259.9	53.47	38.44	27.47	12,497	262.6	9165	13,524
	3	251.7	60.41	41.10	26.31	13,846	265.2	8787	12,489
	6	277.4	81.55	58.05	38.85	19,364	332.1	9571	16,162
	10	251.1	64.75	47.99	35.92	15,348	335.3	9073	13,242
	20	251.6	70.21	50.05	35.28	16,370	309.6	9640	14,970

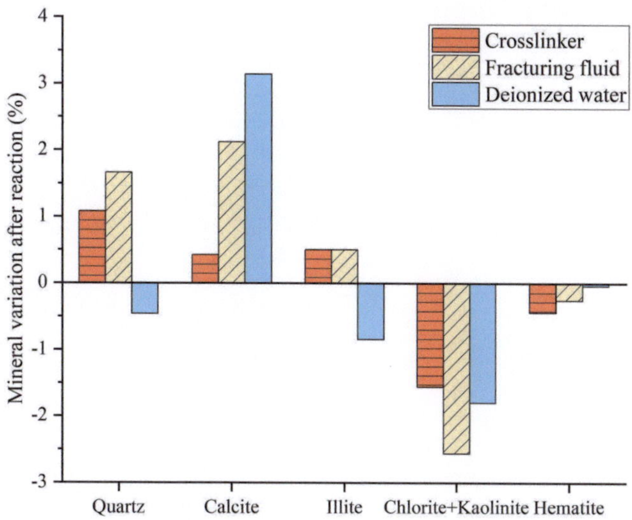

Fig. 5.9 Mineral variation after crosslinker-rock reactions

Moreover, higher F^- concentrations promoted partial fluorite (CaF_2) precipitation, limiting additional Ca^{2+} release. Compared to the pH = 3 environment, the concentrations of major ions were lower at pH = 5. Furthermore, a substantial HCO_3^- concentration was observed, partly due to CO_3^{2-} transformation from the fracturing fluid and partly from calcite dissolution in the sandstone.

At pH = 7, as the system shifted from acidic to neutral, the concentrations of major ions further decreased relative to acidic conditions. In the original fracturing fluid environment (pH = 10), Ca^{2+} and Mg^{2+} concentrations decreased further, as the alkaline condition inhibited carbonate mineral dissolution (Fig. 5.10). Concentrations of other ions showed little change compared with neutral conditions. At pH = 12, CO_3^{2-} content increased markedly, with nearly all HCO_3^- converted to CO_3^{2-} in the strong alkaline environment. Simultaneously, Ca^{2+} dissolution decreased further, and carbonate mineral dissolution was strongly inhibited.

Elemental analysis of the rock cores before and after reaction (Table 5.6) revealed that reactions between the fracturing fluid and the rock proceeded rapidly during the first 4 h, then stabilized, with subsequent interactions slowing down. Comparisons across pH levels clearly indicated that acidic environments facilitated greater element release from the rock. At pH = 3, element release from the rock core was

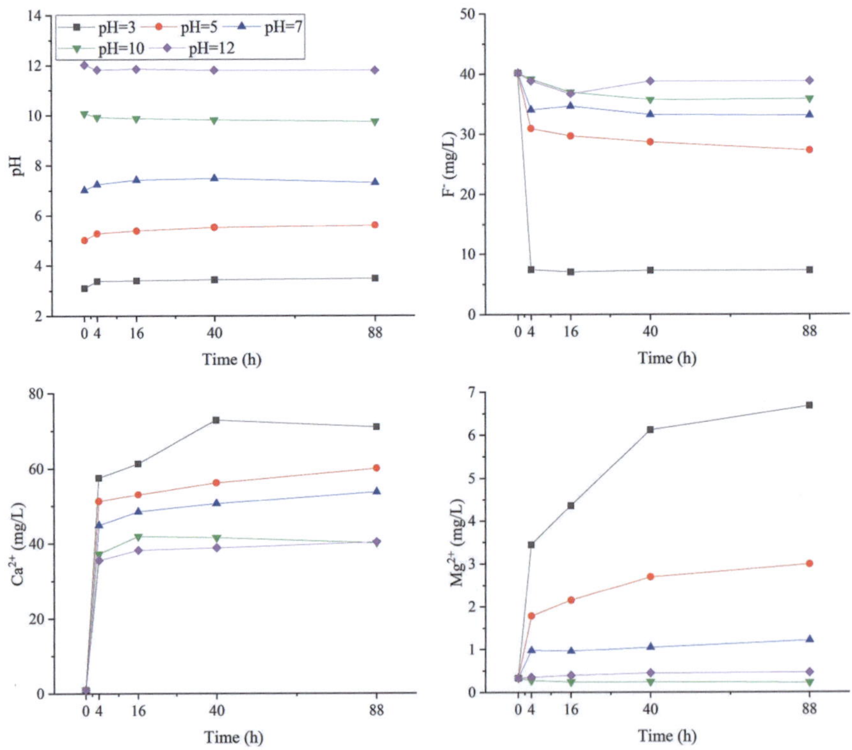

Fig. 5.10 Typical parameters variation in different pH environments

Table 5.6 Elemental variation of the rock in different pH environments (mg/kg)

Condition	Reactive time (h)	Zr	Sr	Rb	Zn	Fe	Mn	V	Ca	K
Natural rock	/	426.0	112.3	97.93	42.34	31,273	350.0	114.6	2316	32,037
pH = 3	4	386.4	53.76	72.54	48.70	22,425	256.5	74.39	798.9	19,531
	16	402.1	46.24	68.60	39.48	20,886	212.5	53.41	629.9	18,644
	40	372.0	39.76	57.10	34.75	16,382	165.3	61.92	512.6	16,062
	88	361.8	40.27	64.00	39.53	17,791	185.7	59.59	1,048	16,395
pH = 5	4	389.1	57.20	66.21	42.52	20,051	246.0	67.72	903.9	17,508
	16	433.9	62.37	72.71	46.02	24,121	232.7	82.52	930.7	19,734
	40	379.8	51.57	70.63	39.34	19,300	219.5	60.01	736.8	16,971
	88	370.3	52.17	72.50	43.47	19,309	210.0	56.46	710.8	18,491
pH = 7	4	323.7	61.03	65.03	43.93	21,657	249.5	47.17	1059	19,928
	16	314.3	59.01	64.64	40.31	20,715	246.0	60.13	1124	18,789
	40	356.9	60.35	81.05	43.51	24,735	233.9	81.09	1012	20,428
	88	348.5	57.76	67.12	39.80	19,708	242.3	62.96	822.4	16,223
pH = 10	4	460.6	75.78	76.05	45.21	24,787	313.0	55.10	1349	22,004
	16	418.3	74.69	76.44	44.58	24,121	234.8	86.38	1331	21,683
	40	388.8	68.36	69.55	38.63	20,800	257.5	81.93	928.1	18,096
	88	460.5	70.76	78.39	48.57	23,689	271.0	59.78	1061	20,991
pH = 12	4	339.6	80.71	79.34	47.93	24,931	316.2	75.25	1399	20,652
	16	391.1	65.38	67.09	43.37	19,102	237.3	55.88	1019	18,094
	40	395.3	77.03	69.20	54.62	21,407	275.8	69.12	1027	17,126
	88	353.6	65.46	68.05	38.19	20,281	235.9	61.59	1109	18,292

generally high and continued to increase with reaction time. Conversely, at pH = 12, representing strong alkalinity, element dissolution from the core was suppressed to some extent.

5.4.2 *Temperature*

In water–rock interaction systems, temperature variations influence the dissolution and precipitation of mineral phases. However, different minerals exhibit varying sensitivities to temperature changes. For instance, sulfate minerals typically become more soluble as temperature rises, while calcite and dolomite exhibit reduced solubility under elevated temperatures. Experimental results indicate that as temperature increases, the rate of pH decline in fracturing fluid accelerates markedly (Fig. 5.11).

In addition, rising temperature facilitates element release from the rock, resulting in elevated solution conductivity. At room temperature (20 °C), conductivity remains nearly constant throughout the leaching process. At 50 °C, conductivity increases, and at 90 °C, the increasing trend becomes more pronounced, though the difference from 50 °C is limited. The behavior of individual ionic species shows distinct variations. For Ca^{2+}, influenced by carbonate mineral solubility, concentrations are lower at higher temperatures than at lower ones. As for Cl^- and SO_4^{2-}, their concentrations rise with temperature, but only to a modest extent. Overall, sustained high temperatures promote element release from cores, raising ion concentrations, but the short-term increases remain limited.

Analyses of residual elements indicate that under 20, 50, and 90 °C, water–rock reactions proceed most rapidly at 50 °C, with greater elemental release during the initial stage (Table 5.7). Although the strength of water–rock interactions generally rises with temperature, the total elemental release from cores remains relatively unchanged among the three tested conditions. Taking K and Sr as examples, the mean leaching amounts of K were 11,343, 12,526, and 13,604 mg/kg, while those of Sr were 39.90, 37.68, and 37.15 mg/kg across the three temperatures. Increasing

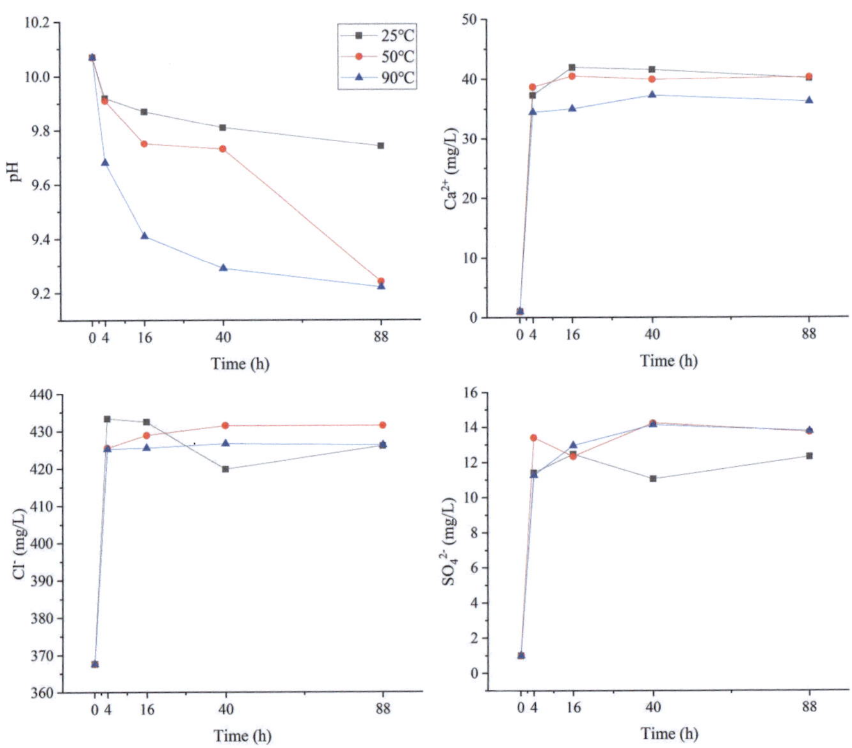

Fig. 5.11 Typical parameters variation in different temperature environments

Table 5.7 Elemental variation of the rock in different temperature environments (mg/kg)

Condition	Reactive time (h)	Zr	Sr	Rb	Zn	Fe	Mn	V	Ca	K
Natural rock	/	426.0	112.3	97.93	42.34	31,273	350.0	114.6	2316	32,037
20 °C	4	460.6	75.78	76.05	45.21	24,787	313.0	55.10	1349	22,004
	16	418.3	74.69	76.44	44.58	24,121	234.8	86.38	1331	21,683
	40	388.8	68.36	69.55	38.63	20,800	257.5	81.93	928.1	18,096
	88	460.5	70.76	78.39	48.57	23,689	271.0	59.78	1061	20,991
50 °C	4	419.2	74.72	76.20	38.39	22,506	256.9	84.78	1430	21,050
	16	404.4	70.86	74.23	45.97	22,130	220.2	87.58	1046	18,403
	40	400.7	73.96	75.55	45.17	22,154	254.1	74.99	1050	18,538
	88	423.6	78.93	74.80	41.76	22,683	251.7	66.81	1525	20,052
90 °C	4	355.7	81.10	79.61	45.14	22,342	263.4	73.15	1095	18,199
	16	338.4	77.00	74.11	41.44	22,614	258.4	92.97	1398	20,335
	40	381.5	70.21	71.44	41.84	21,193	242.5	55.53	970.4	16,746
	88	376.4	72.30	77.52	41.88	22,830	264.6	57.49	1323	18,451

temperature promotes K release in tight sandstone, while exerting a suppressive effect on the dissolution of Sr- and Ca-bearing carbonates.

5.4.3 Solid–Liquid Ratios

During hydraulic fracturing, fracturing fluids mainly occupy fracture channels created within the reservoir. In these fractures, the fluid surrounds the rock matrix, generally with a low solid–liquid ratio, resulting in strong fluid-rock interactions dominated by reservoir dissolution. In contrast, in pores with relatively low fluid-rock ratios, fluids are enclosed by the rock matrix, leading to higher solid–liquid ratios and limited dissolution. The experiments reveal that at high solid–liquid ratios, pH decreases faster, and greater rock abundance intensifies the neutralization of alkaline fluids. At high solid–liquid ratios, ion concentrations in solution are higher than under low ratios (Fig. 5.12). For core residual elements, amounts are lower at low solid–liquid ratios (Table 5.8), indicating that greater contact area enhances reaction intensity. In terms of the rock solid phase, lower ratios promote more extensive elemental leaching. For solution ions, higher ratios often result in higher post-reaction concentrations. Therefore, mineral dissolution tends to occur in fractures with low solid–liquid ratios, whereas precipitation is more likely in high-ratio pores, potentially blocking seepage channels.

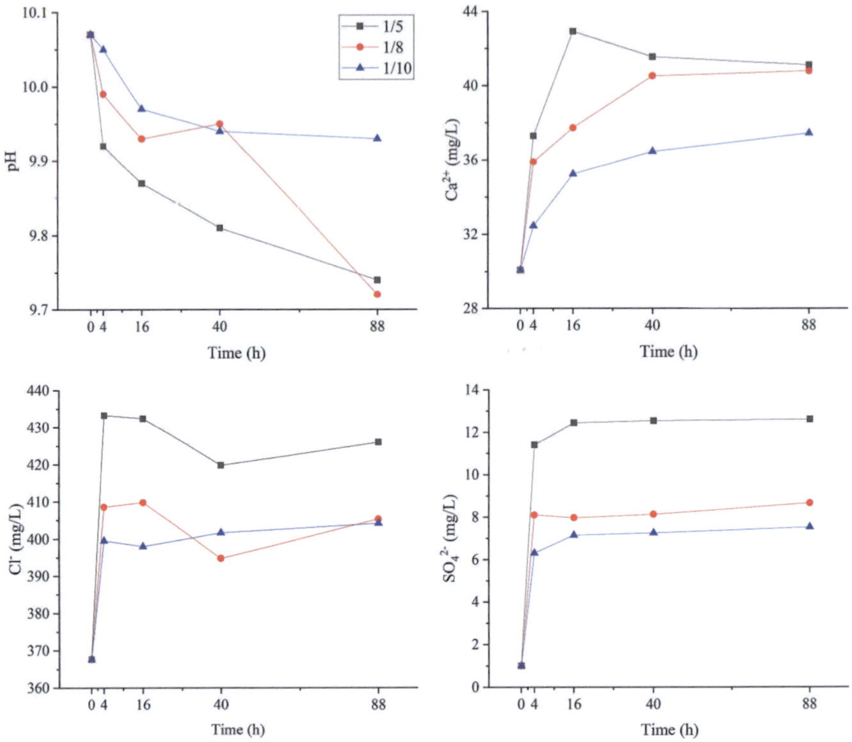

Fig. 5.12 Typical parameters variation in different solid–liquid ratios (g/mL)

In summary, temperature and pH play more critical roles in controlling water–rock reactions in hydraulic fracturing. The injection of large volumes of low-temperature fracturing fluid into fractures lowers reservoir temperature locally, enhancing carbonate mineral dissolution. Previous studies indicate that compared with temperature and pH, pressure changes exert relatively minor effects on water–rock interactions between fracturing fluids and formation waters (Zolfaghari et al. 2016; Huang et al. 2020). Given the uncontrollable nature of temperature and pressure in hydraulic fracturing, the pH of fracturing fluid emerges as the critical factor governing water–rock interactions.

Table 5.8 Elemental variation of the rock in different solid–liquid ratios (mg/kg)

Condition	Reactive time (h)	Zr	Sr	Rb	Zn	Fe	Mn	V	Ca	K
Natural rock	/	426.0	112.3	97.93	42.34	31,273	349.96	114.6	2316	32,037
1/10	4	312.3	55.70	59.48	39.17	17,914	216.28	73.38	1020	18,086
	16	357.9	50.60	62.65	36.28	17,006	191.03	58.72	761.3	16,771
	40	429.8	54.72	66.09	38.09	17,721	197.02	61.49	707.0	16,109
	88	342.2	48.45	58.79	33.03	16,349	185.82	76.48	759.1	17,145
1/8	4	402.4	57.18	63.14	40.07	17,750	208.49	63.68	850.1	17,156
	16	348.1	55.72	62.09	33.17	17,215	193.34	53.17	838.9	16,256
	40	394.3	51.12	60.01	33.06	16,002	184.37	47.47	712.5	14,568
	88	312.3	53.02	63.85	35.21	18,639	221.23	41.11	968.2	18,520
1/5	4	460.6	75.78	76.05	45.21	24,787	313.04	55.10	1349	22,004
	16	418.3	74.69	76.44	44.58	24,121	234.75	86.38	1331	21,683
	40	388.8	68.36	69.55	38.63	20,800	257.53	81.93	928.1	18,096
	88	460.5	70.76	78.39	48.57	23,689	270.96	59.78	1061	20,991

References

He X, Li P, Qian H et al (2026) Biogeochemistry during hydraulic fracturing: a critical review of reservoirs, fluids, processes, and implications. Geoenergy Sci Eng 256:214143. https://doi.org/10.1016/j.geoen.2025.214143

Herz-Thyhsen RJ, Kaszuba JP, Dewey JC (2019) Dissolution of minerals and precipitation of an aluminosilicate phase during experimentally simulated hydraulic fracturing of a mudstone and a tight sandstone in the Powder River Basin, WY. Energy Fuels 33:3947–3956. https://doi.org/10.1021/acs.energyfuels.8b04443

Huang T, Li Z, Mayer B et al (2020) Identification of geochemical processes during hydraulic fracturing of a shale gas reservoir: a controlled field and laboratory water-rock interaction experiment. Geophys Res Lett 47(20):e2020GL090420. https://doi.org/10.1029/2020GL090420

Zolfaghari A, Dehghanpour H, Noel M, Bearinger D (2016) Laboratory and field analysis of flow-back water from gas shales. J Unconv Oil Gas Resour 14:113–127. https://doi.org/10.1016/j.juogr.2016.03.004

Chapter 6
Geochemical Processes During Hydraulic Fracturing

As humanity continues to explore deep Earth, human activities are significantly changing deep-earth environments. Hydraulic fracturing has opened new avenues for studying water–rock interactions in deep reservoirs. In tight sandstone reservoirs enriched with formation water, what interactions occur between the fracturing fluid, tight sandstone, and formation water during hydraulic fracturing? Based on field fracturing test flowback fluid analysis results and laboratory experiments, this chapter characterizes and reveals the chemical processes of fracturing fluid-tight sandstone-formation water interactions in tight sandstone reservoirs during hydraulic fracturing. Focusing on the water-rich characteristics of tight sandstone reservoirs, a detailed study was first conducted on the interaction between fracturing fluid and formation water. Moreover, by simulating the mixing process of fracturing fluid and formation water, the interference of formation water mixing on water–rock interactions was eliminated, enabling the identification of processes such as mineral dissolution, mineral precipitation, redox reactions, and ion exchange during hydraulic fracturing.

6.1 Fluids Mixing

In unconventional natural gas, shale gas, tight sandstone gas, and coalbed methane all originate from coal-bearing reservoirs. Shale gas mainly consists of dry gas, characterized by low water saturation, and generates little or no water during production. In contrast, tight sandstone gas generally has a higher water saturation, with gas–water two-phase flow during development (Zou 2017). As a result, the interaction between fracturing fluid and formation water is significant and plays a crucial role in gas production. Formation water residing in the pores and fractures of gas-bearing reservoirs generally does not exhibit flowability. During hydraulic fracturing, the fractures created under high pressure connect the natural pores and fractures where formation water is stored. The artificial fractures formed during hydraulic fracturing typically

X. He and P. Li, *Geochemical Processes in Tight Gas Hydraulic Fracturing Stimulation*, Environmental Earth Sciences, https://doi.org/10.1007/978-3-032-15180-3_6

have much larger apertures than natural fractures, resulting in a higher conductivity than natural fractures. Once hydraulic fractures are formed, formation water from natural pores and fractures will flow into the fractures under the combined drive of reservoir pressure and natural gas, mixing with the fracturing fluid, and eventually flowing to the wellbore and returning to the surface along with the fracturing fluid (Rowan et al. 2015).

6.1.1 Mixing Pattern

Tight sandstone gas reservoirs are featured by overlapping layers, where multiple sand bodies intersect and merge to create composite sand bodies arranged in strip-like formations (Fig. 6.1). The P_2h_8 Member is widely developed in the Sulige area, characterized by substantial and expansive tight sandstone reservoirs ideally suited for horizontal well exploitation. The horizontal section is divided into multiple stages to facilitate fracturing reconstruction (Fig. 6.1). During the fracturing process, the initial stage closest to the well's bottom is fractured first and isolated by a packer. Subsequently, each subsequent stage is fractured until the final stage near the wellhead is also treated. Ultimately, all fracturing stages undergo flowback concurrently. As a result, each stage contains a blend of fracturing fluid and formation water during the fracturing process, and there is an intermingling of flowback fluid between stages during the flowback phase. Directional well (including vertical well) provides the possibility to extract the tight gas situated in vertical layers, where horizontal well fracturing is not feasible. However, there is significant variation in the chemical compositions of formation water across different vertical formations (Table 4.4, Fig. 4.4). The mixing between fracturing fluid and formation water in different formations have a substantial impact on the hydrochemistry of the flowback fluid. For the directional well in this study, the hydraulic fracturing operations are categorized into five distinct layers, including the P_1s_1 Member layer, two P_2h_8 Member layers, the P_2h_6 Member layer, and the P_2h_4 Member layer. Consequently, each layer contains a blend of fracturing fluid and formation water, and during the flowback phase, there is an intermingling of flowback fluid between different layers, rendering the mixing processes more intricate.

6.1.2 Influencing Factors of the Mixing Process

The mixing processes are dominated by fracturing fluid and formation water, so the types and compositions of fracturing fluid and formation water have an important influence on the hydrochemistry of the flowback fluid. Additionally, the duration of well shut-in, fracturing mode, formation water occurrence, as well as the utilization of perforating fluid and acidic preflush during hydraulic fracturing, exerts a partial influence on the mixing processes. When flowback is initiated immediately after

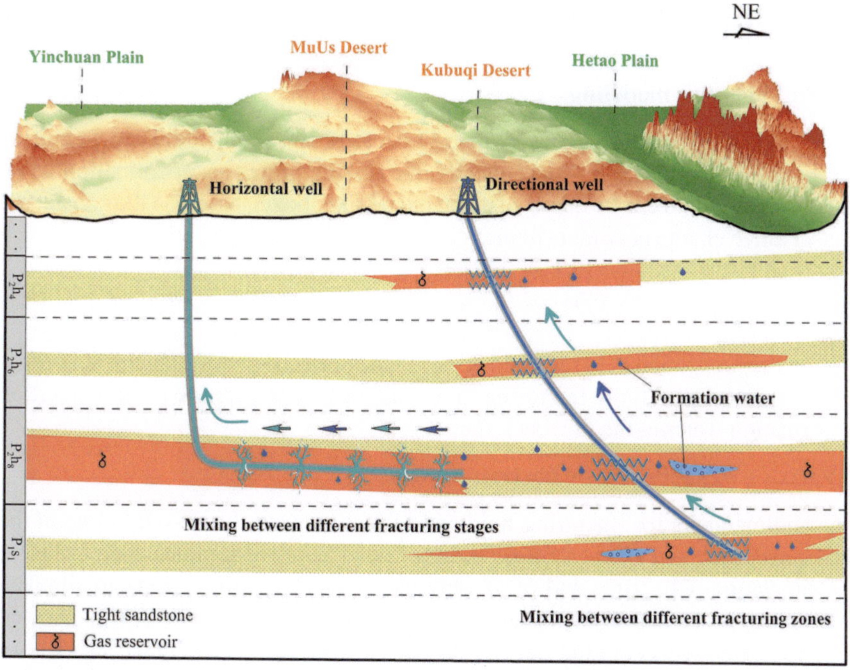

Fig. 6.1 Conceptual map indicating mixing processes and modes during hydraulic fracturing

fracturing, the fracturing fluid and formation water are often insufficiently mixed, and the flowback fluid at the beginning is primarily composed of fracturing fluid. A prolonged well shut-in period will lead to a swift rise in the ratio of formation water in the flowback fluid. Fracturing fluids with specific viscosities typically require gel breaking in the formation. Once fully broken, these fluids mix more readily with formation water. The mixing processes during hydraulic fracturing primarily involve advective mixing between fracturing fluid and formation water within the hydraulic fractures. Additionally, formation water may be released from the rock matrix via reverse imbibition and ion diffusion processes, resulting in the mingling of the fluid in the hydraulic fracture with various proportions of formation water (Osselin et al. 2018; Owen et al. 2020). Therefore, the mixing process is also affected by the distribution of formation water and fracturing fluid in the reservoir, leading to an often uneven mixing process.

6.1.3 Mixing Processes Modeling

(1) End-member modeling

There is a significant disparity in the chemical compositions between fracturing fluid and formation water. Consequently, the mixing processes of formation water and fracturing fluid can bc depicted using the mixing model, based on the considered conservative chemical compositions:

$$c_{FFF} = c_{FF} \times (1 - R) + c_{FW} \times R \tag{6.1}$$

where, c_{FFF}, c_{FF}, and c_{FW} represented the values of conservative chemical compositions for flowback fluid, fracturing fluid, and formation water; R is the proportion of formation water in the flowback fluid.

(2) Mixing modeling in the horizontal well

The end members for fracturing fluid and formation water were well-define in the horizontal well, and extensive time series of flowback fluid samples were available for simulation. Cl^-, Br^-, and $\delta^{18}O$ were used as conservative tracers to simulate the mixing processes of fracturing fluid and formation water. The results showed a progressive increase in the ratio of formation water within the flowback fluid as flowback time extended. Specifically, the ratio of formation water calculated by these tracers was similar in the first 17 samples (within 42 h after flowback), increasing from 0% to approximately 50%. Subsequently, the ratio of formation water indicated via Cl^- and Br^- increased rapidly, while that indicated by $\delta^{18}O$ gradually increased with prolonged flowback time (Table 6.1).

The mixing model using Cl^-, Br^-, and $\delta^{18}O$ as tracers effectively represents the concentration variations of Ca^{2+}, Na^+, Sr^{2+}, and Ba^{2+} of the flowback fluid (Fig. 6.2). However, the model proved inadequate for explaining the concentration changes of other ions (e.g., HCO_3^-, B, and K^+), suggesting that these ions were influenced to a greater extent by water–rock interactions (Fig. 6.3). Three mixing lines closely match with the measured K^+ concentrations during the early stage of flowback. However, the measured K^+ concentrations diverge from the mixing lines, surpassing them. This divergence may be attributed to the influence of formation water from outside the reservoirs (from 51 to 190 h). Subsequently, the primary influence appears to be ion exchange, leading to measured K^+ concentrations higher than the simulated mixing concentrations. HCO_3^- concentrations show a similar variation pattern with K^+, and the mixing model does not effectively depict the variation in HCO_3^- concentrations. Nearly all samples deviate from the mixing lines, demonstrating that the water–rock interactions dominate the variation of HCO_3^- concentrations in the flowback fluid. The mixing process of B is effectively characterized by the mixing lines during the initial and late stages of flowback. However, measured B concentrations during other periods of flowback typically exceed the mixing lines, possibly due to adsorption and desorption effects (He et al. 2022; Huang et al. 2022).

Table 6.1 Ratio of formation water in flowback fluid using different tracers (Unit: %)

Sample number	Mixing-Cl^-	Mixing-Br^-	Mixing-$\delta^{18}O$	Sample number	Mixing-Cl^-	Mixing-Br^-	Mixing-$\delta^{18}O$
1	0	0	0	34	100	100	81.48
2	3.38	0.60	3.82	35	100	84.64	75.17
3	5.92	4.51	16.77	36	100	97.70	80.35
4	13.80	3.84	23.80	37	100	96.16	67.74
5	14.93	13.89	24.28	38	100	91.55	71.44
6	20.56	19.65	35.26	39	65.63	43.46	42.33
7	26.20	28.63	40.16	40	90.42	67.97	78.16
8	40.85	38.85	39.81	41	100	80.80	86.10
9	30.70	30.56	42.80	42	99.44	62.36	85.90
10	31.83	31.71	48.13	43	100	100	77.14
11	36.34	36.78	43.58	44	100	60.13	96.12
12	40.85	38.47	48.91	45	94.93	77.72	78.97
13	40.85	40.08	43.69	46	100	29.56	93.78
14	49.86	39.54	47.82	47	100	45.69	86.76
15	52.11	52.37	49.54	48	97.18	93.85	75.49
16	52.11	40.70	46.78	49	96.06	85.40	75.54
17	54.37	45.69	50.25	50	94.93	100	81.43
18	74.65	44.54	55.73	51	94.93	100	92.26
19	84.79	57.37	55.19	52	92.68	97.70	83.02
20	81.41	61.28	58.59	53	90.42	100	86.09
21	100	56.29	61.62	54	88.17	93.09	88.74
22	100	81.56	67.29	55	90.42	100	91.21

(continued)

Table 6.1 (continued)

Sample number	Mixing-Cl^-	Mixing-Br^-	Mixing-$\delta^{18}O$	Sample number	Mixing-Cl^-	Mixing-Br^-	Mixing-$\delta^{18}O$
23	100	83.87	65.03	56	85.92	97.70	89.33
24	100	85.40	68.13	57	90.42	93.85	91.23
25	100	97.70	59.84	58	88.17	100	92.96
26	100	90.78	69.12	59	97.18	86.94	92.57
27	100	66.81	68.85	60	92.68	80.03	96.31
28	100	95.39	64.46	61	92.68	100	100
29	100	100	76.87	62	92.68	83.87	96.10
30	100	75.72	64.20	63	97.18	88.48	97.13
31	100	83.87	65.22	64	99.44	90.01	97.92
32	100	100	79.21	65	100	100	100
33	98.31	96.93	73.18				

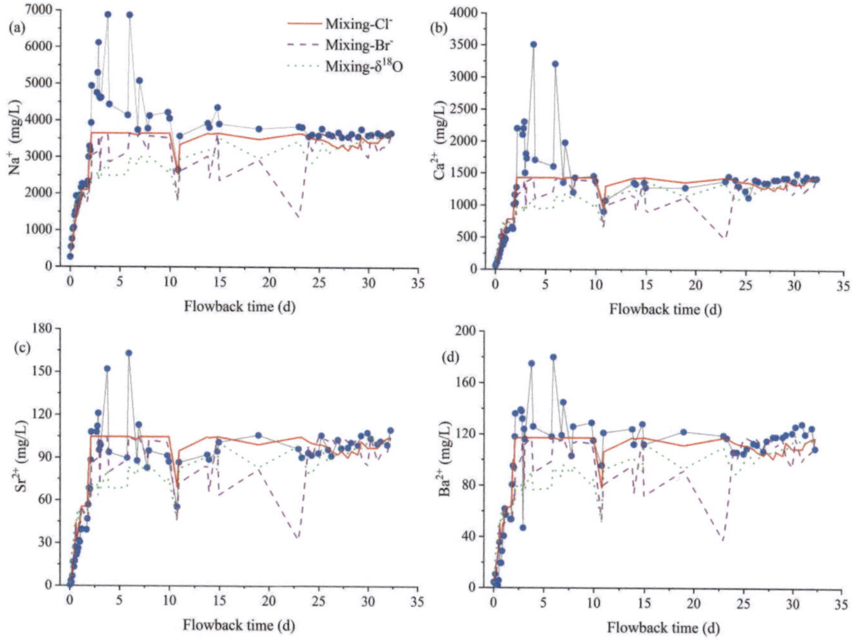

Fig. 6.2 Evolution of measured and simulated ionic concentrations in the horizontal well: **a** Na^+, **b** Ca^{2+}, **c** Sr^{2+}, **d** Ba^{2+}

Given the significant fluctuation in Br^- concentration in the flowback fluid, the calculated mixing concentrations using Br^- tracer exhibits pronounced fluctuations as well. $\delta^{18}O$ and δ^2H also serve as effective conservative tracers, but isotope samples are susceptible to contamination by organic components, leading to reduced accuracy (Owen et al. 2020). The use of Cl^- as tracer yields a more effective mixing simulation. During hydraulic fracturing, the hydraulic fractures exert higher pressure (46.5–60.3 MPa) than the reservoir rock (30.7 MPa). As a result, certain fracturing fluids infiltrate and occupy the micropores of the reservoir under high pressure, with some fluid subsequently draining from the formation due to formation pressure during the flowback process. In the meanwhile, the intricate hydraulic fractures and heterogeneous formation water contribute to uneven mixing processes (Fig. 6.4a). During the initial flowback stage, the flowback fluid primarily consists of fracturing fluid that remained within the wellbore and hydraulic fractures. Subsequently, the fracturing fluid is gradually displaced by formation water. The proportion of formation water steadily increases over time. Ultimately, driven by reservoir pressure and natural gas, the fracturing fluid is entirely displaced by formation water (Fig. 6.4a, He et al. 2022).

(3) Mixing modeling in the directional well

Cl^- was used as a tracer to simulate the mixing processes of ionic concentrations based on the characteristics of fracturing fluid and formation water end members.

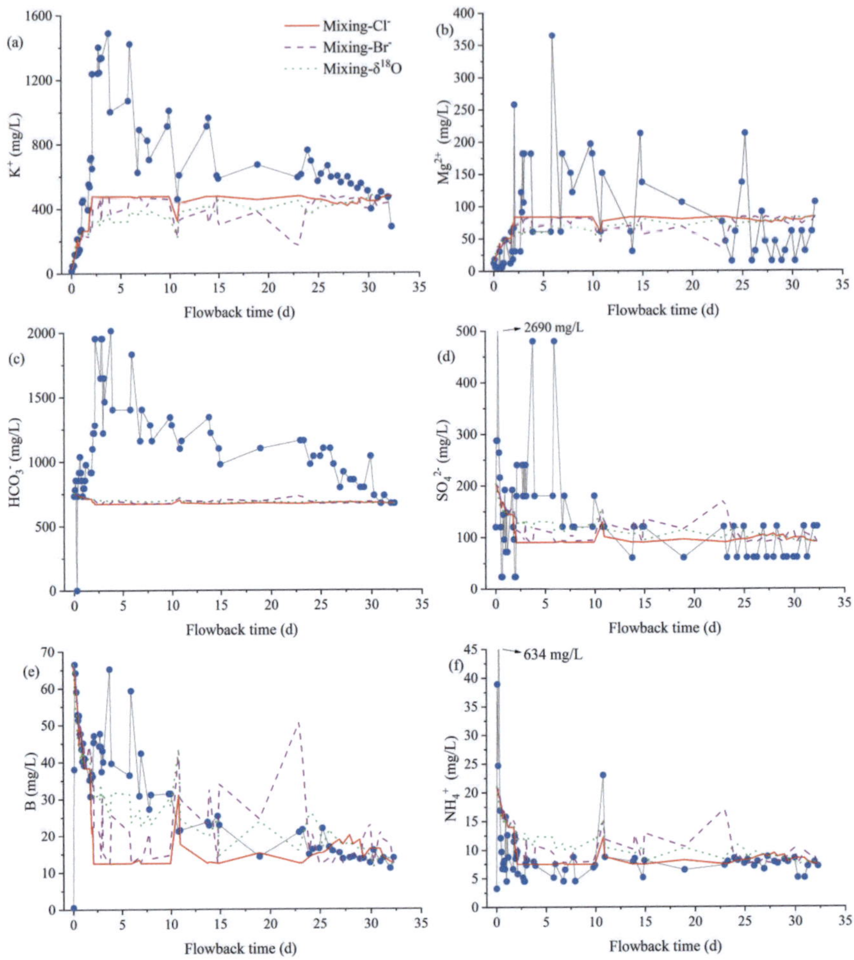

Fig. 6.3 Evolution of measured and simulated ionic concentrations in the horizontal well: **a** K^+, **b** Mg^{2+}, **c** HCO_3^-, **d** SO_4^{2-}, **e** B, **f** NH_4^+

The results indicated that the effectiveness of Cl^- as a mixing tracer was notably poorer in the directional well when compared to its performance in the horizontal well (Fig. 6.5). The mixing line exhibited significant deviation from the measured ionic concentrations. Taking Ca^{2+} as an example, the measured Ca^{2+} concentrations were fell below the mixing line, demonstrating that the fracturing fluid firstly mixed with the formation water characterized by relatively low ionic concentrations in the upper fracturing layer (P_2h_4). The Ca^{2+} concentration in the sample collected at 107 h exhibited a significant decrease compared to the preceding and subsequent samples. Conversely, the HCO_3^- concentration suddenly increased. Given the elevated HCO_3^- concentration in the fracturing fluid used in the directional well, this sample could

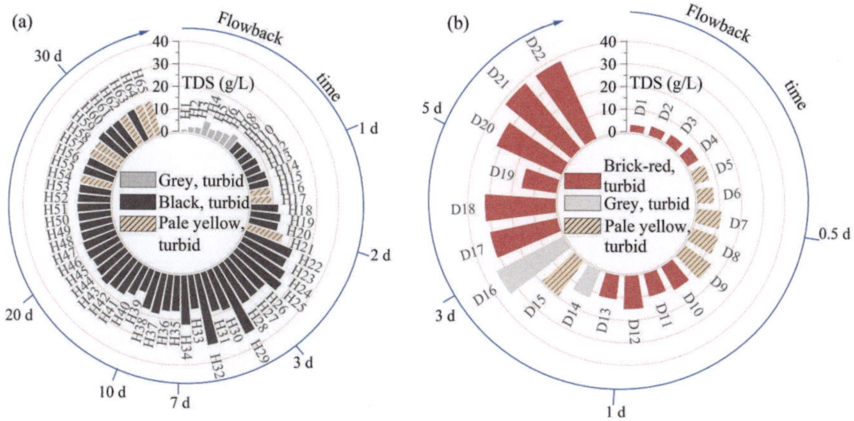

Fig. 6.4 Variation of flowback fluid color and TDS: **a** horizontal well; **b** directional well

represent a mixture of the fracturing fluid retained in the lower fracturing layer's wellbore and the flowback fluid from the upper fracturing layer (Fig. 6.1). Consequently, the TDS values will significantly decrease as the flowback fluid from the lower fracturing layer's wellbore dilutes the water from the upper fracturing layer. Subsequently, there was a general increase in ionic concentrations, indicating the mixing of the flowback fluid and high-salinity formation water from P_1s_1 Member (Fig. 6.4b). The mixing processes in the directional well are notably more intricate in comparison with the horizontal well, due to the interaction of different fracturing layers with varying formation water compositions. It is necessary to acquire the chemical compositions of formation water from each fracturing layer.

6.2 Mineral Dissolution

6.2.1 Halite Dissolution

Unlike the high-ionic-strength brines in formations, water-based fracturing fluids prepared with freshwater generally have a low ionic concentration, which induces mineral dissolution upon entering tight sandstone reservoirs. Halite is highly soluble. Although XRD analysis of tight sandstone cores in this study did not detect halite, scanning electron microscopy revealed minor NaCl crystals within intergranular pores of clay minerals in the core samples from Well *R3* (Fig. 6.6). Similarly, Zhao et al. (2022) documented the halite occurrence of tight sandstones in the Permian Taiyuan Formation (Ordos Basin). In this study, laboratory experiments revealed that with increasing water–rock reaction time, Cl^- and Na^+ in solutions resulting from reactions between tight sandstones from five wells and fracturing fluid exhibited an

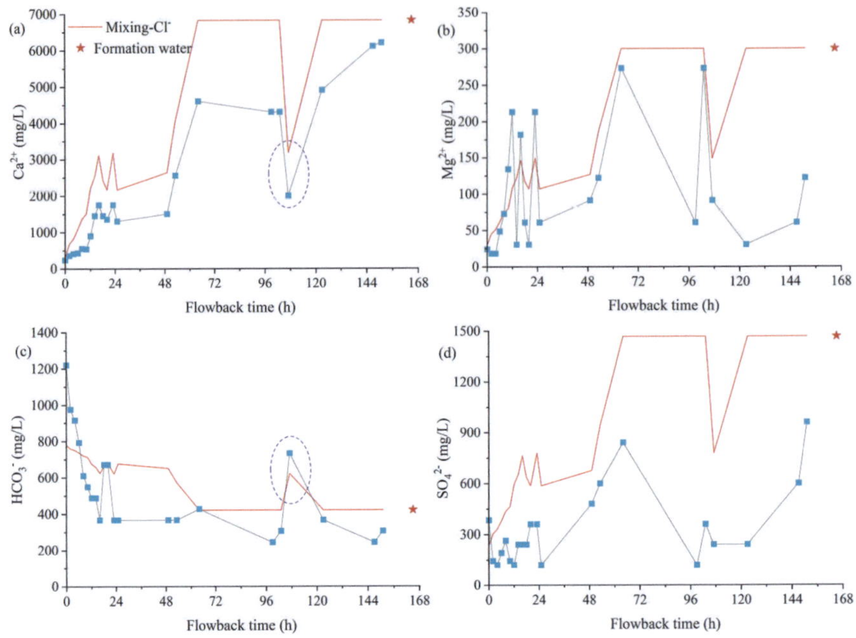

Fig. 6.5 Evolution of measured and simulated ionic concentrations in the directional well: **a** Ca^{2+}, **b** Mg^{2+}, **c** HCO_3^-, **d** SO_4^{2-}

increasing trend (Fig. 6.7). In the case of Well *R3*, where halite crystals were identified, the increases in Cl^- and Na^+ concentrations after water–rock interactions were significantly greater than those in the four wells without halite crystals. Furthermore, SEM analyses showed no residual halite crystals in any of the core samples following solution treatment. At a water–rock ratio of 4/1 (mL: g), fracturing fluid treatment of core samples from Well *R3* resulted in maximum increases of 166.19 mg/L in Cl^- and 58 mg/L in Na^+. In contrast, the maximum increases in Cl^- and Na^+ concentrations in the other four wells were generally below 80 mg/L and 40 mg/L, respectively, providing additional evidence for halite dissolution during hydraulic fracturing.

In actual reservoir fracturing environments, fracturing fluid is typically confined to fractures and pores, where it reacts with the minerals on fracture and pore surfaces. Consequently, the contact area between fracturing fluid and rock minerals is limited, making it difficult to achieve the water–rock ratios of laboratory settings. Compared with mixing between fracturing fluid and high-salinity formation brines, the contribution of halite dissolution to Cl^- and Na^+ concentrations in flowback fluid is relatively limited (Zeng et al. 2020; He et al. 2025). The interaction between fracturing fluid and shale may contribute less than 3% to the salinity of the flowback fluid (Zeng et al. 2020).

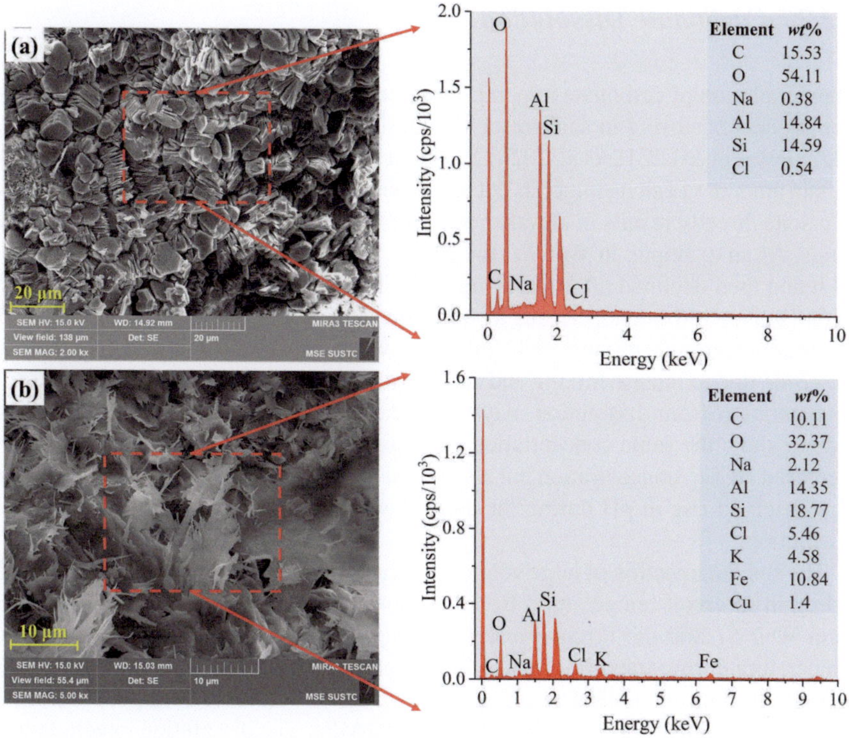

Fig. 6.6 SEM and EDS of halite crystals in the reservoir rock of Well *R3*

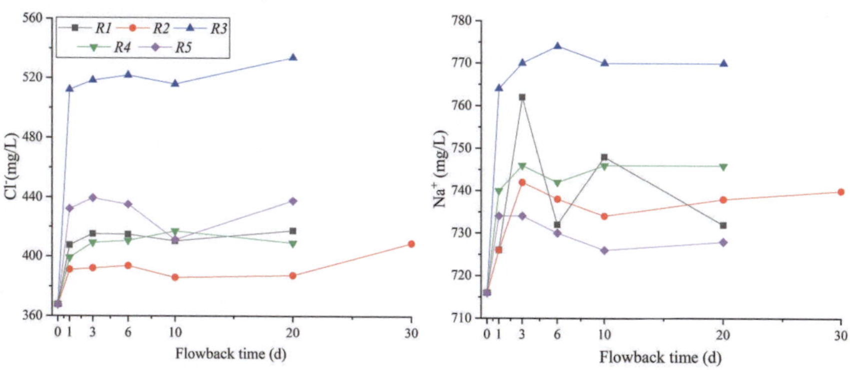

Fig. 6.7 Evolution of Cl^{-}, Na^{+} concentrations after water–rock reactions

6.2.2 Carbonate Dissolution

The dissolution of carbonate minerals represents the most frequent mineral dissolution process occurring in shale reservoirs during hydraulic fracturing (Huang et al. 2022; Jew et al. 2022; He et al. 2026). The tight sandstones in the Sulige area contain a certain amount of calcite minerals but are generally deficient in dolomite. Dissolution of calcite directly results in elevated concentrations of Ca^{2+} and HCO_3^- in flowback fluids. As an example, in Well *S1*, the HCO_3^- concentration rose within the first 0–68 h and then declined after 68 h (Fig. 4.8). This suggests that the fracturing fluid initially dissolved calcite, causing a rapid increase in HCO_3^- concentration, followed by dilution through mixing with formation water of lower HCO_3^- concentration. However, due to intense mixing and cation exchange processes, the increase in Ca^{2+} concentration from dissolution may be masked, and thus Ca^{2+} in flowback fluids did not show the same concentration trend as HCO_3^-. Calcite dissolution not only increased HCO_3^- concentration but also raised the pH of flowback fluids (Fig. 4.11). The observed rise in pH during the early flowback period further supports calcite dissolution.

The sudden injection of large volumes of fracturing fluid into the reservoir caused a drop in reservoir temperature. Based on flowback fluid temperatures (40–50 °C) from Well *S1* and the original reservoir temperature (104–109 °C), the reservoir temperature during fracturing was estimated at approximately 70 °C. The operating pressure during hydraulic fracturing ranged from 46.5 to 60.3 MPa, with an estimated average fracturing pressure of around 50 MPa. The dissolution–precipitation behavior of calcite in fracturing fluid under reservoir conditions (70 °C, 50 MPa) was modeled using PHREEQC. The simulation showed that calcite in fracturing fluids had a saturation index of − 1.96 under these conditions, indicating undersaturation and continued dissolution capacity for reservoir calcite. Without mixing with formation water, fracturing fluid can dissolve calcite until saturation, increasing Ca^{2+} concentration from 64 mg/L to 580 mg/L and HCO_3^- from 638 mg/L to 2422 mg/L. Additional modeling revealed that during the first 12 h of flowback, calcite in all samples remained undersaturated (Fig. 6.8). Therefore, during early fracturing when formation water mixing is limited, calcite dissolution in the reservoir is consistently triggered.

In field operations, gas wells rich in carbonate minerals are commonly acidized with preflush acids (such as HCl) prior to fracturing to dissolve carbonates, increasing pore and fracture networks for subsequent fracturing. Carbonate minerals such as calcite are more susceptible to dissolution under acidic conditions. In the field hydraulic fracturing experiment, two wells utilizing a variable-viscosity slickwater system showed high initial HCO_3^- concentrations in flowback fluids. This resulted both from the relatively high HCO_3^- content of the fracturing fluid and from pre-acid treatment that dissolved calcite, allowing subsequent fracturing fluids to mix with dissolution products, thereby producing elevated HCO_3^- concentrations in early flowback (Figs. 4.8 and 6.5).

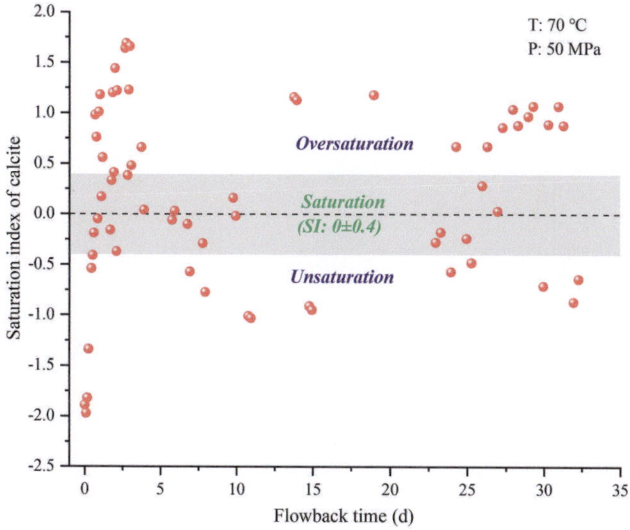

Fig. 6.8 Saturation index of calcite in the flowback fluids from Well *S1* under simulated reservoir fracturing conditions

Laboratory experiments revealed that calcite minerals in all five core samples significantly decreased or even disappeared after fracturing fluid treatment. For instance, in Well *R1*, all calcite minerals (0.4%) dissolved completely within the first three days of reaction, showing a pronounced decrease in the remaining calcium content in the core and a rapid rise in Ca^{2+} concentration in the solution (Fig. 6.9). However, after six days of reaction, XRD results detected the reappearance of some calcite, accompanied by an increase in Ca content in the core and a corresponding decrease in Ca^{2+} concentration in the solution. These observations suggest that, in the early stages of hydraulic fracturing, low-ion fracturing fluids promote the dissolution of calcite minerals in tight sandstones. In the control experiment using deionized water, calcite minerals were completely dissolved during the entire water–rock reaction. For Well *R3*, characterized by a higher initial calcite content, the calcite proportion decreased from 18% to about 10% after exposure to fracturing fluid. The Ca content in the core was reduced from 50,787 mg/kg in the untreated rock to about 32,869 mg/kg after reaction. These results demonstrate that the injection of fracturing fluid into tight sandstone formations induces calcite dissolution. However, because the fracturing fluid used was an alkaline guar-based fluid, the degree of calcite dissolution was lower than that observed in the deionized water environment.

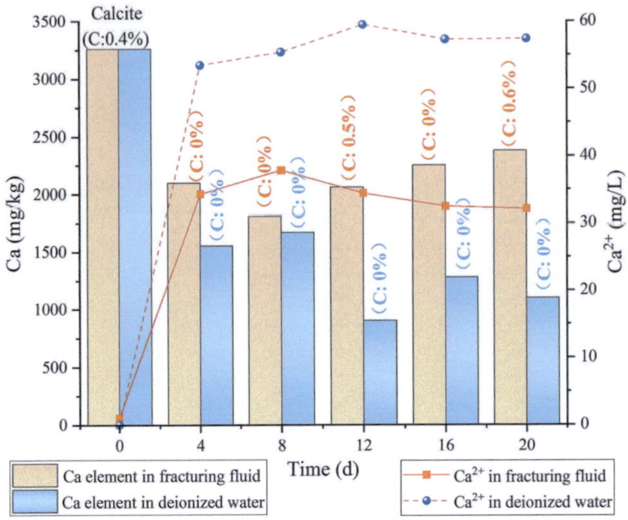

Fig. 6.9 Ca element, Ca^{2+}, and calcite evolution in deionized water and fracturing fluid environments

6.2.3 Dissolution of Other Minerals

XRF results show that elemental levels in the cores decreased after fracturing fluid exposure, suggesting that in addition to dissolving halite and calcite, the fluid also mobilized other soluble minerals, leading to elements release (e.g., Sr, Mn) from the rock (Fig. 6.10). SEM images showed that detrital minerals were corroded after fracturing fluid treatment, resulting in the formation of abundant dissolution pores (Fig. 6.11). This indicates that, besides carbonate cement dissolution, other soluble detrital minerals also dissolve during hydraulic fracturing, increasing K^+, Mg^{2+}, and SO_4^{2-} in the reacted fluid. A experiment on tight siltstone-shale interactions with fracturing fluid revealed that Al-bearing detrital minerals started dissolving immediately upon contact (Herz-Thyhsen et al. 2019). In this study, the clay mineral content in some core samples decreased after fracturing fluid treatment. With prolonged water–rock interaction, some detrital clay minerals were corroded, the rock became more brittle, and clay minerals attached to quartz grains detached, further reducing the clay content.

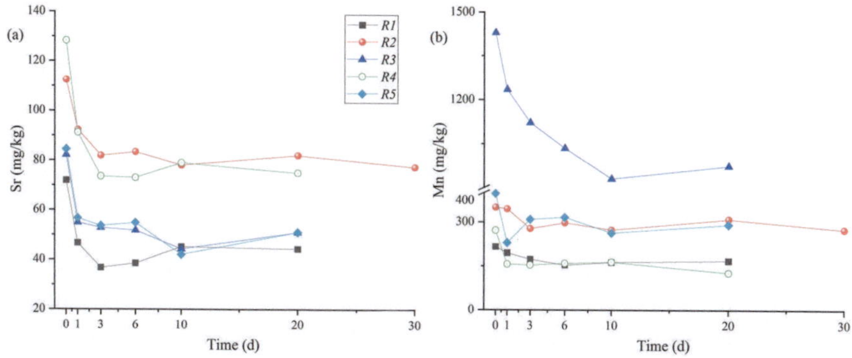

Fig. 6.10 Evolution of Sr (**a**) and Mn (**b**) in the rock after water–rock reaction

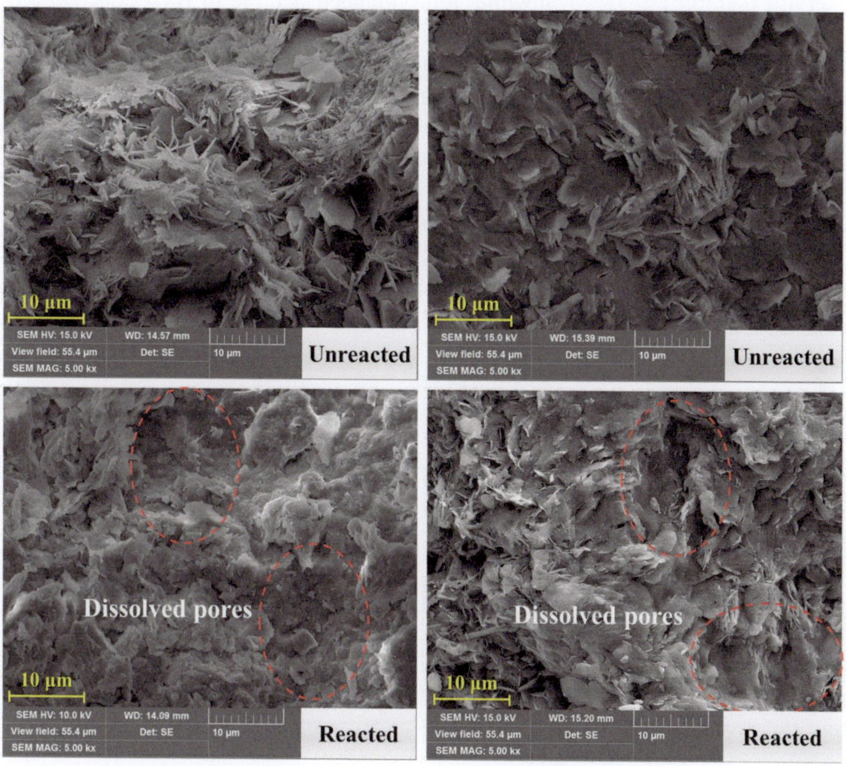

Fig. 6.11 Microscopic surface morphology changes of core samples before and after fracturing fluid treatment

6.3 Mineral Precipitation

6.3.1 *Barite Precipitation*

When fracturing fluid interacts with tight sandstone, its low ionic concentration typically does not trigger mineral precipitation. Nevertheless, hydraulic fracturing not only generates artificial fractures but also connects natural pores and fractures in the formation, facilitating the mixing of formation water with fracturing fluid that triggers mineral precipitation. Barite ($BaSO_4$) is one of the most common precipitates formed during hydrocarbon production (Paukert Vankeuren et al. 2017; Xiong et al. 2020). Taking Well *S1* as an example, the addition of ammonium persulfate breaker and the oxidation of pyrite during fracturing introduced abundant SO_4^{2-} into the reservoir (samples 2–7). Meanwhile, a pronounced decline in Ba^{2+} concentration (2.25–10.6 mg/L) during the early flowback stage (Fig. 6.12) indicates the formation of barite precipitation (Eq. 6.2). Similarly, Matecha et al. (2022) reported that sulfate release from pyrite oxidation during shale hydraulic fracturing rapidly induced barite precipitation, resulting in a significant decrease in Ba^{2+} concentration in the reservoir solution.

$$Ba^{2+} + SO_4^{2-} = BaSO_4 \text{ (Barite)} \tag{6.2}$$

A comparison between simulated and observed Ba^{2+} concentrations indicates that barite precipitation reduced Ba^{2+} in early flowback fluids by as much as 28 mg/L

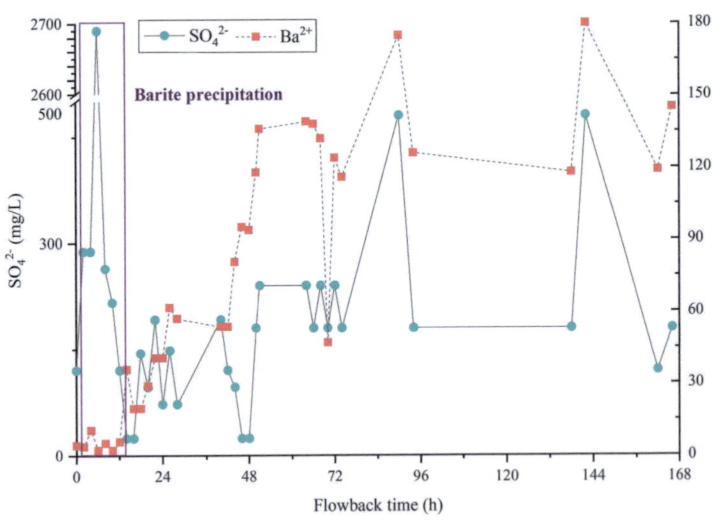

Fig. 6.12 Evolution of Ba^{2+} and SO_4^{2-} concentrations of the flowback fluid in Well *S1*

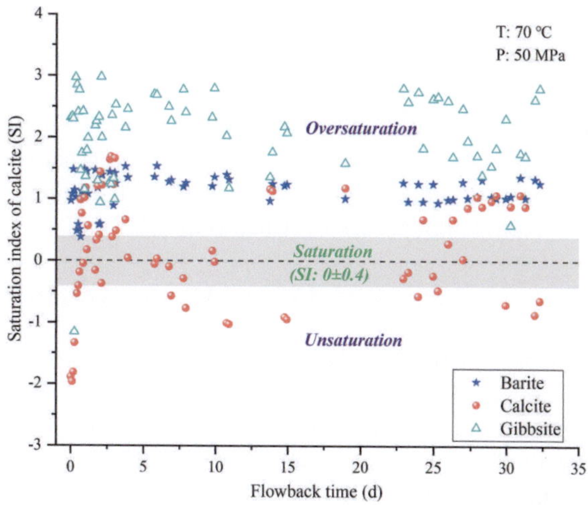

Fig. 6.13 Saturation index of barite, calcite, and gibbsite in the flowback fluid from Well *S1*

(Fig. 6.2d). Subsequently, the decline in SO_4^{2-} concentration and the influx of high-Ba^{2+} formation water led to an evident increase in Ba^{2+} concentration, with SO_4^{2-} and Ba^{2+} exhibiting a similar variation trend (Fig. 6.12). Because of the high-Ba^{2+} formation water in tight sandstone reservoirs, the SO_4^{2-} concentration essentially governs barite precipitation during hydraulic fracturing. Further simulations under reservoir fracturing conditions (70 °C, 50 MPa) reveal that barite minerals remain in a supersaturated state during the entire flowback process (Fig. 6.13), implying a persistent tendency toward precipitation. Once the flowback fluid accumulates in zones favorable for crystallization, severe barite precipitation can occur. Additionally, the injection of fracturing fluid decreases the reservoir temperature, and the solubility of barite diminishes with decreasing temperature. Therefore, barite scale is more likely to occur during the low-temperature stage of early fracturing and in the cooler regions near the wellbore.

6.3.2 Calcite Precipitation

Formation water typically contains a high concentration of Ca^{2+}, while alkaline fracturing fluids (e.g., guar-based fluids) exhibit elevated CO_3^{2-} concentrations. The mixing of fracturing fluid with formation water under near-neutral pH conditions may induce secondary calcite precipitation (Eq. 6.3). Unlike sulfate minerals (e.g., barite and gypsum), the solubility of calcite in aqueous solution decreases with increasing temperature. Following hydraulic fracturing, the higher temperatures in distal fracture zones of the reservoir promote calcite precipitation. Simulation results indicate that the flowback fluid samples exhibit high calcite saturation indices during the early-mid and late flowback stages, suggesting a supersaturated state (Fig. 6.13). During the early to middle stages, mineral dissolution and mixing with formation water increase Ca^{2+} and HCO_3^- concentrations, whereas in the later stage, depletion of fracturing fluid, temperature rise, and pressure decline reduce the solubility of calcite in the solution. Therefore, calcite precipitation primarily occurs during the early-mid and late stages of hydraulic fracturing. Given that fracturing fluids typically contain high HCO_3^- and low Ca^{2+}, whereas formation waters in the study area show the opposite trend (high Ca^{2+}, low HCO_3^-), mixing between the two often triggers co-precipitation processes (Chen et al. 2013).

$$Ca^{2+} + CO_3^{2-} = CaCO_3 \text{ (Calcite)} \tag{6.3}$$

Under laboratory conditions without formation water, minerals display pronounced dissolution when exposed to the fracturing fluid. In controlled laboratory environments, the later phase of the water–rock interaction favors the formation of calcite (Fig. 6.9). For instance, in Well *R1* fracturing fluid–tight sandstone experiment, no calcite minerals were detected after 1 or 3 days of reaction, while small amounts of calcite were identified after 6 and 20 days. Additionally, fluoride present in the guar-based fracturing fluid leads to elevated F^- concentrations, which may react with high Ca^{2+} concentrations to form fluorite (CaF_2) precipitates. A simulation based on the laboratory-measured chemical composition of the guar fracturing fluid under Well *S1* reservoir conditions showed that the saturation index of fluorite (CaF_2) was -0.87, implying an undersaturated state. However, when the Ca^{2+} concentration increases from 1 mg/L to 3 mg/L, the solution reaches saturation with respect to fluorite. Therefore, after the fracturing fluid breaks gel, the dissolution of calcite and influx of Ca^{2+}-rich formation water are likely to induce fluorite precipitation within the reservoir.

6.3.3 Precipitation of Other Minerals

Formation water contributes Fe and Al ions, and additional Fe (from pyrite oxidation and chlorite dissolution) and Al (from clay mineral dissolution and oxidation) are released through water–rock interaction processes, leading to high Fe and Al concentrations in the flowback fluid. In Well *S1*, the concentrations of Fe and Al in the flowback fluid reach maximum values of 79.33 mg/L and 14 mg/L, respectively. Precipitation of $Fe(OH)_3$ and $Al(OH)_3$ are likely to form during fracturing stimulation. Additional simulations show that the gibbsite mineral phase in the Well *S1* flowback fluid is supersaturated, with a high saturation index (Fig. 6.13), implying a significant likelihood of precipitation.

Laboratory observations reveal that after water–rock interaction between fracturing fluid and tight sandstone, newly formed chlorite and Ti-bearing mineral precipitates appear on the sample surfaces (Fig. 6.14). Herz-Thyhsen et al. (2019) also found in laboratory experiments that aluminosilicate precipitates began to form about 27 h after tight mudstone was treated with fracturing fluid. Overall, in the absence of formation water, fracturing fluids primarily induce mineral dissolution within the

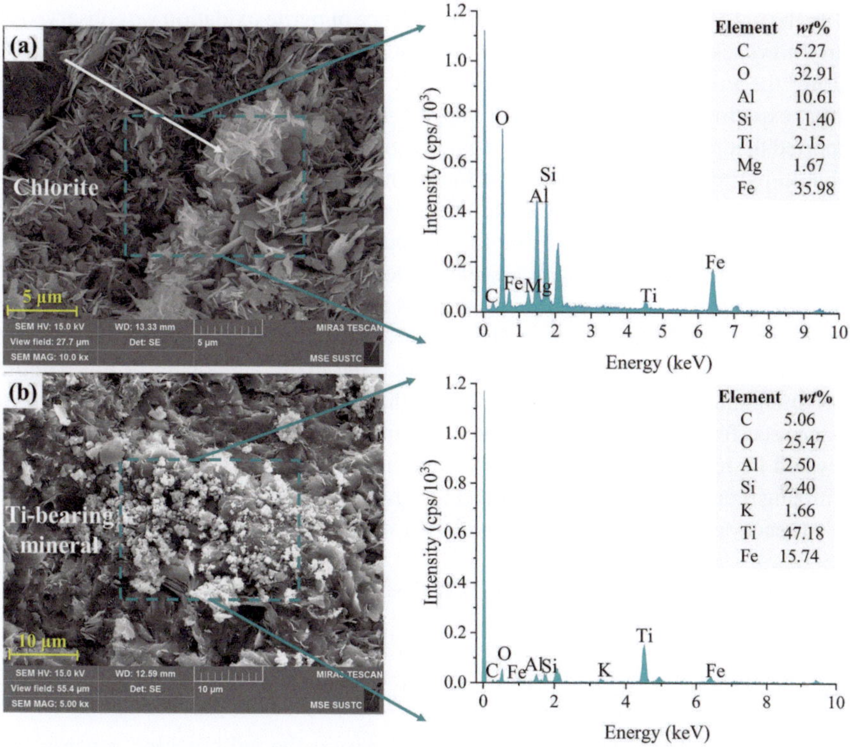

Fig. 6.14 Microscopic morphology of newly formed mineral precipitates in the water-rock experiment (**a** Chlorite; **b** Ti-bearing mineral)

reservoir. However, with continued dissolution and as mineral phases in the solution approach saturation, the subsequent mixing with saline formation water can readily trigger secondary mineral precipitation during the later stages of fracturing.

6.4 Pyrite Oxidation

In low-permeability reservoirs, pyrite typically occurs in association with organic matter, and its oxidation is regarded as a key water–rock interaction process during shale gas production (Harrison et al. 2017; Xu et al. 2018). Owing to the typically low abundance of pyrite in tight sandstone reservoirs, its oxidation process during hydraulic fracturing is frequently neglected. This study documented significant pyrite oxidation during tight gas fracturing stimulation of the study area. Although pyrite was not detected in mineral composition analyses due to its low abundance and reservoir heterogeneity, SEM observations revealed localized pyrite morphologies (Figs. 3.6 and 3.7).

In the field-fractured *Well S1*, distinct evidence of pyrite oxidation was observed in early flowback samples, where both Fe and SO_4^{2-} concentrations increased markedly during the initial flowback stage. Huang et al. (2020) conducted hydraulic fracturing experiments using river water in pyrite-rich shale reservoirs (3.8% pyrite) and reported that pyrite oxidation was insignificant. Notably, pyrite oxidation was negligible in pyrite-rich shale formations but pronounced in pyrite-deficient tight sandstone reservoirs. This difference can be attributed to the use of clean water in shale fracturing tests, while Well *S1* employed a guar-based fracturing fluid containing a breaker. In the fourth flowback sample, collected six hours after fracturing, NH_4^+ and SO_4^{2-} concentrations were as high as 634 mg/L and 2690 mg/L, respectively, far exceeding those of other samples. The sample also exhibited a pH of 2.17, much lower than other samples. These results suggest that the main component of this sample was ammonium persulfate ($(NH_4)_2S_2O_8$), the oxidizing breaker typically used in guar-based fracturing. Samples collected before and after this interval also showed elevated Fe levels (13.53–47.29 mg/L) and acidic conditions (pH 2.17–5.31). The oxidizing breaker not only facilitated gel degradation but also significantly enhanced redox reactions, particularly pyrite oxidation within the reservoir (Eq. 6.4, Fig. 6.15).

$$2FeS_2\ (\text{Pyrite}) + 15(NH_4)_2S_2O_8 + 16H_2O = 2Fe^{3+} + 34SO_4^{2-} + 30NH_4^+ + 32H^+ \quad (6.4)$$

$$2Mn^{2+} + 5(NH_4)_2S_2O_8 + 8H_2O = 2MnO_4^- + 10SO_4^{2-} + 10NH_4^+ + 16H^+ \quad (6.5)$$

The strong acidic environment generated during the gel-breaking process can promote the acidification of chlorite minerals, accompanied by the dissolution of calcite (Khan et al. 2021). However, during this stage, no significant increase in Ca^{2+} or Mg^{2+} concentrations was observed in the flowback fluid. This indicates that chlorite experienced limited acidification within a short time and contributed minimally to Fe variations in the flowback fluid. Moreover, pre-existing dissolution of fracture-surface calcite before gel breaking explains the limited calcite dissolution observed during this phase. Further mixing simulations revealed that pyrite oxidation

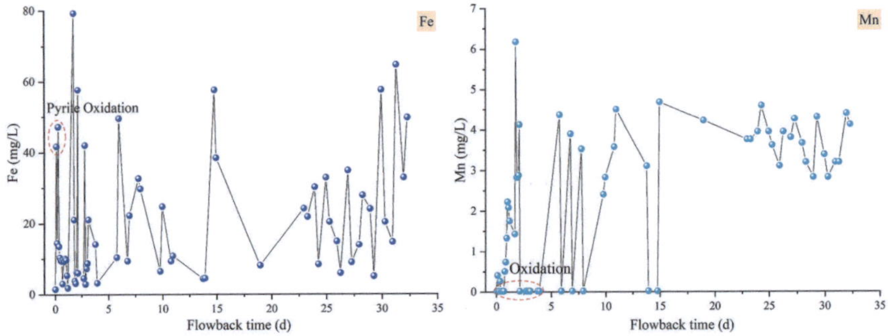

Fig. 6.15 Evolution of Fe and Mn concentrations of the flowback fluid in Well *S1*

could increase Fe concentration in the flowback fluid by up to 40 mg/L. During this period, Mn concentrations were mostly below detection (<0.05 mg/L), implying Mn^{2+} oxidation (Eq. 6.5, Fig. 6.15) with up to 1.7 mg/L oxidized. In contrast, elevated Al concentrations (9.3–14.0 mg/L) were detected, suggesting that the strong acid and oxidation conditions during gel breaking promoted the oxidative dissolution of aluminosilicates, releasing Al into the flowback fluid. In Well *S3* utilizing variable-viscosity slickwater, Al and Mn concentrations were markedly lower than in guar-based wells (Well *S1*, Well *S2*), suggesting stronger redox-driven metal release in the guar-based system.

Laboratory water–rock experiments showed that pyrite morphology in tight sandstone cores remained largely intact after deionized water treatment, with no obvious signs of oxidation (Fig. 6.16). Pyrite was also identified in one deionized-water sample, comprising 0.7% of the whole rock. This confirms that pyrite oxidation does not occur significantly in the absence of oxidizing agents. In contrast, SEM images after fracturing-fluid exposure showed no discernible pyrite features. In *R1*, *R3*, and *R5* wells where pyrite morphology was observed, SO_4^{2-} concentrations increased by 14.7, 15.3, and 10.2 mg/L after fracturing-fluid treatment, compared with 13.3, 15.0, and 10.8 mg/L after deionized-water exposure. Overall, SO_4^{2-} concentrations after fracturing-fluid treatment were slightly higher than in the deionized-water environment; however, due to the low pyrite content in tight sandstone, the increase in SO_4^{2-} from pyrite oxidation was limited. Li (2021) conducted pyrite oxidation experiments in shale reservoirs and found that, under oxygen-rich conditions, both deionized water and fracturing fluids reacted with shale powders to induce a certain degree of pyrite oxidation. In conclusion, pyrite oxidation is expected to occur during hydraulic fracturing under oxygen-rich or oxidizing conditions.

Fig. 6.16 The microscopic morphology of pyrite after deionized water treatment

6.5 Ion Exchange and Adsorption

Ion exchange/adsorption–desorption is a fundamental geochemical interactions within aqueous-mineral systems. Since the solid phase of rocks generally carries negative surface charges, cations in the fluid (e.g., Ca^{2+}, Na^+, Sr^{2+}, B) participate in ion exchange/adsorption–desorption processes to neutralize the mineral surface charge. Given that clay minerals exhibit extensive surface negative charges, ion exchange/adsorption between aqueous solutions and clays is especially significant (Tipper et al. 2021; Huang et al. 2022). Different cations exhibit distinct affinities toward clay surfaces, with higher-valence cations showing greater adsorption potential. The commonly observed order of cation adsorption affinity is $Fe^{3+} > Al^{3+} > Sr^{2+} > Ca^{2+} > Mg^{2+} > Cs^+ > NH_4^+ > K^+ > Na^+ > Li^+$ (Qian et al. 2012). Tight sandstone reservoirs contain abundant clay minerals. The injection of external fluids during hydraulic fracturing inevitably triggers a series of ion exchange and adsorption processes. Moreover, hydraulic fracturing provides an excellent setting for investigating ion exchange and adsorption–desorption processes.

6.5.1 *Ion Exchange Between $Na^+(K^+)$ and $Ca^{2+}(Mg^{2+})$*

In natural aquifer systems, ion exchange processes typically occur between Na^+ (K^+) and Ca^{2+} (Mg^{2+}). Simulation results show that within 2–40 h after flowback (when the formation water mixing ratio is below 50%), the measured $Na^+ + K^+$ concentrations in Well *S1* were higher than the simulated end-member mixing concentrations, whereas $Ca^{2+} + Mg^{2+}$ values were lower. The molar ratio of concentration changes, $\Delta(Ca + Mg)/\Delta(Na + K)$ approached $-$ 1:2 (Fig. 6.17). This suggests that substantial cation exchange occurred at formation water proportions below 50%, decreasing Ca^{2+} and Mg^{2+} concentrations while increasing Na^+ and K^+ levels in the flowback fluid (Eq. 6.6). Although significant calcite dissolution occurred after the fracturing fluid entered the reservoir, increasing Ca^{2+} concentrations in the solution. The presence of abundant clay minerals, particularly kaolinite with higher affinity for divalent cations (e.g., Ca^{2+}, Mg^{2+}) than for monovalent ones (e.g., Na^+, K^+), resulted in rapid ion exchange between Ca^{2+} in the fluid and Na^+ (K^+) on the rock surface (Civan 2015). This accounts for the observed deviation in the mixing simulation, where Ca^{2+} concentrations fell below and HCO_3^- concentrations rose above the predicted mixing line. Similar ion exchange phenomena have also been observed in shale gas and coalbed methane fracturing stimulation (Baublys et al. 2015; Li et al. 2020; Phan et al. 2020; Huang et al. 2022).

$$2Na(K)X + Ca^{2+}(Mg^{2+}) \longrightarrow 2Na^+(K^+) + Ca(Mg)X_2 \tag{6.6}$$

Laboratory experiments revealed that after water–rock interactions, the molar increase of Na^+ in the fracturing fluid was significantly greater than that of Cl^-. As

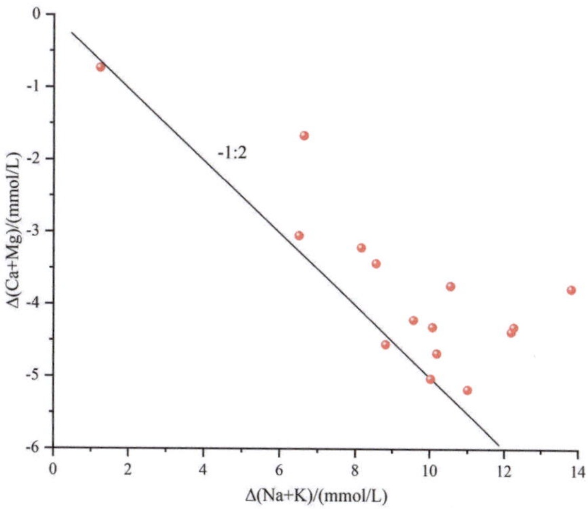

Fig. 6.17 Relationship between the increment of (Ca + Mg) and (Na + K) of flowback fluid within 2 to 40 h for Well *S1*

no albite or sodium montmorillonite was detected in the tight sandstone reservoir, the increased Na^+ and Cl^- concentrations in the post-reaction solution primarily originated from the dissolution of residual halite within the tight sandstone. The dissolution of halite should yield equal or nearly equal increases in the molar concentrations of Na^+ and Cl^- (Eq. 6.7). However, experimental results showed that most post-reaction samples deviated from the 1:1 line in the ΔCl^- and ΔNa^+ diagram, plotting above and to the left of the line, indicating an excess of Na^+ in the solution (Fig. 6.18a). Further analysis of post-reaction solutions showed that the increase in Ca^{2+} concentration was generally lower than that of CO_3^{2-}. Similarly, since anorthite, Ca-montmorillonite, and dolomite were absent in the tight sandstone, the increased Ca^{2+} and CO_3^{2-} concentrations primarily originated from calcite dissolution. The dissolution of calcite should theoretically produce equal molar increases of Ca^{2+} and CO_3^{2-}. Experimental results indicated that most samples deviated from the 1:1 line in the $\Delta\ CO_3^{2-}$ and ΔCa^{2+} diagram, plotting below and to the right of the line, suggesting a Ca^{2+} deficit in solution (Fig. 6.18b). These observations collectively indicate that pronounced ion exchange occurred during the water–rock interaction between the fracturing fluid and the tight sandstone, driven by the abundance of clay minerals. Ion exchange between aqueous Ca^{2+} and solid-phase Na^+ resulted in elevated Na^+ and reduced Ca^{2+} concentrations post-reaction (Eq. 6.8), in agreement with field-scale hydraulic fracturing observations. Because K^+ and Mg^{2+} concentrations in post-reaction solutions were low, no significant ion exchange involving these ions was identified.

$$\text{NaCl (Halite)} \longrightarrow Na^+ + Cl^- \tag{6.7}$$

$$2NaX + Ca^{2+} \longrightarrow 2Na^+ + CaX_2 \tag{6.8}$$

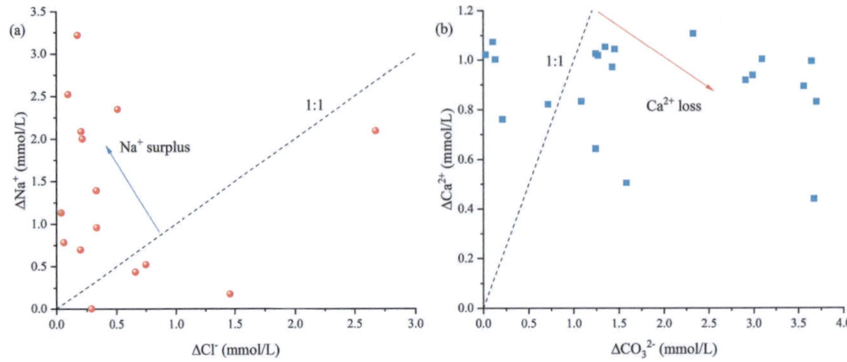

Fig. 6.18 ΔNa^+ and ΔCa^{2+} relationship of the fracturing fluid solution after the water–rock reaction

6.5.2 *Ion Exchange of Sr^{2+}*

The similar ionic radii and charge of Sr^{2+} and Ca^{2+} result in their comparable geochemical behaviors (Appelo and Postma 2005; Huang et al. 2022). Sr^{2+} readily adsorbs onto exchange sites in clay minerals and exhibits exchange behavior similar to that of Ca^{2+} during cation exchange processes. Similar to Ca^{2+}, the measured Sr^{2+} concentrations in Well *S1* during the first 2–50 h of flowback were lower than the simulated mixing concentrations (Fig. 6.19). Simulations of mineral saturation indices showed that strontianite ($SrCO_3$) and celestite ($SrSO_4$), the main Sr-bearing minerals, were undersaturated during this stage, indicating that the decrease in Sr^{2+} concentration in the flowback fluid was not due to mineral precipitation. The $^{87}Sr/^{86}Sr$ ratio serves as a highly sensitive indicator for detecting fluid mixing processes (Zheng et al. 2021). During the identified calcite dissolution period (0–12 h), the measured $^{87}Sr/^{86}Sr$ ratio was slightly higher than the simulated value, whereas from 14–50 h it was significantly lower. Calcite generally possesses a lower $^{87}Sr/^{86}Sr$ ratio (< 0.711). Hence, its dissolution leads to a reduction in the solution's isotopic ratio. However, during calcite dissolution, some Sr-bearing minerals with higher $^{87}Sr/^{86}Sr$ ratios also dissolved, and the concurrent mixing with formation water of high $^{87}Sr/^{86}Sr$ (0.714979) slightly increased the overall measured isotopic ratio above the simulated value. Between 14 and 50 h, as ion exchange intensified, Sr^{2+} in the solution was adsorbed onto the solid phase, resulting in significantly lower $^{87}Sr/^{86}Sr$ ratios than predicted by mixing models (Fig. 6.19). Quantitative modeling further suggests that, within 2–50 h of flowback, cation exchange transferred about 20–60% of Sr^{2+} from the fluid to the solid phase, resulting in a maximum Sr^{2+} reduction of approximately 22 mg/L.

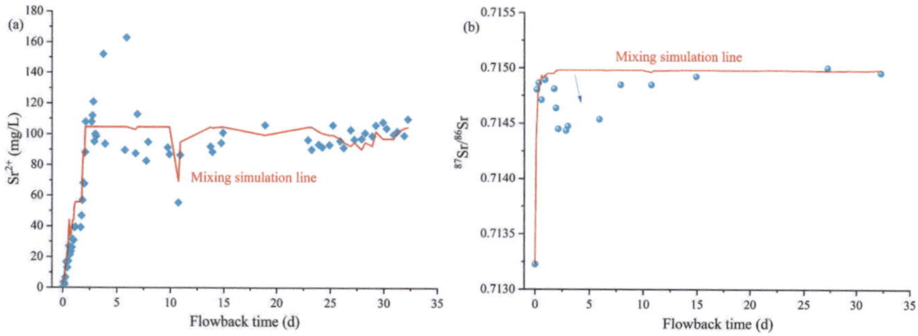

Fig. 6.19 Evolution of measured Sr^{2+}, simulated Sr^{2+} concentrations (**a**), and $^{87}Sr/^{86}Sr$ values (**b**) of the flowback fluid in Well *S1*

6.5.3 Adsorption of Boron (B)

Guar gum fracturing fluids typically employ borate cross-linkers, resulting in elevated B concentrations in flowback fluids from the *S1* and *S2* wells, with peak levels of 66.5 mg/L and 57.3 mg/L, respectively. Simulation of the mixing process in Well *S1* showed that measured B levels during certain intervals exceeded those expected from fracturing-fluid and formation-water mixing, indicating other B sources (Fig. 6.20a). The use of borate cross-linkers results in higher B concentrations in fracturing fluids compared to formation water. After gel breaking, part of the B is adsorbed by reservoir minerals, causing the measured B concentrations in early flowback (2–42 h) to be slightly lower than the mixed-endmember values, consistent with the cation exchange processes involving Ca^{2+} and Sr^{2+}.This may be attributed to the preferential cation exchange of Ca^{2+} and Sr^{2+} over B, resulting in a relatively low overall B adsorption during this stage, averaging only 1.95 mg/L. As formation water continued to mix with the fracturing fluid, the increasing proportion of formation water in the flowback led to further dilution of B, reducing its concentration. When the B concentration in the flowback fluid dropped to approximately 30 mg/L (a decline exceeding 50%), desorption of previously adsorbed B from reservoir rocks occurred, causing a subsequent rise in B concentration. Simulation results indicate that during the flowback period between 44 to 238 h, the average desorbed B amount reached 26.6 mg/L. Tight-reservoir clay minerals commonly adsorb large B quantities, and similar B desorption has been observed in shale fracturing systems (Huang et al. 2022). B desorption in tight sandstones explains the rising B concentration with flowback time in Well *S3*.

High-boron stimulation fluids significantly alter the boron adsorption and desorption equilibrium in subsurface environments. During hydraulic fracturing, fluid movement induced by pressure fluctuations causes continuous mixing between fracturing fluid and formation water, leading to dynamic changes in B adsorption–desorption states. Consequently, B adsorption and desorption may occur throughout the entire fracturing and flowback processes (Fig. 6.20b). For low- or non-B-containing fracturing fluids, the hydraulic fracturing process is primarily characterized by B

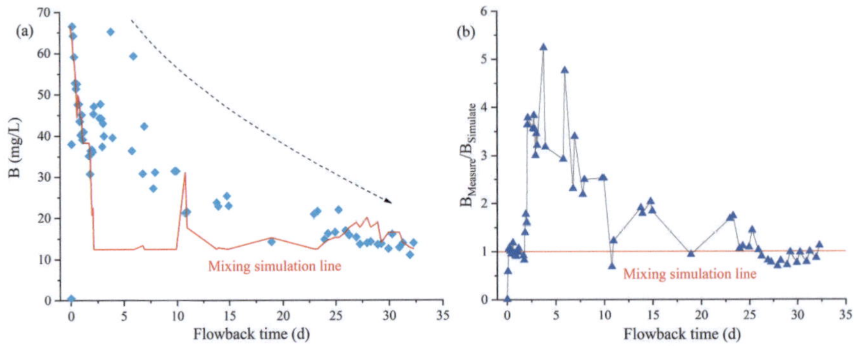

Fig. 6.20 Evolution of measured B, simulated B concentrations (**a**), and $B_{Measure}/B_{Simulate}$ values (**b**) of the flowback fluid in Well *S1*

desorption, resulting in a progressive increase in B concentration with flowback time (Fig. 4.10).

6.6 Geochemical Processes During Hydraulic Fracturing in Low-Permeability Reservoirs

Integrating field-scale fracturing experiments with laboratory water–rock interaction tests, this study systematically characterizes and reveals the geochemical processes during hydraulic fracturing in tight sandstone reservoirs. The fracturing flowback fluid is a complex multiphase mixture dominated by the liquid phase, containing gases and suspended solids. The mixing process controls the chemical composition of the flowback fluid, while fracturing fluid-tight sandstone interactions such as mineral dissolution and precipitation, redox reactions, and ion exchange, also play crucial roles (Fig. 6.21). Hence, both the mineralogical composition of gas-bearing reservoir rocks and the geochemical nature of formation waters exert substantial influence on the effectiveness of hydraulic fracturing operations. The complexity of stimulation fluids, heterogeneous reservoir conditions, and varying fracturing and flowback strategies further complicate the composition of flowback fluids. Nevertheless, by performing high-frequency sampling of flowback fluids and integrating laboratory water–rock interaction simulations, it is possible to elucidate in detail the geochemical processes occurring within the reservoir during hydraulic fracturing, offering critical information for optimizing fracturing operations and improving oil and gas production.

Field analysis of flowback fluids identified multiple geochemical reactions during hydraulic fracturing, including calcite dissolution, pyrite oxidation, barite precipitation, and cation exchange involving Ca^{2+}, Sr^{2+}, Na^{+}, and B. Laboratory experiments revealed the dissolution of halite, calcite, and detrital cements, the precipitation of secondary minerals such as calcite, fluorite, and aluminosilicates, and cation

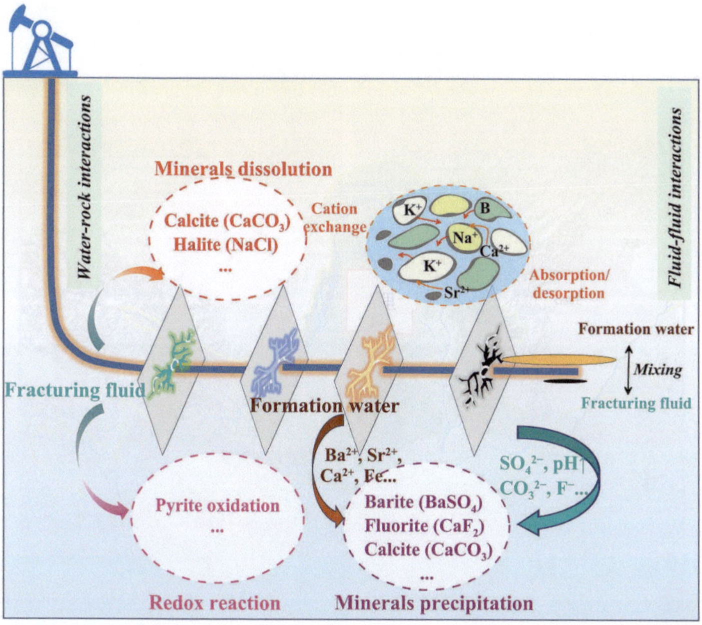

Fig. 6.21 Illustration of the geochemical processes during tight gas fracturing stimulation

exchange between Na^+ and Ca^{2+}, further confirming pyrite oxidation in the presence of oxidants. Integrated field and laboratory results indicate that upon entering the reservoir, fracturing fluid first reacts with the rock, dissolving soluble minerals such as halite, calcite, and detrital cements, thereby creating abundant dissolution pores. Concurrently, interphase ion exchange involving Ca^{2+}, Sr^{2+}, Na^+, and B takes place between the solid and aqueous phases. In the presence of oxygen and oxidants, pyrite oxidation occurs, increasing SO_4^{2-} and Fe concentrations in the flowback fluid. When fracturing fluid enters pores and fractures containing formation water, high concentrations of Ca^{2+}, Ba^{2+}, and Al mix into the fluid, transforming minerals such as calcite, barite, and gibbsite from undersaturated to supersaturated states, leading to secondary mineral precipitation (Fig. 6.21). Because fracturing fluid significantly reduces the local formation temperature, and the solubility of sulfate minerals decreases with decreasing temperature, barite precipitation tends to occur more readily during the early fracturing stage and in near-wellbore low-temperature zones. Calcite tends to precipitate during early and middle fracturing stages due to fluid mixing, while in the later flowback phase, increasing temperature and decreasing pressure reduce solubility, facilitating additional secondary precipitation.

Varied reservoir conditions and fracturing fluid compositions yield distinct geochemical responses across tight gas, coalbed methane, and shale gas systems (He et al. 2026). Deeply buried gas-bearing reservoirs typically contain brine-like formation water enriched in Ca^{2+}, Ba^{2+}, Sr^{2+}, Fe, and Al in tight sandstone and shale formations, facilitating the precipitation of carbonate, barite, and Fe/Al-bearing secondary

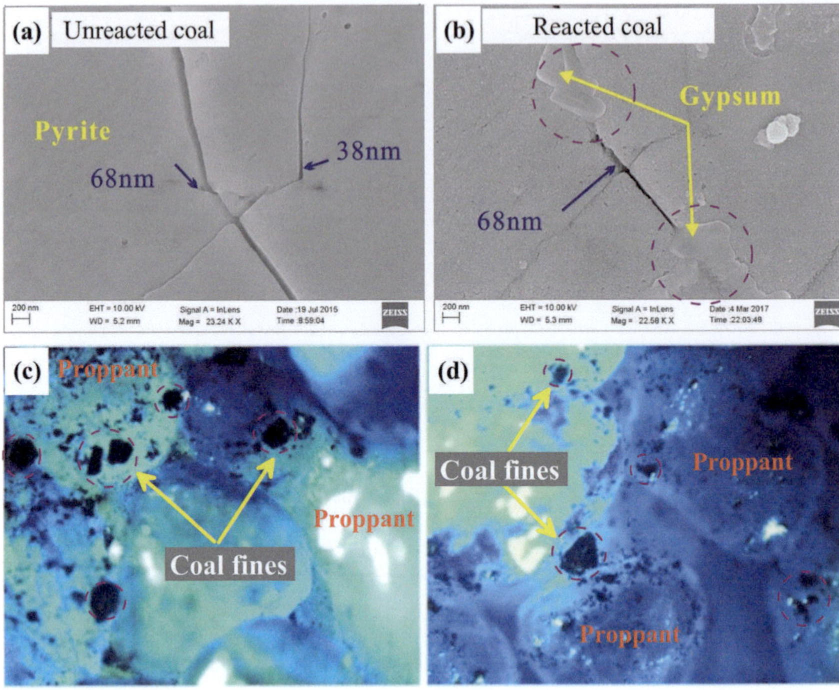

Fig. 6.22 Gypsum precipitation (**a**, **b** Du et al., 2020) and coal fines migration (**c**, **d** Zou et al. 2014)

minerals during fracturing stimulation. In coalbed methane reservoirs, which are generally shallowly buried, formation water exhibits lower TDS and ionic concentrations than those in tight sandstone and shale, resulting in a lower likelihood of secondary mineral precipitation. However, due to strong hydraulic connectivity with overlying aquifers, coal seam water contains higher Ca^{2+} concentrations. Additionally, pyrite-rich coal seams are prone to pyrite oxidation during fracturing, which can readily induce gypsum precipitation (Fig. 6.22a, b). Similarly, in the shale-fracturing fluid laboratory experiment, significant pyrite oxidation and the induced barite precipitation were observed (Fig. 6.23). The low mechanical strength of coal leads to the easy generation of coal fines during the fracturing process. The migration and aggregation of coal fines can block proppant packs and pore throats (Fig. 6.22c, d). Due to the presence of a large amount of clay minerals, significant ion exchange has also been observed in shale gas fracturing flowback fluid (Fig. 6.24), similar to tight sandstone reservoirs (Figs. 6.19 and 6.20). The oxidation and dissolution of organic matter in shale reservoirs may be an important source of heavy metal pollutants in the flowback fluid (Fig. 6.23).

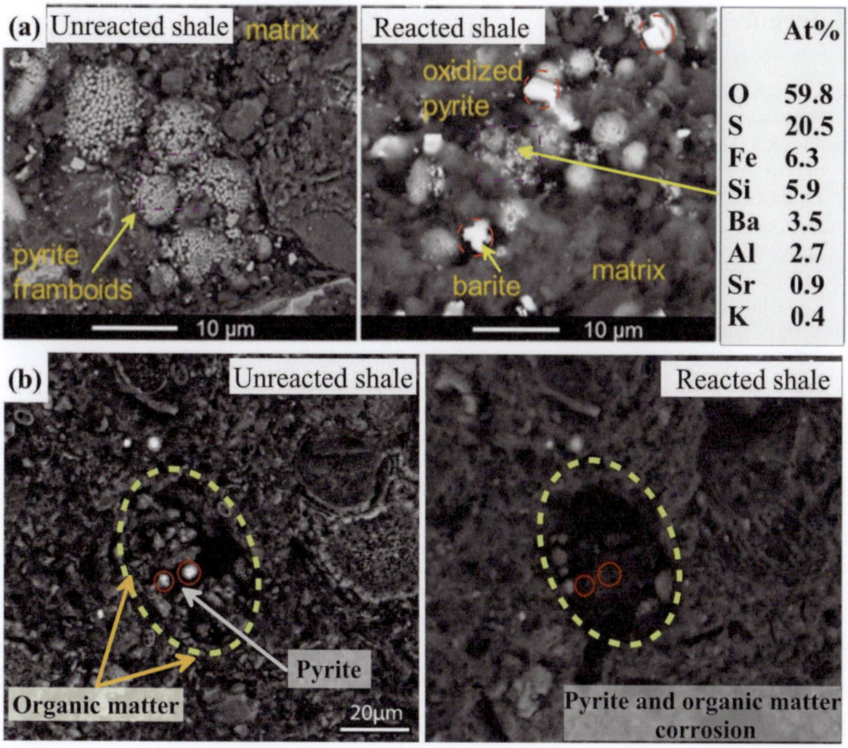

Fig. 6.23 Pyrite oxidation and barite precipitation in shale (**a** Xiong et al. 2020); Pyrite and organic matter oxidation in shale (**b** Raynes et al. 2024)

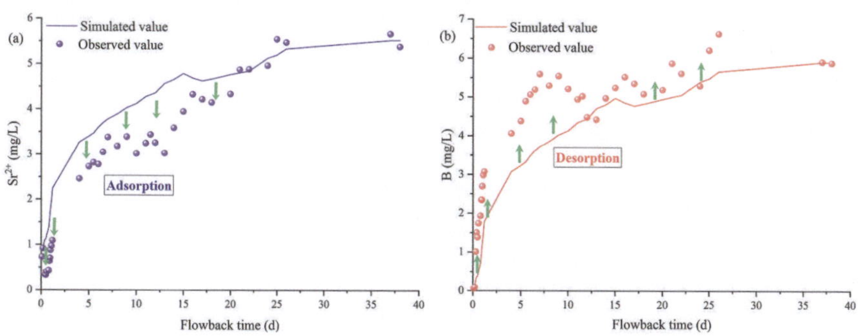

Fig. 6.24 Temporal variations in Sr^{2+} (**a**) and B (**b**) levels in shale gas flowback fluid (*Data source* Huang et al. 2022)

References

Appelo CAJ, Postma D (2005) Geochemistry, Groundwater and Pollution, 2nd edn. Balkema Publishers, Amsterdam

Baublys KA, Hamilton SK, Golding SD et al (2015) Microbial controls on the origin and evolution of coal seam gases and production waters of the Walloon Subgroup; Surat Basin, Australia. Int J Coal Geol 147–148:85–104. https://doi.org/10.1016/j.coal.2015.06.007

Chen J, Qian H, Li P (2013) Mixing precipitation of $CaCO_3$ in natural waters. Water 5(4):1712–1722. https://doi.org/10.3390/w5041712

Civan F (2015) Reservoir formation damage: fundamentals, modeling, assessment, and mitigation (3rd edition). Gulf Professional Publishing, Elsevier. https://doi.org/10.1016/C2014-0-01087-8

Du Y, Fu C, Pan Z et al (2020) Geochemistry effects of supercritical CO_2 and H_2O on the mesopore and macropore structures of high-rank coal from the Qinshui Basin China. Int J Coal Geol 223:103467. https://doi.org/10.1016/j.coal.2020.103467

Harrison AL, Jew AD, Dustin MK et al (2017) Element release and reaction-induced porosity alteration during shale-hydraulic fracturing fluid interactions. Appl Geochem 82:47–62. https://doi.org/10.1016/j.apgeochem.2017.05.001

He X, Li P, Ning J et al (2022) Geochemical processes during hydraulic fracturing in a tight sandstone reservoir revealed by field and laboratory experiments. J Hydrol 612:128292. https://doi.org/10.1016/j.jhydrol.2022a.128292

He X, Li P, Qian H et al (2025) Impacts of additives in hydraulic fracturing technology: an experimental investigation in the Sulige Gas Field China. Environ Earth Sci 84:558. https://doi.org/10.1007/s12665-025-12606-5

He X, Li P, Qian H et al (2026) Biogeochemistry during hydraulic fracturing: a critical review of reservoirs, fluids, processes, and implications. Geoenergy Sci Eng 256:214143. https://doi.org/10.1016/j.geoen.2025.214143

Herz-Thyhsen RJ, Kaszuba JP, Dewey JC (2019) Dissolution of minerals and precipitation of an aluminosilicate phase during experimentally simulated hydraulic fracturing of a mudstone and a tight sandstone in the Powder River Basin, WY. Energy Fuels 33:3947–3956. https://doi.org/10.1021/acs.energyfuels.8b04443

Huang T, Li Z, Mayer B et al (2020) Identification of geochemical processes during hydraulic fracturing of a shale gas reservoir: a controlled field and laboratory water-rock interaction experiment. Geophys Res Lett 47:e2020GL090420. https://doi.org/10.1029/2020GL090420

Huang T, Li Z, Long Y et al (2022) Role of desorption-adsorption and ion exchange in isotopic and chemical (Li, B, and Sr) evolution of water following water-rock interaction. J Hydrol 610:127800. https://doi.org/10.1016/j.jhydrol.2022.127800

Jew AD, Druhan JL, Ihme M et al (2022) Chemical and reactive transport processes associated with hydraulic fracturing of unconventional oil/gas shales. Chem Rev 122(9):9198–9263. https://doi.org/10.1021/acs.chemrev.1c00504

Khan HJ, Spielman-Sun E, Jew AD et al (2021) A critical review of the physicochemical impacts of water chemistry on shale in hydraulic fracturing systems. Environ Sci Technol 55:1377–1394. https://doi.org/10.1021/acs.est.0c04901

Li Y, Tang S, Zhang S et al (2020) Biogeochemistry and water-rock interactions of coalbed methane co-produced water in the Shizhuangnan Block of the southern Qinshui Basin China. Water 12(1):130. https://doi.org/10.3390/w12010130

Li Z (2021) Tracing water-rock interactions of a low-permeability reservoir using multiple isotopes. Ph.D. thesis, Chinese Academy of Sciences. (In Chinese with English abstract)

Matecha RM, Xiong W, Heck WF et al (2022) Experimental investigation of barium sources and fluid-rock interaction in unconventional Marcellus shale wells using Ba isotopes. Energy Fuels 36:4470–4478. https://doi.org/10.1021/acs.energyfuels.2c00118

Osselin F, Nightingale M, Hearn G et al (2018) Quantifying the extent of flowback of hydraulic fracturing fluids using chemical and isotopic tracer approaches. Appl Geochem 93:20–29. https://doi.org/10.1016/j.apgeochem.2018.03.008

Owen J, Bustin RM, Bustin AMM (2020) Insights from mixing calculations and geochemical modeling of Montney Formation post hydraulic fracturing flowback water chemistry. J Petrol Sci Eng 195:107589. https://doi.org/10.1016/j.petrol.2020.107589

Paukert Vankeuren AN, Hakala A, Jarvis K et al (2017) Mineral reactions in shale gas reservoirs: barite scale formation from reusing produced water as hydraulic fracturing fluid. Environ Sci Technol 51:9391–9402. https://doi.org/10.1021/acs.est.7b01979

Phan TT, Hakala JA, Sharma S (2020) Application of isotopic and geochemical signals in unconventional oil and gas reservoir produced waters toward characterizing in situ geochemical fluid-shale reactions. Sci Total Environ 714:136867. https://doi.org/10.1016/j.scitotenv.2020.136867

Qian H, Ma Z, Li P et al (2012) Hydrochemistry, 2nd edn. Geological Publishing House, Beijing (In Chinese)

Raynes D, Zhang H, Emmanuel S (2024) Carbonate dissolution enhances particulate iron sulfide mobilization during water-shale interaction. J Hydrol 636:131257. https://doi.org/10.1016/j.jhydrol.2024.131257

Rowan EL, Engle MA, Kraemer TF et al (2015) Geochemical and isotopic evolution of water produced from Middle Devonian Marcellus shale gas wells, Appalachian basin. Pennsylvania. AAPG Bulletin 99(2):181–206. https://doi.org/10.1306/07071413146

Tipper ET, Stevenson EI, Alcock V et al (2021) Global silicate weathering flux overestimated because of sediment-water cation exchange. PNAS 118:e2016430118. https://doi.org/10.1073/pnas.2016430118

Xiong W, Gill M, Moore J et al (2020) Influence of reactive flow conditions on barite scaling in Marcellus shale during stimulation and shut-in periods of hydraulic fracturing. Energy Fuels 34(11):13625–13635. https://doi.org/10.1021/acs.energyfuels.0c02156

Xu M, Binazadeh M, Zolfaghari A et al (2018) Effects of dissolved oxygen on water imbibition in gas shales. Energy Fuels 32(4):4695–4704. https://doi.org/10.1021/acs.energyfuels.7b03955

Zeng L, Reid N, Lu Y et al (2020) Effect of the fluid–shale interaction on salinity: implications for highsalinity flowback water during hydraulic fracturing in shales. Energy Fuels 34:3031–3040. https://doi.org/10.1021/acs.energyfuels.9b04311

Zhao S, Wang Y, Li Y et al (2022) Tight sandstone petrophysics and flow rate/water/alkali sensitivities variations: a cross section of Permian in the northeastern of Ordos Basin, China. J Petrol Sci Eng 208:109443. https://doi.org/10.1016/j.petrol.2021.109443

Zheng Z, Cui X, Zhu P et al (2021) Sensitivity assessment of strontium isotope as indicator of polluted groundwater for hydraulic fracturing flowback fluids produced in the Dameigou Shale of Qaidam Basin. J Groundwater Sci Eng 9(2):93–101. https://doi.org/10.19637/j.cnki.2305-7068.2021.02.001

Zou C (2017) Unconventional Petroleum Geology, second ed. Elsevier. https://www.sciencedirect.com/book/9780128122341/unconventional-petroleum-geology

Zou Y, Zhang S, Zhang J (2014) Experimental method to simulate coal fines migration and coal fines aggregation prevention in the hydraulic fracture. Transp Porous Media 101:17–34. https://doi.org/10.1007/s11242-013-0228-9

Chapter 7
Formation Damage During Hydraulic Fracturing

A series of water–rock interactions occur during hydraulic fracturing. The key questions concern how these interactions affect the tight reservoir and influence natural gas production. For unconventional hydrocarbon reservoirs, the effects of water–rock interactions on the reservoir begin during diagenesis, where compaction and cementation alter and degrade the reservoir, leading to formation damage. Formation damage also occurs during production stages of unconventional reservoirs, including drilling, hydraulic fracturing, and well completion. Compared with other processes, reservoir damage is more pronounced during hydraulic fracturing. The incompatibility between fracturing fluids and reservoir rocks often leads to formation damage through water–rock interactions, impeding hydrocarbon migration and production. As an external fluid, the injection of large volumes of fluid disrupts the equilibrium of the formation environment, leading to reservoir damage. Clay minerals, second only to quartz in abundance within tight sandstone reservoirs, play a critical role in determining the stability and effectiveness of hydraulic fracturing. Based on laboratory and field experiments, this chapter investigates the impacts and damage mechanisms of fracturing fluid-tight sandstone interactions during hydraulic fracturing, focusing on clay mineral instability and scale-induced pore blockage.

7.1 Potential Formation Damage in Gas Reservoirs

The mineralogical composition and distribution patterns of tight reservoirs influence both the primary porosity and permeability as well as gas production behavior during fracturing stimulation. Quartz remains chemically stable, while the Lower He 8 Member contains only minor soluble carbonate and sulfate minerals. Thus, the relatively high clay content emerges as the primary factor controlling the hydraulic fracturing performance of tight sandstones (Li et al. 2023). The abundance of clay minerals (e.g., kaolinite, illite, and chlorite) in tight sandstones can easily cause clay

X. He and P. Li, *Geochemical Processes in Tight Gas Hydraulic Fracturing Stimulation*, Environmental Earth Sciences, https://doi.org/10.1007/978-3-032-15180-3_7

instability during hydraulic fracturing. Clay minerals, composed of hydrous aluminosilicates, exhibit limited solubility and chemical reactivity under natural conditions. When external fracturing fluids are introduced into gas-bearing formations, the crystalline stability of clay minerals becomes susceptible to alteration (Khan et al. 2021). Interaction between montmorillonite and freshwater leads to hydration swelling, a process frequently observed in shale reservoirs (Zhao et al. 2022). Unlike shale, tight sandstone reservoirs generally contain less montmorillonite, so clay swelling is not the main factor affecting clay stability. Furthermore, the distribution patterns and structural characteristics of clays exhibit greater sensitivity to fracturing fluids than their abundance alone.

SEM results show that some kaolinite crystals in the lower He 8 tight sandstone of the Sulige Gas Field occur in dispersed aggregates (Fig. 7.1), which can easily migrate during hydraulic fracturing, leading to clay dispersion. Although kaolinite is a two-layer non-expanding clay, it tends to disperse and migrate under alkaline conditions. Guar-based fracturing fluids, widely applied in tight reservoir stimulation, generally require alkaline environments for crosslinking. Therefore, when interacting with guar-based fracturing fluids, kaolinite stability is easily disturbed (Li et al. 2023). Illite, with its interlayered structure, exhibits both dispersive and expansive tendencies, resulting in low structural stability. Illite in the studied tight sandstone predominantly occurs as flake-like or filamentous aggregates within intergranular pores or coating grain surfaces. Fibrous illite creates numerous micropores but also makes pore channels tortuous, leading to high water saturation and potential water blocking. Migrating particles can accumulate at illite pore throats, causing blockage and impeding gas flow (Fig. 7.2). Chlorite, characterized by a three-layer structure, typically shows minimal swelling upon hydration. However, Mg- and Fe-enriched chlorite is highly acid-sensitive and readily forms Fe oxide/hydroxide secondary precipitates when in contact with acidic pre-fluids (e.g., HCl) or oxidizing agents such as ammonium persulfate (Khan et al. 2021; Jew et al. 2022). Consequently, chlorite precipitation commonly occurs during the initial fracturing stages, resulting in matrix plugging and a decline in permeability. In summary, kaolinite in the study area demonstrates pronounced rate and alkali sensitivities, whereas illite exhibits strong water and rate sensitivities, making it particularly susceptible to water-lock effects.

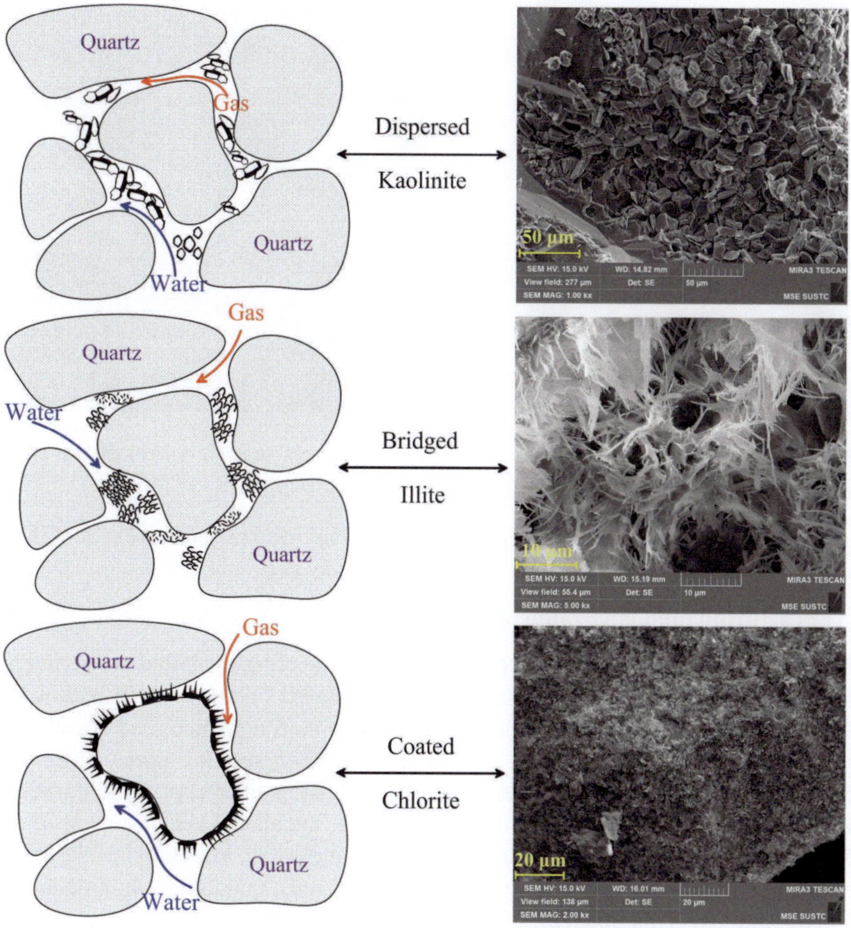

Fig. 7.1 The occurrence morphology of clay minerals in the tight sandstone of the study area

7.2 The Influence of Water–Rock Interaction on the Micro-area Morphology

7.2.1 Morphological Changes in Micro-areas on the Rock Surface

The morphology of tight sandstone samples after reactions between tight sandstones with various solutions (fracturing fluid, deionized water, biocide, surfactant, clay stabilizer, pH adjustment agent, and crosslinker) was tested and analyzed to compare the effects of different solution environments on the micro-scale surface morphology of rocks. For consistent comparison of surface micro-morphological changes before

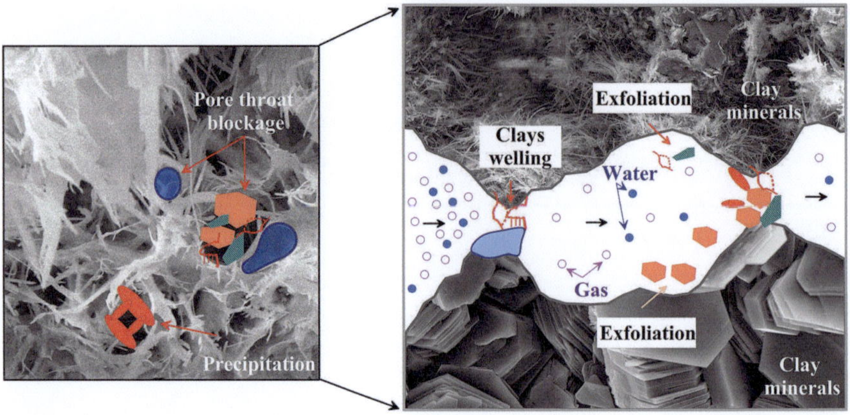

Fig. 7.2 Schematic diagram of the potential impact of clay mineral structure on hydraulic fracturing

and after water–rock interaction, SEM images at a uniform magnification of 5000 × were selected for discussion.

(1) Fracturing fluid environments

Laboratory experiments revealed that tight sandstone cores exhibited pore enlargement when immersed in low-ionic-strength guar-based fracturing fluid. Following fracturing fluid treatment, the detrital matrix became fragmented, and dissolution of fine-grained detrital minerals generated enlarged dissolution pores (Fig. 7.3). Some large mineral grains further disintegrated into smaller crystalline particles. Kaolinite underwent widespread dissolution under the alkaline environment of the fracturing fluid, resulting in crystal fragmentation and dispersion. Booklet-like kaolinite aggregates disaggregated into platy single crystals, undergoing fragmentation and migration that disrupted the original intercrystalline pores. After fracturing fluid exposure, fibrous illite crystals aggregated, obliterating the original intergranular micropores and forming abundant illite pore throats. Chlorite minerals were generally stable, showing minimal morphological change, though some detrital chlorite matrices exhibited localized dissolution and fragmentation. Overall, the fracturing fluid environment exhibited strong dissolution capacity, capable of partially dissolving carbonate minerals and detrital cements in the reservoir, thereby increasing porosity but also altering the micro-morphology and stability of clay minerals, especially kaolinite.

(2) Deionized water environments

In comparison with the fracturing fluid, deionized water treatment produced a more prominent increase in pore size on tight sandstone core surfaces. In deionized water, detrital minerals attached to quartz grains and clay minerals were prone to dissolution and hydration, forming larger pores. After treatment with deionized water, some illite minerals underwent hydration expansion, transforming from platy to fibrous

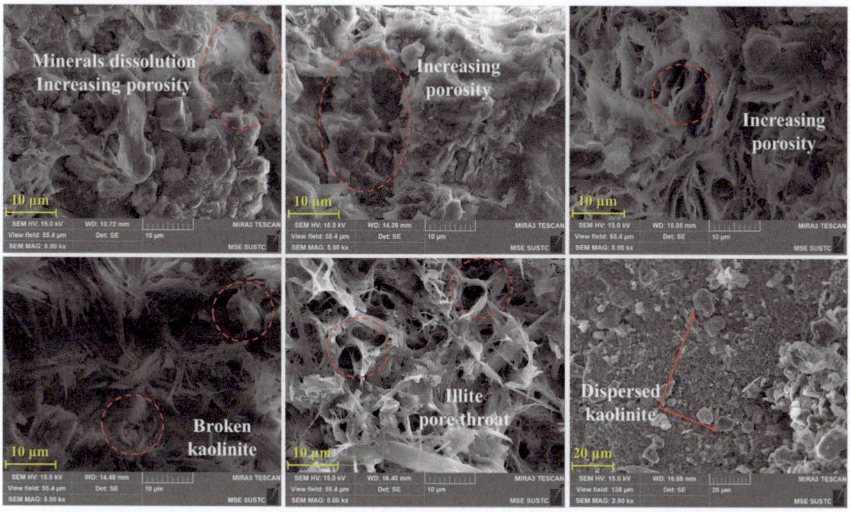

Fig. 7.3 Changes in the microscopic morphology of the rock surface after fracturing fluid treatment

forms. Well-cemented chlorite, kaolinite, and pyrite crystals remained morphologically stable after exposure to deionized water, showing no substantial fragmentation, deformation, or migration (Fig. 7.4). Compared with the fracturing fluid environment, deionized water increased rock porosity to a greater extent but also made illite and mixed-layer illite–montmorillonite in tight sandstone more susceptible to hydration swelling.

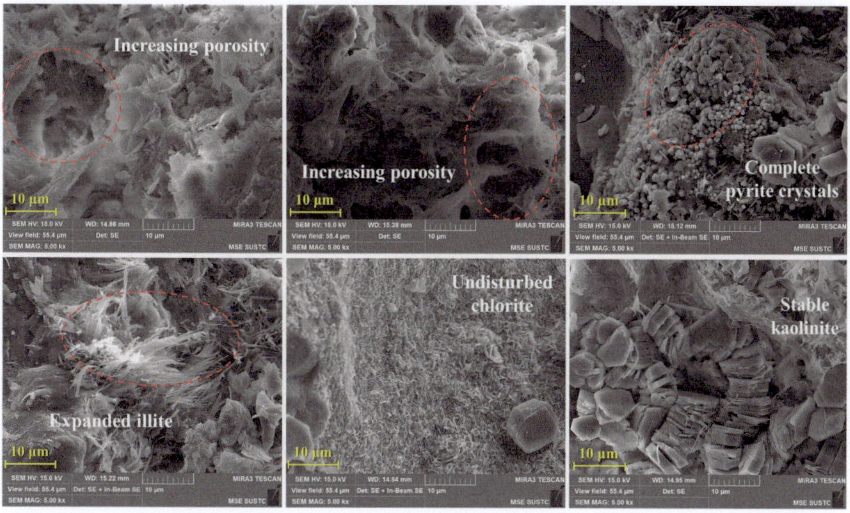

Fig. 7.4 Changes in the microscopic morphology of the rock surface after deionized water treatment

(3) Fracturing fluid additives environments

After treatment with bactericide, detrital minerals within the core underwent fragmentation, with flaky clastic minerals being dissolved and decomposed into smaller detrital grains, generating extensive secondary pores through dissolution processes. Some platy illite minerals exhibited swelling and deformation, and their crystal edges reassembled to form intercrystalline pores. Meanwhile, after long-term immersion (30 days), kaolinite minerals showed almost complete dissolution of the intergranular cement, generating new intercrystalline pores, and some loosely attached kaolinite single crystals were broken and transported (Fig. 7.5). Chlorite minerals remained generally stable. The bactericide treatment effectively increased the surface porosity of the tight sandstone, and its impact on surface morphology was comparable to that of deionized water.

After treatment with the clay stabilizer, the crystal morphology of kaolinite showed no significant change, indicating that the stabilizer helped maintain the stability of clay minerals to some extent. However, with prolonged contact between the stabilizer solution and the rock, noticeable dissolution and fragmentation of detrital minerals occurred, and some disintegrated particles detached from the rock surface (Fig. 7.6). As these detrital minerals and easily dissolvable clay matrix materials were dissolved and removed, the surface porosity of the rock samples increased significantly. Nevertheless, during the late stage of reaction between the clay stabilizer and tight sandstone, deposition of these fragmented particles was observed. Once such dispersed particles deposit and accumulate within pore throats or microfractures, they inevitably affect the reservoir's porosity and permeability. It was also found that the clay stabilizer exhibits a time-dependent effect. Prolonged immersion of tight sandstone samples in the stabilizer caused brittleness and deterioration of

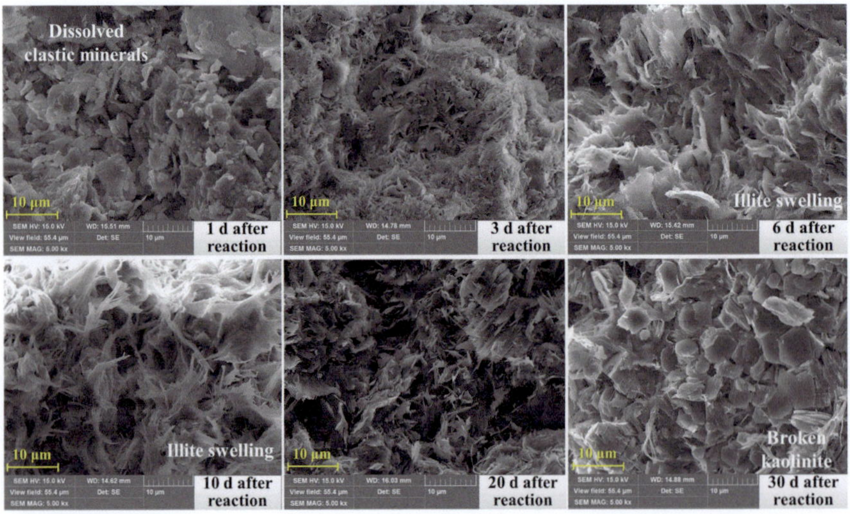

Fig. 7.5 Changes in the microscopic morphology of the rock surface after bactericide treatment

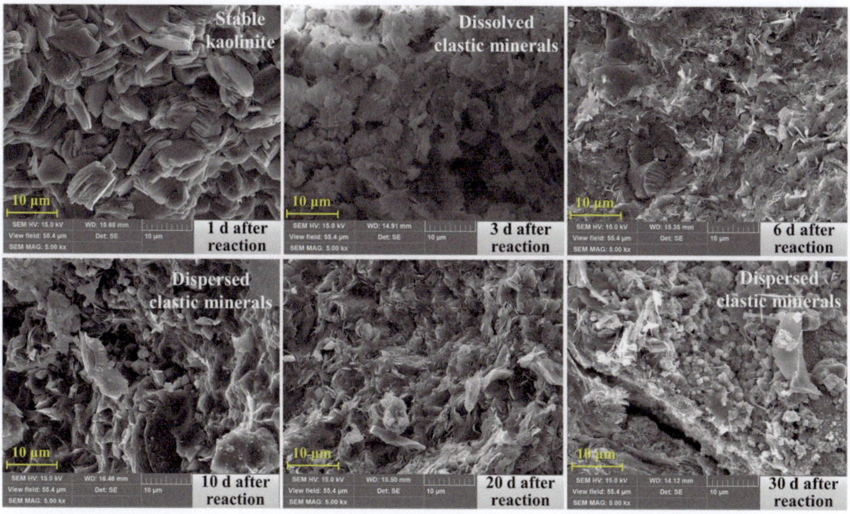

Fig. 7.6 Changes in the microscopic morphology of the rock surface after clay stabilizer treatment

mechanical properties, as well as a reduction in clay crystals adhesion. Such conditions may trigger clay particle dislodgment, gradually undermining clay minerals stability.

In the presence of a surfactant, some illite minerals expanded and dispersed as the reaction progressed. The flaky illite swelled into filamentous forms and dispersed. Similar to other additive environments, dissolution of detrital mineral cements occurred, generating dissolution pores and fissures, and increasing the pore and fracture size on the core surface. During the late reaction stage, some kaolinite crystals were observed to break and disperse (Fig. 7.7), suggesting that prolonged contact of the surfactant fluid and tight sandstone also adversely affects the clay minerals stability.

After treatment with the pH regulator, alkaline-sensitive kaolinite lost stability and showed pronounced fragmentation and detachment. Under scanning electron microscopy, the instability of kaolinite minerals was more apparent than under other additive conditions. In contrast, chlorite, which is sensitive to acidic conditions, maintained its morphology and distribution, remaining stably attached to quartz grain surfaces even after 20 days of soaking in the alkaline environment (Fig. 7.8). The effect of the pH regulator on kaolinite in tight sandstone was similar to that observed under fracturing fluid conditions, both unfavorable to kaolinite stability.

After treatment with the crosslinker, flaky illite minerals expanded, with their crystals stretched, thinned, and elongated into hair-like forms. The original fibrous illite flocculated, forming numerous bundled pore throats (Fig. 7.9). The edges of incompletely developed laminated chlorite were corroded after immersion, and the original platy crystals were fragmented and detached. This phenomenon became

Fig. 7.7 Changes in the microscopic morphology of the rock surface after surfactant treatment

Fig. 7.8 Changes in the microscopic morphology of the rock surface after pH control treatment

more pronounced as the reaction time increased, particularly in the later stage of reaction.

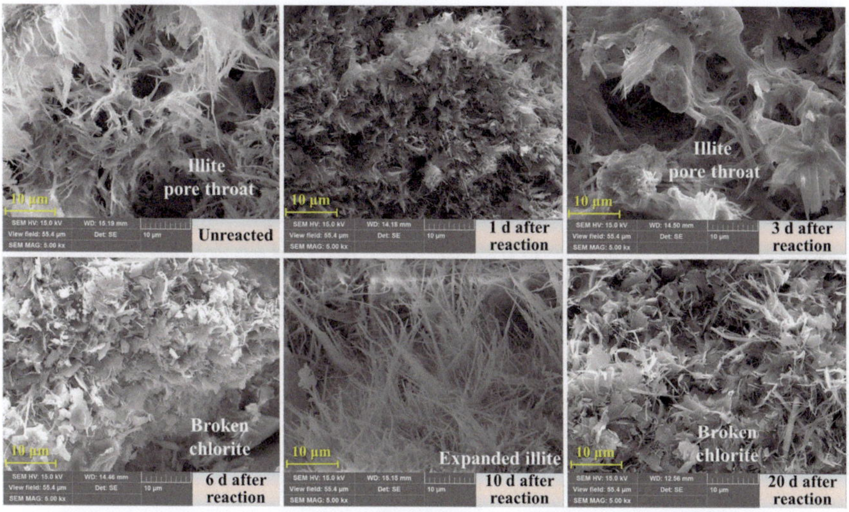

Fig. 7.9 Changes in the microscopic morphology of the rock surface after crosslinker treatment

7.2.2 Potential Influence of Water-rock Chemical Reactions on Reservoirs

After fracturing fluid enters the reservoir, water–rock interactions modify the rock surface morphology, which in turn influences reservoir porosity and permeability. Mineral dissolution favors pore and fracture expansion, while precipitation may cause pore blockage. Herz-Thyhsen et al. (2019) reported that the dissolution of calcite after fracturing fluid treatment increased the porosity of tight mudstone samples by 2–2.9%, and that of sandstone samples by 5–5.8%. Comparative analysis of mineral compositions of tight sandstones from different regions (Table 7.1) shows that these rocks are primarily composed of quartz, clay minerals, feldspars (K-feldspars and plagioclase), and carbonate minerals (calcite and dolomite). Quartz, together with clay minerals such as kaolinite, illite, and chlorite, are common and abundant across all tight sandstone samples, while feldspar and carbonate minerals exhibit regional variability. The tight sandstones from the Xujiahe Formation in the Sichuan Basin and the Montney Formation in the Canadian Deep Basin contain higher contents of K-feldspar and plagioclase, while those from the Tarim Basin and the Wall Creek Formation in the Powder River Basin, USA, are richer in carbonate minerals. Moreover, regions with higher calcite contents generally have lower dolomite contents, and vice versa. Feldspar and carbonate minerals in tight sandstones exhibit strong heterogeneity, while sulfate minerals are generally absent. Gypsum occurs only in the Montney Formation of the Canadian Deep Basin and is absent elsewhere. Pyrite occurs in tight sandstones but in much lower abundance than in organic-rich tight reservoirs such as shales and coal seams. Additionally, minor iron oxide minerals such as siderite and hematite are present in low levels. Based on the above mineralogical analysis, soluble and easily dissolved minerals are generally scarce in tight sandstone reservoirs, implying that porosity increase due to mineral dissolution during hydraulic fracturing is limited.

Laboratory water–rock interaction experiments revealed that treatment with low-ionic-strength fracturing fluid caused dissolution, fragmentation, and dispersion of weakly cemented detrital and clay matrix minerals, forming large dissolution pores in original filling zones and thereby enhancing surface porosity and overall reservoir porosity. Simulation of field fracturing data for Well *S1* shows that each liter of fracturing fluid dissolved about 397 mg of calcite. With 2481 m^3 of injected fluid, total calcite dissolution reached approximately 985 kg, equivalent to a pore volume

Table 7.1 Mineral composition of tight sandstone in representative tight sandstone gas basins (wt%, (average value))

Region	Jurassic Ahe Formation, Tarim Basin, China[a]	Xujiahe Formation, Sichuan Basin, China[b]	Montney Formation, Deep Basin, Canada[c]	Wall creek formation, Powder River Basin, USA[d]	Shihezi Formation, Ordos Basin, China[e]	Shihezi Formation, Ordos Basin, China[f]
Quartz	54–82 (72.5)	43.2–96.8 (81.3)	31.2–53.8 (45.3)	66.4	36.0–98.0 (72.0)	45.9–81.6 (63.8)
K-feldspar	0–12.0 (6.3)	0–14.1 (5.3)	3.3–9.2 (6.9)	/	0–12.8 (1.6)	0–0.2 (0.04)
Plagioclase	0–8.0 (3.7)	0–21.4 (11.2)	2.6–8.9 (5.8)	4.8	/	0–0.2 (0.04)
Calcite	0–30.0 (5.8)	0–0.8 (0.1)	/	20.1	0–28.5 (6.7)	0.4–18.0 (4.0)
Dolomite	0–6.0 (0.4)	0–3.0 (0.5)	7.6–40.7 (17.0)	/	0–1.4 (0.05)	/
Pyrite	/	/	1.0–4.0 (1.8)	1.0	/	/
Gypsum	/	/	0–5.5 (0.8)	/	/	/
Kaolinite	3–21 (10.2)	0–5.8 (0.9)	0–0.6 (0.2)	1.5	0–12.4 (4.6)	10.9–31.8 (21.1)
Chlorite		0–7.3 (2.0)	0.2–4.9 (2.6)	0.7	0.7–22.7 (8.7)	
Illite		0–10.3 (2.4)	8.4–42.8 (18.2)	6.3	0.3–4.4 (2.3)	3.9–27.3 (10.2)
Illite/smectite (I/S)		0–1.9 (0.1)	/	/	0.6–19.6 (3.8)	/
Others	0–14 (1.2)	/	/	/	0–7.0 (0.3)	0.4–1.0 (0.8)

Note [a]Guo et al. (2018); [b]Luo et al. (2019); [c]Yang and Lee (2018); [d]Herz-Thyhsen et al. (2020); [e]Wang et al. (2020); [f]This study increase of about 0.36 m^3. In real reservoir conditions, the actual calcite dissolution may be lower than the theoretical value due to mineral heterogeneity and limited fracturing fluid distribution. Overall, calcite and cement dissolution increase porosity, but long-term fluid retention can induce illite swelling and kaolinite fragmentation. Although dissolution by fracturing fluid can enhance porosity, migration and dispersion of clay minerals may block pore throats and microfractures. Consequently, despite the increase in porosity after fracturing fluid treatment, the permeability may decrease. Given the limited dissolution capacity of fracturing fluids in tight sandstone reservoirs, preventing permeability reduction caused by clay mineral instability is particularly important.

7.3 Reservoir Micro-damage Caused by Water–Rock Chemical Reactions

7.3.1 Stability of Clay Minerals

(1) Clay swelling

Clay swelling reflects the reservoir's sensitivity to water and involves crystalline and osmotic mechanisms (Khan et al. 2021; Jew et al. 2022). Crystalline swelling can occur in all types of clay minerals and is caused by the hydration of exchangeable cations within the clay structure. It develops gradually through the formation of hydration layers on the surfaces of clay minerals. Differential cation concentrations at clay-solution interfaces drive interlayer water adsorption, inducing mineral expansion. Osmotic swelling occurs only in specific clay minerals; for instance, sodium montmorillonite tends to form crystalline hydrates more readily and exhibits greater swelling than calcium montmorillonite. Compared with crystalline swelling, osmotic swelling results in a more significant increase in mineral volume (Civan 2015).

Kaolinite, a 1:1 layer clay, exhibits a CEC of 3–15 meq/100 g, and its surface charge strongly depends on pH. Higher pH results in more negative surface charge (Fig. 7.10). Thus, drastic pH variations during fracturing may destabilize kaolinite. Illite, which has an interlayer structure, possesses a CEC ranging from 10 to 40 meq/100 g. Variations in pH have limited influence on its surface charge, allowing it to maintain relatively high surface charge across different pH conditions. Montmorillonite, a typical swelling clay with a 2:1 layered structure, has a specific surface area of 700–850 m^2/g and a high CEC of 80–120 meq/100 g (Tipper et al. 2021). The surface charge of montmorillonite is also weakly affected by pH. Its large surface area and high CEC enable strong adsorption of water and cations between layers, leading to high water sensitivity. When montmorillonite contacts low-salinity fracturing fluid, ion exchange occurs, increasing the net repulsive force between clay layers, expanding clay volume, and occupying pore and fracture spaces, which reduces reservoir permeability (Bishop 1997).

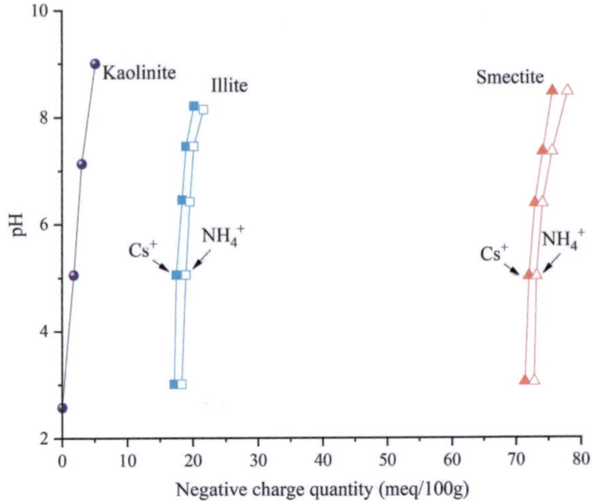

Fig. 7.10 Variation of surface negative charges in different clay minerals with pH (Modified from Qian et al. 2012)

Unlike shale reservoirs, tight sandstone reservoirs typically lack well-developed montmorillonite and thus do not exhibit large-scale montmorillonite hydration swelling. However, laboratory experiments revealed that some illite minerals still undergo swelling after fracturing fluid treatment, with fibrous illite showing pronounced expansion (Fig. 7.11). Wei (2019) reported that the He 8 Member of the Sulige Gas Field lacks montmorillonite but contains 0.96–3.13% of illite–smectite mixed-layer (I/S) minerals. Therefore, tight sandstone reservoirs still exhibit a certain degree of clay swelling during hydraulic fracturing, which adversely affects reservoir properties.

(2) Fine migration

Fine migration is an inherent process that affects reservoir permeability. Fine migration is most typical in coalbed methane production. Because of coal's brittleness, fracturing produces abundant coal fines that are mainly composed of clay minerals (Wei et al. 2019). Therefore, clay mineral dispersion is the primary factor contributing to particle migration in tight reservoirs. While kaolinite maintains non-expansive properties as an acid-formed clay, it demonstrates pronounced dispersion instability in alkaline systems such as crosslinked gel fracturing fluids prepared under alkaline conditions (Maley et al. 2013; Sruthi and Reddy 2019). This study found that under the alkaline guar-based fracturing fluid-rock system, kaolinite crystals commonly exhibit fragmentation, where regular hexagonal crystals break into irregular flakes, destroying the original intergranular pores (Fig. 7.12). Kaolinite crystals are weakly bonded and adhere poorly to quartz grains (Anderson et al. 2010; Khan et al. 2021). Consequently, kaolinite tends to fragment and detach under the high flow velocities of fracturing and flowback fluids. The detached fragments migrate through fractures with fracturing and flowback fluids, as well as natural gas. Kaolinite crystals are typically larger than other clay mineral particles. In the study area, single kaolinite grains range from 5–15 μm, whereas illite-related pore throat diameters are concentrated

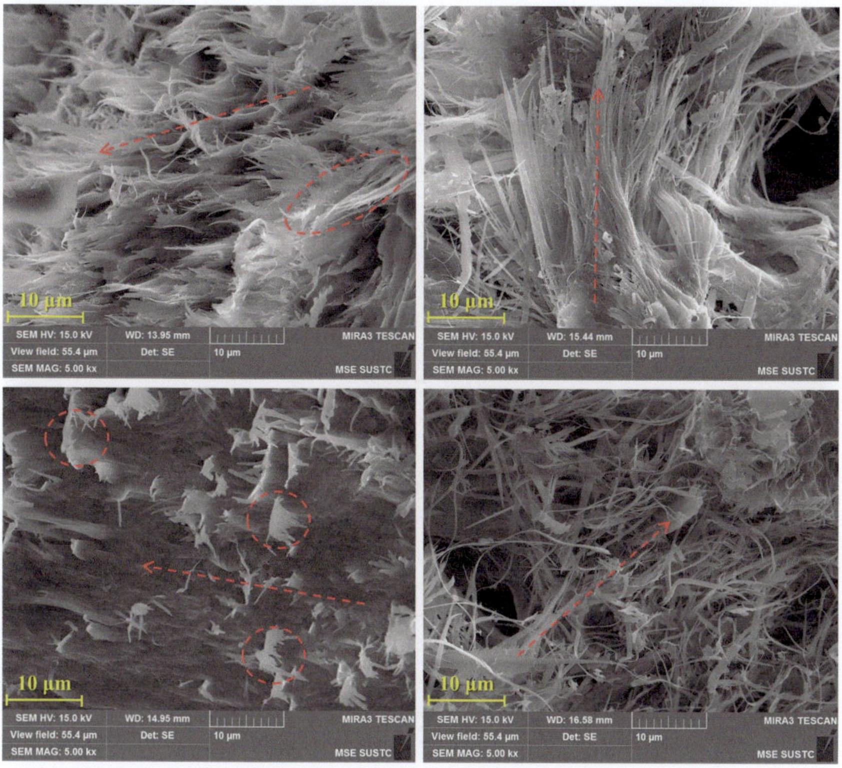

Fig. 7.11 Illite expansion in tight sandstone after fracturing fluid treatment

between 2.7 and 7.4 μm. Thus, once dispersed, kaolinite particles are more likely to block pore throats (Fig. 7.12). As the dominant clay mineral in tight sandstone reservoirs, kaolinite exhibits high velocity and alkali sensitivity after water–rock interaction, requiring careful attention during hydraulic fracturing.

Illite, with its sandwich-like 2:1 structure, has poor stability and readily disperses. Flaky chlorite adhering to grains is likewise susceptible to dispersion and migration. Laboratory observations revealed directional dispersion of fibrous illite and flaky chlorite (Fig. 7.13). However, their dispersibility in fracturing fluid environments is generally weaker than that of kaolinite.

Geochemical modeling of K-feldspar, kaolinite, and illite in the *S1* flowback fluid shows that at 70 °C and 50 MPa, most samples fall within the K-feldspar stability field, suggesting that fracturing conditions destabilize kaolinite and illite (Fig. 7.14). Besides clay dispersion, dissolution of cements, mineral flaking from fracture surfaces, and proppant failure also produce fines (Khan et al. 2021). Migrating fines tend to flocculate in high-salinity conditions, where they accumulate or redeposit in pore throats, forming larger aggregates that reduce permeability (Bishop 1997; Maley et al. 2013).

Fig. 7.12 Broken and dispersed kaolinite crystals after fracturing fluid treatment

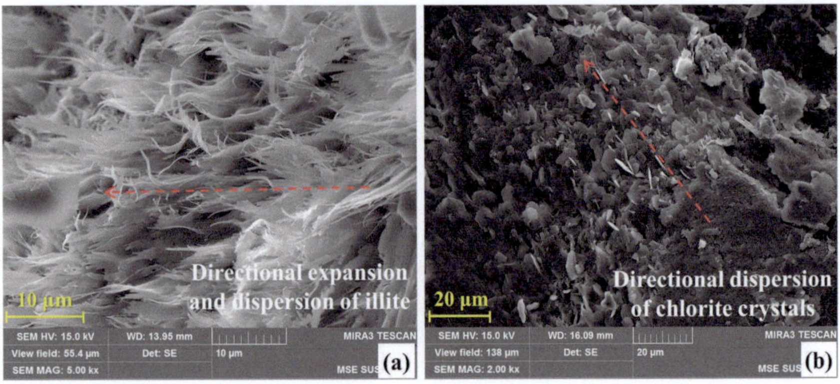

Fig. 7.13 Directional dispersion of illite (**a**) and chlorite (**b**) after fracturing fluid treatment

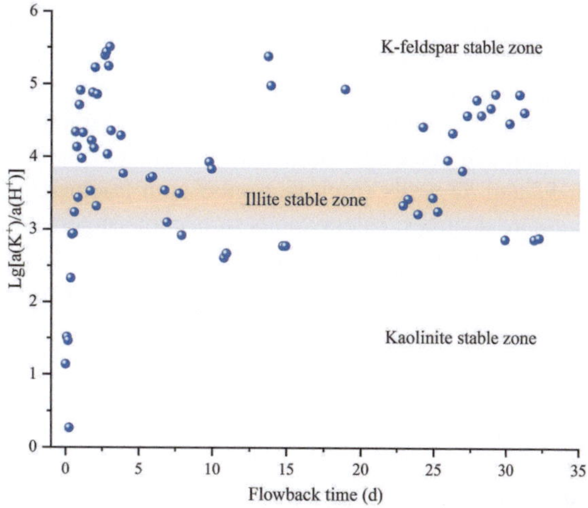

Fig. 7.14 Stability zoning map of clay minerals in Well *S1* based on ion activity

7.3.2 Blocking Problems of Scaling

Mineral precipitation and scaling are significant water–rock interactions during hydraulic fracturing and constitute major factors affecting natural gas output in reservoirs. Secondary mineral deposits can irreversibly block microfracture systems and flow pathways, decrease porosity and permeability, and thereby impair the overall fracturing performance (Jew et al. 2017). Scaling can be categorized into inorganic and organic types, both representing major formation damage mechanisms that threaten oil and gas production. The principal scale-forming ions originate from multivalent cations such as Ca^{2+}, Ba^{2+}, Sr^{2+}, and Fe^{3+}, and anions, including SO_4^{2-} and CO_3^{2-}, released through water–rock interactions, formation water, and fracturing fluids. Ionic interactions yield secondary mineralization including carbonate (calcite/strontianite), sulfate (gypsum/barite/celestite), and iron oxide-hydroxide phases. The scaling process typically proceeds through four stages: solution supersaturation, nucleation, crystalline development, and particle aggregation. Thermodynamically, scale initiates when ionic activities surpass mineral-specific saturation thresholds. The precipitating ions subsequently form micro-aggregates on rock surfaces, which act as crystallization nuclei for crystallization. The crystals continue to grow on the surfaces, developing into precipitate layers that further adsorb ions from the solution, ultimately forming a dense scale layer (Menefee et al. 2020). Several parameters, such as water incompatibility, pH, pressure, salinity, and temperature, can trigger mineral precipitation. Inorganic scaling is governed not only by hydrochemical conditions but also by hydrodynamic factors. Variations in fracture aperture during hydraulic fracturing also affect precipitation dynamics: larger fracture widths correspond to lower flow rates along fracture surfaces, higher saturation, and thus a greater likelihood of mineral precipitation (Li et al. 2017). Secondary mineral precipitation irreversibly clogs pores, decreases natural gas yield, and impairs well productivity.

According to the geochemical data of flowback fluids, the formation water in the Sulige region is enriched in Ca^{2+}, Ba^{2+}, Sr^{2+}, and Fe ions, facilitating the precipitation of secondary carbonates and sulfates during fracturing operations. Modeling of mineral saturation indices under reservoir fracturing conditions (temperature and pressure) in six wells indicated that calcite oversaturation occurred intermittently in *S1* and *S2* wells employing guar gum fracturing fluids (Fig. 7.15). As reservoir pressure declines, calcite tends to become supersaturated, leading to the formation of secondary calcite precipitates. In contrast, wells employing other fracturing fluid systems (e.g., variable-viscosity slickwater, slickwater, or biopolymer fluids) generally exhibited unsaturated or near-saturated calcite states, suggesting a lower likelihood of secondary calcite precipitation compared to guar-based systems. Additional modeling of barite and gibbsite saturation in *S1*, *S2*, and *S3* wells demonstrated that all water samples were supersaturated with respect to barite, highlighting the substantial risk of barite scale formation (Fig. 7.16). For instance, modeling of Well *S1* showed concentrated barite precipitation between 2 and 48 h after flowback initiation, with about 11.7 mg of barite forming per liter of fluid. Given the 140 m^3 flowback volume during sampling, total barite precipitation was estimated at 1.63 kg. Once deposited within pore throats and microfractures, barite minerals can clog gas flow pathways, severely reducing gas productivity. For gibbsite, almost all simulated data points from *S1* and *S2* wells (guar systems) fell within the supersaturated region, while in *S3* (variable-viscosity slickwater), only a few samples showed oversaturation, with most being undersaturated and saturation indices below − 1 (Fig. 7.16). Simulation results for Well *S1* indicated an average precipitation of 2.5 mg of gibbsite per liter of flowback fluid throughout the process (1048 m^3), corresponding to a total of about 2.62 kg of gibbsite. The above analysis demonstrates that guar gum fracturing fluids are more prone than other systems to trigger secondary mineral precipitation, leading to scaling and pore blockage in tight sandstone reservoirs.

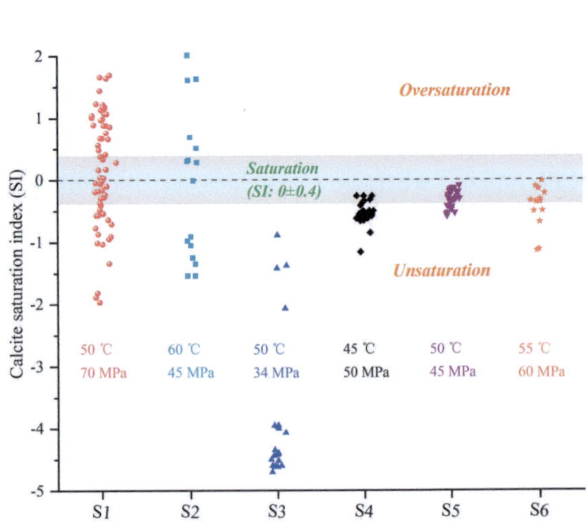

Fig. 7.15 Calcite saturation index of flowback fluids in different fracturing wells

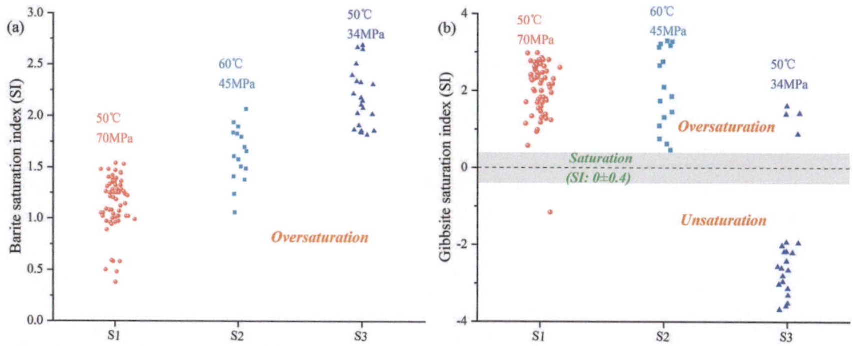

Fig. 7.16 Barite (**a**) and gibbsite (**b**) saturation indexs of flowback fluids in different fracturing wells

Paraffin, asphaltenes, and emulsions are the main organic scales. Paraffin crystallization generally occurs in zones of temperature decline, whereas asphaltene deposition results from pressure depletion and flocculation (Adams 2014; Orem et al. 2014). The fracturing flowback fluids in the Sulige Gas Field commonly exhibit high COD concentrations, containing dispersed condensate oil, which can lead to organic scaling during fracturing and flowback processes due to variations in temperature and pressure. In addition, microbial metabolites represent another significant scaling component in hydrocarbon production. These metabolites can form biofilms on rock surfaces, clogging pore spaces within the reservoir (Sharma et al. 2020). Scale inhibitors are widely applied in conventional hydrocarbon production to prevent precipitation. Common inhibitors include chelating agents, phosphate- or phosphonate-based inhibitors that suppress crystal growth, and polymer inhibitors that impede nucleation. These agents can effectively prevent the formation of common scale minerals such as calcite, gypsum, celestite, barite, and iron-bearing compounds. However, their ability to inhibit precipitation reactions under unconventional reservoir conditions remains uncertain, and the chemical composition of formation water can significantly influence inhibitor performance (Khan et al. 2021; Jew et al. 2022).

7.3.3 Other Potential Reservoir Damage

Hydraulic fracturing can also alter the wettability of reservoir rocks. Prolonged exposure to fracturing fluids often results in sandstone softening and reduced mechanical strength. Wettability refers to the solid phase's capacity to retain fluids, determining the nature of interactions between fluid and solid surface (Song et al. 2019; Wang et al. 2021). It exerts a direct influence on water saturation and the distribution of gas and water, leading to substantial impacts on gas permeability, which, in turn, affects gas flow and production (Civan, 2015; Wang et al. 2021). Based on contact angles

(θ), wettability can be categorized into water-wet ($\theta < 90$), mixed-wet ($\theta \approx 90$), and oil-wet ($\theta > 90$). Wettability is contingent on fluid composition, the mineral composition and roughness of the rock surface, as well as temperature and pressure (Khan et al. 2021; Zhou et al. 2021). Alterations in wettability can occur when fracturing fluids are injected into reservoirs due to changes in the reservoir environment. For example, the surface of organic matter tend to be predominantly hydrophobic, while inorganic mineral exhibits hydrophilic characteristics. Once wettability shifts, it leads to changes in gas–water distribution, the magnitude and direction of capillary forces, thus affecting gas migration (Kang et al. 2016; Wang et al. 2021). Wettability alteration has the potential to confine the water phase within micropores, forming a water block in the pore throat that obstructs gas flow. The interactions between fracturing fluids and the rock matrix may also modify the functional groups on mineral surfaces, leading to wettability alteration (Dustin et al. 2018). The dissolution of minerals on fracture surfaces during fracturing exposes less soluble minerals, generating distinct surface properties that alter the original wettability and hinder fluid flow. Illite, rich in hydrophilic elements such as Si and Al, generally exhibits water-wet behavior. In the tight sandstones of the study area, illite commonly occurs as fibrous aggregates between grains or on grain surfaces, forming abundant micropores. Fluids migrating into these pores can readily cause localized water blockage. Additionally, the interaction between fracturing fluid and organic matter could induce shifts in functional group composition and wettability (Dustin et al. 2018; Liu and Gao 2018).

In addition, incomplete gel breaking during fracturing can lead to colloidal residues that cause reservoir plugging. Fracturing gels have certain viscosities, and when trapped in pore throats, they significantly hinder fluid flowback and gas production. Tight sandstones with high clay content are prone to rock matrix softening, which prevents effective proppant support. Consequently, proppants become embedded in the softened rock, partially closing fractures and drastically reducing fracture conductivity, thereby impairing hydrocarbon production (Fig. 7.17, He et al. 2026). Laboratory experiments reveal that prolonged exposure to fracturing fluids makes tight sandstone noticeably more brittle, with minerals detaching from quartz surfaces and intergranular fillings, ultimately reducing the overall mechanical strength of the rock.

In summary, water–rock interactions during hydraulic fracturing exert multiple influences on tight sandstone reservoirs. Initially, low-ionic-strength fracturing fluids dissolve easily soluble minerals such as carbonates and detrital cements, effectively enhancing porosity. The addition of strongly oxidizing breakers further oxidizes minerals like pyrite, creating additional pore spaces. However, pyrite oxidation releases Fe and SO_4^{2-}, which promote scaling. As fractures propagate, high-concentration formation waters rich in Ca^{2+}, Ba^{2+}, Sr^{2+}, and Fe mix with the injected fluids, triggering the precipitation of secondary minerals such as barite, calcite, gibbsite, and iron oxides/hydroxides (Fig. 7.17). Disruption of the original water–rock equilibrium induces ion exchange and adsorption reactions between fracturing fluids and clay minerals, leading to swelling, dispersion, and migration of clays. These processes cause pore blockage, modify local wettability, and ultimately reduce permeability. Although mineral dissolution during fracturing can temporarily

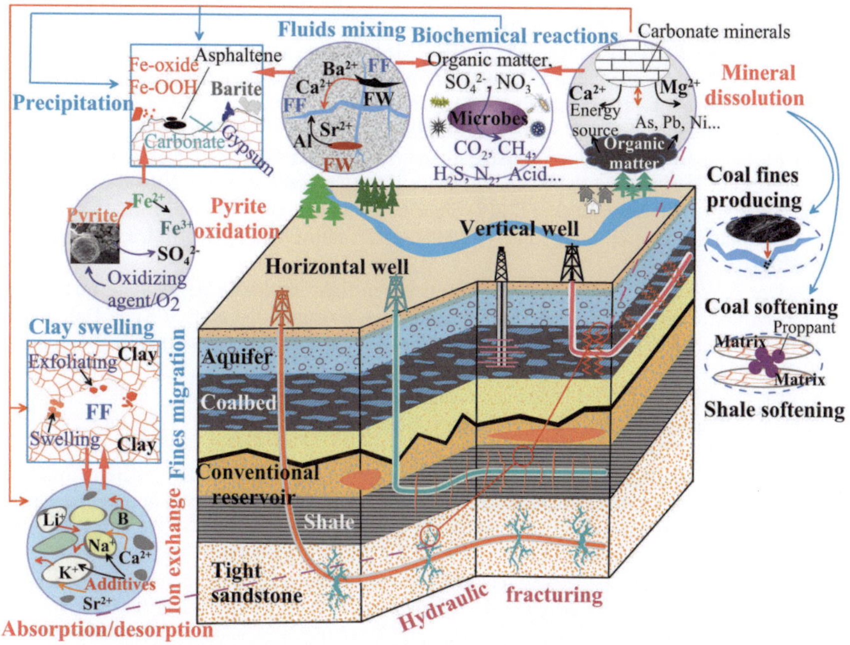

Fig. 7.17 Schematic diagram of formation damage during fracturing stimulation (FF: Fracturing ffuid; FW: Formation water, He et al. 2026)

increase porosity, continuous dissolution under high water–rock ratios may soften the rock matrix, causing proppant embedment, fracture closure, and consequently reduced gas production.

Due to the presence of high-salinity brines, shale and tight sandstone formations are highly susceptible to scaling, whereas coalbed methane reservoirs, characterized by abundant pores and fractures, are more prone to plugging from fracturing fluid residues. Unlike dry-gas-dominated shale reservoirs, tight sandstones typically have higher water saturations. During hydraulic fracturing, the release of trapped formation water leads to mixing with fracturing fluids, creating a highly saline environment. Non-uniform mixing causes sharp fluctuations in pH and ion concentration, destabilizing kaolinite and illite crystals along fracture surfaces. The mobilized minerals can flocculate and precipitate within pore throats, obstructing gas migration pathways and reducing reservoir permeability.

References

Adams JJ (2014) Asphaltene adsorption, a literature review. Energy Fuels 28:2831–2856. https://doi.org/10.1021/ef500282p

Anderson RL, Ratcliffe I, Greenwell HC et al (2010) Clay swelling—A challenge in the oilfield. Earth Sci Rev 98:201–216. https://doi.org/10.1016/j.earscirev.2009.11.003

Bishop SR (1997) Experimental investigation of formation damage due to the induced flocculation of clays within a sandstone pore structure by a high salinity brine. In: SPE European Formation Damage Conference. https://doi.org/10.2118/38156-MS

Civan F (2015) Reservoir formation damage: fundamentals, modeling, assessment, and mitigation, third ed. Gulf Professional Publishing, Elsevier. https://doi.org/10.1016/C2014-0-01087-8

Dustin MK, Bargar JR, Jew AD et al (2018) Shale kerogen: hydraulic fracturing fluid interactions and contaminant release. Energy Fuels 32:8966–8977. https://doi.org/10.1021/acs.energyfuels.8b01037

Guo S, Lyu X, Zhang Y (2018) Relationship between tight sandstone reservoir formation and hydrocarbon charging: a case study of a Jurassic reservoir in the eastern Kuqa Depression, Tarim Basin, NW China. J. Nat. Gas Sci. Eng. 52:304–316. https://doi.org/10.1016/j.jngse.2018.01.031

He X, Li P, Qian H et al (2026) Biogeochemistry during hydraulic fracturing: a critical review of reservoirs, fluids, processes, and implications. Geoenergy Sci. Eng. 256:214143. https://doi.org/10.1016/j.geoen.2025.214143

Herz-Thyhsen RJ, Kaszuba JP, Dewey JC (2019) Dissolution of minerals and precipitation of an aluminosilicate phase during experimentally simulated hydraulic fracturing of a mudstone and a tight sandstone in the Powder River Basin, WY. Energy Fuels 33:3947–3956. https://doi.org/10.1021/acs.energyfuels.8b04443

Herz-Thyhsen RJ, Kaszuba JP, Dewey JC (2020) Mineral dissolution and precipitation induced by hydraulic fracturing of a mudstone and a tight sandstone in the Powder River Basin, Wyoming, USA. Appl Geochem 119:104636. https://doi.org/10.1016/j.apgeochem.2020.104636

Jew AD, Dustin MK, Harrison AL et al (2017) Impact of organics and carbonates on the oxidation and precipitation of iron during hydraulic fracturing of shale. Energy Fuels 31(4):3643–3658. https://doi.org/10.1021/acs.energyfuels.6b03220

Jew AD, Druhan JL, Ihme M et al (2022) Chemical and reactive transport processes associated with hydraulic fracturing of unconventional oil/gas shales. Chem Rev 122(9):9198–9263. https://doi.org/10.1021/acs.chemrev.1c00504

Kang Y, Huang F, You L et al (2016) Impact of fracturing fluid on multi-scale mass transport in coalbed methane reservoirs. Int J Coal Geol 154–155:123–135. https://doi.org/10.1016/j.coal.2016.01.003

Khan HJ, Spielman-Sun E, Jew AD et al (2021) A critical review of the physicochemical impacts of water chemistry on shale in hydraulic fracturing systems. Environ Sci Technol 55:1377–1394. https://doi.org/10.1021/acs.est.0c04901

Li J, Tang M, Ye Z et al (2017) Scale formation and control in oil and gas fields: a review. J Dispersion Sci Technol 38(5):661–670. https://doi.org/10.1080/01932691.2016.1185953

Li P, He X, Zhou C et al (2023) Mineral compositions and microstructural characteristics of the tight sandstone reservoir in the Sulige area and their potential influence on hydraulic fracturing. Hydrogeol Eng Geol 50(3):1–11 (In Chinese with English abstract)

Liu N, Gao Q (2018) Experimental study on damage mechanism of fracturing fluid in CBM reservoir of eastern Yunnan and western Guizhou. IOP Conf Series: Earth and Environ Sci 189:052075. https://doi.org/10.1088/1755-1315/189/5/052075

Luo L, Meng W, Gluyas J et al (2019) Diagenetic characteristics, evolution, controlling factors of diagenetic system and their impacts on reservoir quality in tight deltaic sandstones: Typical example from the Xujiahe Formation in Western Sichuan Foreland Basin, SW China. Mar Pet Geol 103:231–254. https://doi.org/10.1016/j.marpetgeo.2019.02.012

Maley D, Farion G, Giurea-Bica G et al (2013) Non-polymeric permanent clay stabilizer for shale completions. SPE European Formation Damage Conference and Exhibition. https://doi.org/10.2118/165168-MS

Menefee AH, Welch NJ, Frash LP et al (2020) Rapid mineral precipitation during shear fracturing of carbonate-rich shales. J Geophys Resh-Solid Earth 125:e2019JB018864. https://doi.org/10.1029/2019JB018864

Orem W, Tatu C, Varonka M et al (2014) Organic substances in produced and formation water from unconventional natural gas extraction in coal and shale. Int J Coal Geol 126:20–31. https://doi.org/10.1016/j.coal.2014.01.003

Qian H, Ma Z, Li P et al (2012) Hydrochemistry, 2nd edn. Geological Publishing House, Beijing (In Chinese)

Sharma S, Agrawal V, Akondi RN (2020) Role of biogeochemistry in efficient shale oil and gas production. Fuel 259:116207. https://doi.org/10.1016/j.fuel.2019.116207

Sruthi PL, Reddy PHP (2019) Swelling and mineralogical characteristics of alkali-transformed kaolinitic clays. Appl Clay Sci 183:105353. https://doi.org/10.1016/j.clay.2019.105353

Song X, Qin Y, Ma H et al (2019) A new method to determine wettability of tight sandstone: water imbibition evaporation rate ratio measurements. Energy Fuels 33(3):1998–2007. https://doi.org/10.1021/acs.energyfuels.8b04184

Tipper ET, Stevenson EI, Alcock V et al (2021) Global silicate weathering flux overestimated because of sediment-water cation exchange. PNAS 118:e2016430118. https://doi.org/10.1073/pnas.2016430118

Wang R, Shi W, Xie X et al (2020) Clay mineral content, type, and their effects on pore throat structure and reservoir properties: Insight from the Permian tight sandstones in the Hangjinqi area, north Ordos Basin China. Marine Petroleum Geol 115:104281. https://doi.org/10.1016/j.marpetgeo.2020.104281

Wang Z, Liu S, Qin Y (2021) Coal wettability in coalbed methane production: a critical review. Fuel 303:121277. https://doi.org/10.1016/j.fuel.2021.121277

Wei Y (2019) Study on the spontaneous imbibition effect of tight sandstone gas reservoirs in Sulige. Master's thesis, Xi'an Shiyou University. (in Chinese with English abstract)

Wei Y, Li C, Cao D et al (2019) The effects of particle size and inorganic mineral content on fines migration in fracturing proppant during coalbed methane production. J Petrol Sci Eng 182:106355. https://doi.org/10.1016/j.petrol.2019.106355

Yang IH, Lee HS (2018) Desorbed gas volume estimation using conventional well-log data for the Montney Formation, Deep Basin, Canada. J Petrol Sci Eng 162:633–651. https://doi.org/10.1016/j.petrol.2017.10.082

Zhao S, Wang Y, Li Y et al (2022) Tight sandstone petrophysics and flow rate/water/alkali sensitivities variations: a cross section of Permian in the northeastern of Ordos Basin, China. J Petrol Sci Eng 208:109443. https://doi.org/10.1016/j.petrol.2021.109443

Zhou Z, Li X, Teklu TW (2021) A critical review of osmosis-associated imbibition in unconventional formations. Energies 14:835. https://doi.org/10.3390/en14040835

Chapter 8
Environmental Problems in the Hydraulic Fracturing Areas

Water is involved throughout the entire life cycle of unconventional hydrocarbon production. From pre-fracturing freshwater sourcing (surface and groundwater), fracturing fluid preparation (mixing freshwater with chemical additives), injection during fracturing, and post-fracturing flowback to surface treatment and reuse, water is involved in every stage. Due to the use of chemical additives, the fracturing fluid itself is inherently polluting. During hydraulic fracturing, fracturing fluid not only mixes with formation brines rich in dissolved ions but also reacts with reservoir rocks, leading to the release of various chemical constituents, especially metallic elements, resulting in flowback water with complex water quality. Consequently, hydraulic fracturing consumes vast amounts of freshwater while generating large volumes of toxic and hazardous wastewater with strong pollution potential. Once these fluids infiltrate groundwater aquifers, they can cause severe environmental pollution. Therefore, potential environmental contamination, especially groundwater pollution in fracturing areas, must receive sufficient attention. The Sulige region is rich in groundwater resources, which are highly susceptible to contamination by fracturing fluids, drilling fluids, and flowback wastewater during oil and gas field development. This chapter discussed hydraulic fracturing activities, groundwater hydrochemical characteristics, potential pollution sources, and pathways in the Sulige region. Furthermore, an identification model for groundwater pollution induced by flowback fluids was established and recommendations are proposed for the protection of soil and groundwater in hydraulic fracturing zones.

8.1 Hydraulic Fracturing in the Sulige Gas Field

The Sulige Gas Field is currently in a phase of stable production, projected to maintain this state for at least 15 more years (Dai 2022). However, for tight sandstone gas, production from individual wells declines rapidly, with stable production lasting

X. He and P. Li, *Geochemical Processes in Tight Gas Hydraulic Fracturing Stimulation*, Environmental Earth Sciences, https://doi.org/10.1007/978-3-032-15180-3_8

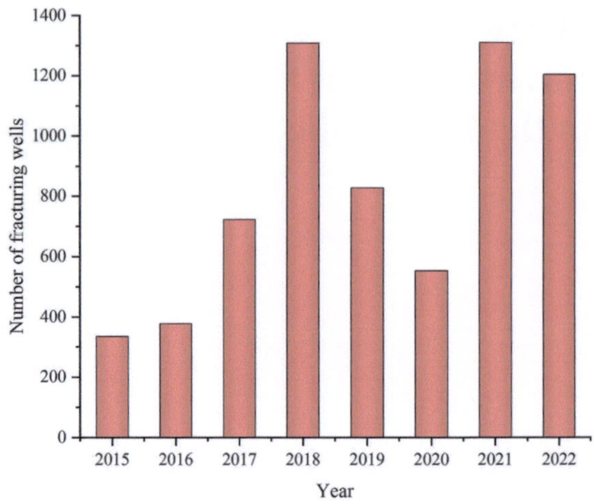

Fig. 8.1 The interannual variation of fractured wells in the Sulige Gas Field

only 2–3 years and annual decline rates exceeding 20% (Li 2021; Jia et al. 2022). To sustain and increase annual natural gas output, a large number of new wells must be developed each year to offset the declining productivity of existing wells. As shown by the historical trend of fracturing wells in the Sulige Gas Field (Fig. 8.1), the number of newly drilled and fractured wells has steadily increased, with a pronounced rise in recent years. In 2020, 1310 wells were completed and 1295 new wells were put into production. In 2022, to achieve a daily gas production exceeding 100 million m^3, 1204 new wells were added (He 2023), nearly four times the number in 2015. With reservoir quality gradually declining, maintaining an annual output of 30 billion m^3 will require an additional 6 billion m^3 of new capacity each year. Thus, in the coming years, the Sulige Gas Field will need to drill over 1000 new fracturing wells annually, possibly more. Clearly, hydraulic fracturing activities in the Sulige region are expected to intensify in the future. Frequent fracturing operations require large volumes of fracturing fluid and generate more wastewater, inevitably imposing tremendous pressure and threats on the local water environment.

8.1.1 Usage of Fracturing Fluid

As the Sulige Gas Field lies in the arid to semi-arid region of northwest China with scarce surface water resources, its fracturing operations primarily rely on local groundwater. In the northern Ordos Desert Plateau, water is mainly obtained by drilling wells near the sites, whereas in the southern Loess Plateau, water is primarily supplied through tank truck transportation. By 2020, nearly 16,500 wells had been developed in the Sulige Gas Field, with a cumulative fracturing fluid consumption of approximately 12 million m^3 (He 2023). As the development scale has expanded, the number of hydraulic fracturing wells and the total consumption of fracturing

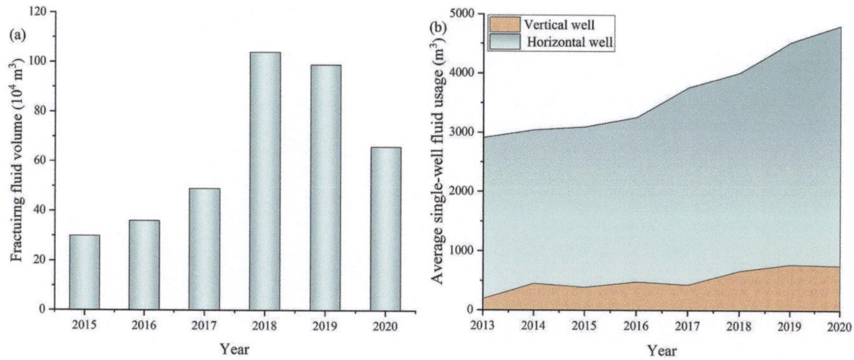

Fig. 8.2 Annual usage of fracturing fluid (**a**) and usage variation of individual well (**b**) in the Sulige Gas Field

fluids have shown a continuous upward trend (Figs. 8.1 and 8.2a). Moreover, with the continuous improvement of multi-layer and multi-stage fracturing technologies and the implementation of the volumetric fracturing concept, the average fracturing fluid volume per well has increased annually, particularly for horizontal wells. In 2020, the mean fracturing fluid consumption per vertical well was 757 m^3, whereas horizontal wells used an average of 4042 m^3, approximately 5–6 times higher than that of vertical wells (Fig. 8.2b). A comparison between the close-cluster fracturing tests and conventional bridge-plug fracturing in 2020 shows that the average number of fractured stages increased from 8.8 to 14, the number of clusters from 20.8 to 37, and the mean per-well fracturing fluid volume from 3971 to 6617 m^3. From 2018 to 2020, the total volume of fracturing fluid used in the Sulige Gas Field was about 2.69 million m^3, 2.5 times higher than that used from 2015 to 2017. Although the field has entered a stable production stage, maintaining an annual output of nearly 30 billion m^3 will require the fracturing of numerous new wells and continued consumption of large volumes of freshwater and fracturing fluid. Fracturing fluids, which contain a variety of chemical additives, already pose notable environmental hazards. After hydraulic fracturing, these fluids are converted into substantial quantities of more heavily polluted flowback wastewater, posing serious risks to the local ecological environment.

8.1.2 *Wastewater Generated from Hydraulic Fracturing*

Unconventional hydrocarbon production, requires substantial freshwater input and simultaneously produces diverse waste fluids such as drilling muds, well-flushing fluids, post-fracturing flowback, and produced water. The fluids returned from drilling and well-flushing operations contain large amounts of drilling mud with complex compositions and high ionic concentrations. Fracturing flowback fluids are complex

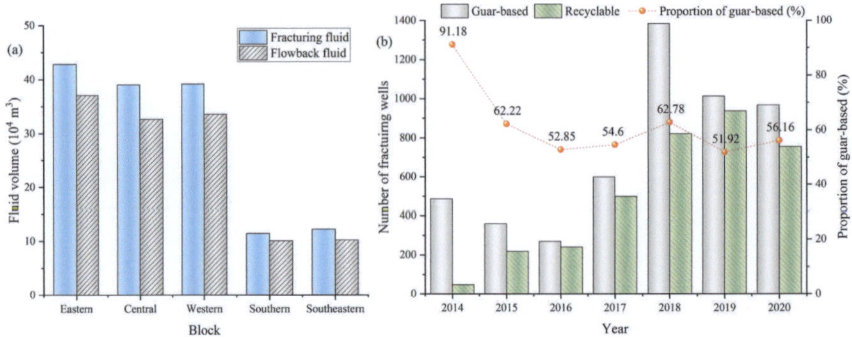

Fig. 8.3 The volume of fracturing fluid and flowback fluid in different blocks (**a**) and the usage of diverse fracturing fluid systems (**b**) in the Sulige Gas Field

mixtures comprising solids, gels, hydrocarbons, and formation brines. Flowback fluids from the Upper Paleozoic tight sandstone reservoirs typically contain formation brines, exhibiting high TDS concentrations. Flowback fluids from acid-fractured Lower Paleozoic carbonate reservoirs have high residual acid content, making on-site treatment and reuse difficult. In recent years, the flowback fluid volumes in the eastern, western, and central fracturing zones of the Sulige Gas Field have each exceeded 300,000 m^3 (Fig. 8.3a). Currently, fracturing fluids used for reservoir stimulation in the Sulige Gas Field are primarily guar-based and polymer recyclable systems (e.g., EM50, biogels, and variable-viscosity slickwater). Guar based systems are challenging to recycle, demanding dedicated treatment units and yielding low overall reuse efficiency. Although recyclable fracturing fluid systems have been widely applied since 2014, with usage increasing from less than 10% to nearly 50%, guar systems remain the dominant fracturing fluid type across Sulige areas (Fig. 8.3b).

Tight sandstone reservoirs in the Sulige Gas Field generally produce water, with severe water production particularly in the western blocks. At present, the Sulige Gas Field has 115 gas gathering stations, with produced liquid volumes averaging 5–117.5 m^3/d, primarily consisting of high-ionic-strength formation brines. Over more than ten years of production, the average liquid-to-gas ratio (water produced per 10^4 m^3 of gas) increased from 0.5 m^3/10^4 m^3 to 0.75 m^3/10^4 m^3, reaching up to 3 m^3/10^4 m^3 in some cases (Zhang et al. 2021). According to 2020 gas production data, the Sulige Gas Field generated roughly 1.95×10^6 m^3 of produced water during that year.

8.2 Groundwater Quality in the Sulige Area

8.2.1 The Northern Desert Plateau Area

Geomorphologically, the Sulige Gas Field lies within the Ordos Desert Plateau and the loess-covered northern fringe of the Loess Plateau. Currently, the tight sandstone gas production areas (east, central, and west) are primarily distributed in the northern desert plateau. The northern desert plateau hosts abundant groundwater within the Quaternary Salawusu Formation and the Cretaceous Huanhe and Luohe Formations, characterized by good water quality and serving as major local water sources (Fig. 8.4). The Salawusu Formation groundwater is mainly of the low-salinity HCO_3-Ca hydrochemical type. The average cation concentrations follow $Ca^{2+} > Na^{+} > Mg^{2+} > K^{+}$, while the anion concentrations follow $HCO_3^{-} > SO_4^{2-} > Cl^{-} > CO_3^{2-}$ (Table 8.1). The pH ranges from 7.10 to 8.97, indicating neutral to weakly alkaline conditions. The total dissolved solids (TDS) are generally low, with an average below 300 mg/L. Nitrate concentrations range from 0 to 228.59 mg/L, with an average of 29.59 mg/L, indicating nitrate contamination in some areas.

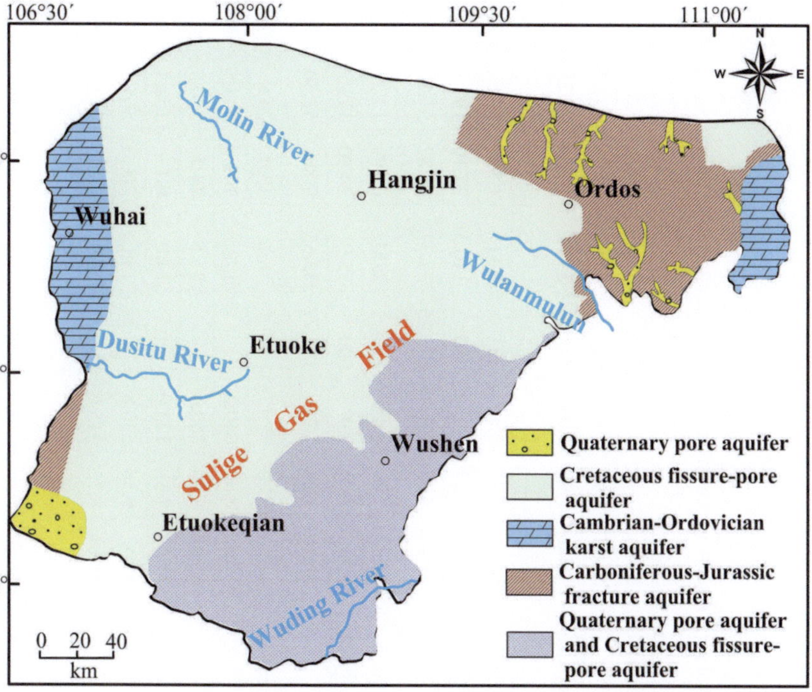

Fig. 8.4 Distribution of aquifer types in the northern Desert Plateau area (Modified from Hou et al. 2008)

Table 8.1 Statistical results of groundwater chemical parameters in the study area (Unit mg/L)

Aquifer		pH	K^+	Na^+	Ca^{2+}	Mg^{2+}	Cl^-	SO_4^{2-}	HCO_3^-	CO_3^{2-}	NO_3^-	TDS	δ^2H	$\delta^{18}O$
Salawusu Formation, northern Desert Plateau area (Yang et al. 2016; Li et al. 2022)	Min	7.10	0.03	6.35	2.00	0.47	3.5	4.81	27.5	0.09	0	93.0	– 86.19	– 11.02
	Max	8.97	15.52	600.5	159.69	77.65	194.99	315.75	445.42	36.31	228.59	1062	– 56.71	– 7.57
	Mean	7.93	2.04	52.96	54.22	19.67	27.25	51.40	214.61	3.55	29.59	267.72	– 68.66	– 8.50
Huanhe Formation, northern Desert Plateau area (Su et al. 2011)	Min	7.00	0.42	22	5.11	1.30	5.16	24	42.7	0	–	272.07	– 91.78	– 12.07
	Max	8.37	2.9	750	140.3	27.10	197	1882.8	252	4.81	–	2860.66	– 48.78	– 7.40
	Mean	7.98	1.64	172.75	38.77	11.87	51.56	294.21	168.78	2.09	–	735.90	– 73.86	– 9.56
Luohe Formation, northern Desert Plateau area (Su et al. 2011)	Min	7.20	0.45	7.64	2.00	0.60	1.72	19.2	73.2	0	–	263.54	– 92.70	– 11.60
	Max	8.40	2.9	600	75.9	24.30	173.31	1306.4	263.59	15	–	2051.1	– 45.88	– 6.50
	Mean	7.94	1.74	165.77	35.63	9.88	52.30	250.52	176.63	2.36	–	697.31	– 75.18	– 9.88
Huanhe Formation, southern Loess Plateau area	Min	7.70	77.1		36.1	13.4	24.8	110.5	67.1	0	0	504	– 103.71	– 13.94
	Max	8.30	640.2		170.3	109.4	744.5	888.6	402.7	0	166.6	2266	– 76.70	– 10.28
	Mean	8.00	303.4		71.1	44.7	320.5	417.0	227.4	0	26.80	1253	– 89.45	– 11.83
Luohe Formation, southern Loess Plateau area	Min	7.88	123.0		14.0	6.1	152.4	144.1	9.2	0	1.80	796.4	– 101.58	– 13.90
	Max	9.38	530.8		158.3	200.3	565.4	1215.2	323.4	15	118.04	2285.7	– 80.04	– 10.84
	Mean	8.32	315.6		61.9	45.7	342.2	367.4	134.8	2.0	61.81	1311.0	– 94.27	– 12.63

Cretaceous groundwater primarily occurs in the Luohe and Huanhe Formations. The absence of a stable aquiclude between these aquifers results in strong vertical hydraulic connectivity and similar hydrochemical characteristics (Su et al. 2011). TDS values are generally higher than those of the overlying Salawusu Formation groundwater but typically remain below 1 g/L, indicating good water quality in most areas. The dominant hydrochemical type is low-salinity $HCO_3{\cdot}SO_4-Ca{\cdot}Na$ water. Controlled by the lithology and material composition of overlying strata, the aeolian sands above the Cretaceous units enhance meteoric recharge, resulting in overall low mineralization in the groundwater (He 2023).

8.2.2 *The Southern Loess Plateau Area*

In the southern Loess Plateau area, the TDS of groundwater in the Huanhe Formation ranges from 564 to 2266 mg/L, with values exceeding 1 g/L in most regions. A general increasing trend in TDS is observed from east to west and from south to north. The hydrochemical facies are diverse, predominantly characterized by $SO_4{\cdot}Cl$–Na, $SO_4{\cdot}Cl-Ca{\cdot}Na$, and HCO_3-Na types. For cations, $Na^+ + K^+$ concentrations range from 77.1 to 640.2 mg/L, with an average of 303.4 mg/L. Ca^{2+} and Mg^{2+} concentrations range from 36.1–170.3 mg/L and 13.4–109.4 mg/L, with averages of 71.1 mg/L and 44.7 mg/L, respectively. Among anions, Cl^- and $SO_4{}^{2-}$ are predominant, ranging from 24.8–744.5 mg/L (mean 320.5 mg/L) and 110.5–888.6 mg/L (mean 417.0 mg/L), respectively. In certain regions, $HCO_3{}^-$ is the dominant anion, with concentrations between 67.1 and 402.7 mg/L, averaging 227.4 mg/L.

Similar to the overlying Huanhe Formation groundwater, the Luohe Formation also exhibits high TDS values (796.4–2285.7 mg/L), mainly of $SO_4{\cdot}Cl-Na$ and HCO_3-Na types. The δ^2H and $\delta^{18}O$ values of the Luohe Formation are slightly lower than those of the Huanhe Formation. Due to evaporative effects, δ^2H and $\delta^{18}O$ values in aquifers of the southern study area are generally lower than those in the northern Desert Plateau area (Table 8.1). Portions of the Cretaceous strata are overlain by Quaternary loess and Neogene mudstone, impeding direct recharge from precipitation and leading to generally inferior groundwater quality (Hou et al. 2017). The joint influence of agricultural practices and geological conditions results in elevated nitrate levels in groundwater across the southern Loess Plateau.

8.3 Identification of Potential Groundwater Pollution

8.3.1 Potential Pollution Sources and Pollution Pathways

(1) **Pollution Sources**

The production cycle of the Sulige Gas Field mainly includes four stages: site preparation, drilling, hydraulic fracturing, and production. Each stage consumes a certain amount of freshwater and generates wastewater. During the wellsite preparation phase, wastewater mainly originates from domestic use by onsite personnel and is typically collected in storage tanks before being transported to treatment facilities. The volume of wastewater produced during this stage is minimal and generally poses negligible environmental risk. Water-based drilling fluids are commonly used in the Sulige Gas Field, with an average water consumption of approximately 0.2 m^3 per meter of drilling. Vertical wells in the study area have an average depth of 3500 m, while horizontal wells range from 4000 to 5000 m (median 4600 m). The average water consumption per vertical well is 700 m^3, and 960 m^3 per horizontal well. A drilling platform usually contains multiple wells. Thus, drilling wastewater from one well is partially recycled for use in other wells on the same platform to conserve freshwater. Consequently, over 85% of drilling wastewater can be recycled and reused in other wells, while about 15% (105–144 m^3) must be stored and transported to treatment plants for further processing (He 2023).

The main procedures during the hydraulic fracturing stage include well site preparation, wellbore preparation, fluid mixing, fracturing operations, and flowback testing. In the Sulige Gas Field, a factory-style clustered well development approach is commonly used. Each well pad generally includes multiple directional (vertical) and horizontal wells, which are often fractured simultaneously to improve operational efficiency. The typical well pad configuration in the Sulige Gas Field is a nine-well pattern, consisting of seven vertical wells and two horizontal wells (Zhang et al. 2019). Vertical wells are generally fractured in one to three layers, with each layer consuming approximately 200–300 m^3 of fracturing fluid and an average total of 757 m^3 per well. Horizontal wells usually contain 5–12 fracturing stages, averaging 7.1 stages per well (Fei et al. 2022). Each stage consumes between 200 and 750 m^3 of fluid, with an average of 434 m^3 per stage. The post-fracturing flowback rate of gas wells in the Sulige Gas Field typically varies between 10 and 50%, with an average value of approximately 40%. As a result, a significant amount of polluted wastewater is produced during the hydraulic fracturing phase.

The production stage mainly involves gas production and produced-water recovery and treatment, with minimal water consumption. Because of differences in geographic location and production strata, the water volume produced from each well varies considerably, with substantial discrepancies observed between different well clusters. The volume of produced water can be estimated using the average water-to-gas ratio of the entire gas field. The Sulige Gas Field exhibits an average water–gas ratio ranging from 0.32 to 1.06 m^3 of water per 10^4 m^3 of gas, with a median value of about 0.75 $m^3/10^4$ m^3 (Zhang et al. 2021).

From the above analysis, it is evident that, relative to other stages of natural gas production, the hydraulic fracturing stage involves large volumes of fracturing fluids and flowback fluids (including produced water), which are highly polluted and represent the primary potential source of groundwater contamination in the Sulige Gas Field.

(2) **Pollution Pathway**

Groundwater contamination associated with unconventional natural gas extraction primarily includes gaseous pollution (such as methane) and contamination from waste fluids produced during operation. Various potential contamination pathways exist, which can generally be divided into three categories: (1) upward migration from deep reservoirs, (2) aquifer leakage, and (3) surface leakage and discharge.

1) Upward migration of deep fluids: During hydraulic fracturing, high-pressure fluids are injected into reservoirs several kilometers deep to create fractures and connect existing natural fissures. This process may induce fractures in the subsurface, connect pre-existing faults, or intersect abandoned wellbores. The high pressure also establishes a sustained upward hydraulic gradient, promoting the migration of deep fluids and fracturing fluids toward shallower zones. However, research indicates that the extent of fractures induced by hydraulic fracturing is limited, which the maximum observed vertical propagation is about 588 m, and statistical data show that the probability of fractures extending beyond 350 m is less than 1% (Davies et al. 2012). In the Sulige Gas Field, hydraulic fracturing is typically conducted at depths of 2800–3700 m (Jia et al. 2022). The field is situated within the Yishan Slope, a structurally stable region with poorly developed faults. Thus, the presence of kilometer-scale faults connecting the fractured reservoirs and shallow aquifers is unlikely, indicating a very low probability of upward migration of deep fluids contaminating shallow groundwater. The gas-bearing formations in the Sulige Field are mainly within the Middle-Lower Permian Shihezi and Shanxi Formations, overlain by Triassic, Jurassic, Cretaceous, and Quaternary strata. The directly overlying Triassic strata consist primarily of mudstone interbedded with fine sandstone and shale, exhibiting very low vertical permeability (< 0.001 mD), forming an effective caprock. The Jurassic strata, composed mainly of mudstone and sandstone-coal interbeds (320–570 m thick), have permeability typically below 0.1 mD, acting as a competent aquitard (Wang et al. 2021). Together, the Triassic and Jurassic caprocks provide effective sealing for the tight sandstone gas reservoirs and significantly hinder the upward migration of gases and fluids from the deep fracturing zones into the overlying Cretaceous and Quaternary aquifers.
2) Aquifer leakage: During intensive hydraulic fracturing operations, inadequate management leading to fluid leakage (e.g., casing and cementing failures, leakage during deep well reinjection) poses a direct contamination risk to underlying aquifers.

3) Surface leakage and discharge: During storage and transportation, fracturing fluids and flowback fluids face multiple leakage risks, including spills from storage ponds, tanks, and pipelines; leakage during transportation, and substandard discharge from wastewater treatment facilities (Fig. 8.5). Compared to the upward migration of deep fluids, wastewater leakage represents a more common and direct threat to groundwater during tight gas production (Zhang et al. 2024). The surface impoundment of significant volumes of flowback fluids for temporary storage prior to treatment and reuse elevates the risk of containment breaches. Furthermore, the uncontrolled release and the discharge of untreated or improperly treated wastewater, along with extreme weather-induced leaching from drilling solids, can release toxic and hazardous substances into groundwater. In the northern desert plateau of the Sulige Gas Field, groundwater levels are shallow, typically within 5 m of the surface and less than 1 m in some areas. The surface is covered by permeable Quaternary unconsolidated deposits (sand and gravel), allowing leaked surface fluids to rapidly infiltrate into underlying aquifers, while the overlying materials offer limited capacity for contaminant degradation. Consequently, improper anti-seepage measures at well sites can easily lead to groundwater contamination. Strict management of fracturing and flowback fluid storage and transport is therefore essential to prevent leakage.

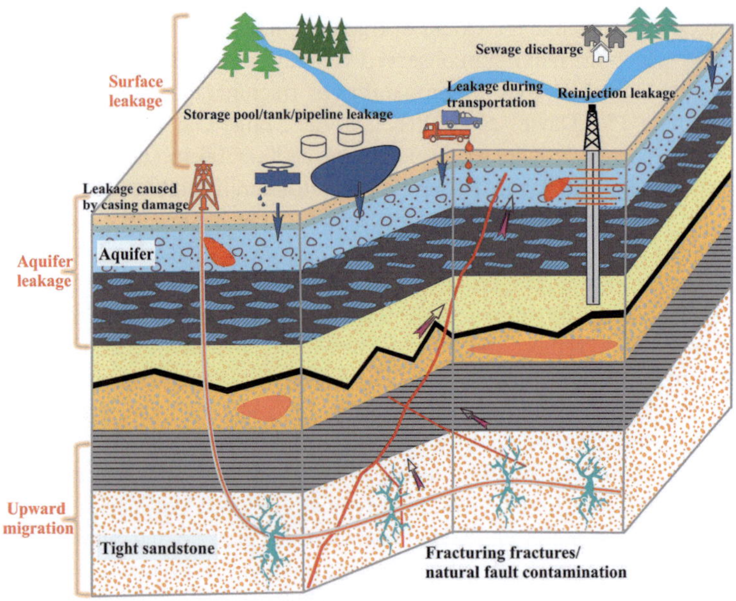

Fig. 8.5 Schematic diagram of potential groundwater pollution pathways in tight gas development area

8.3.2 *Pollution Identification Model*

Flowback fluid from hydraulic fracturing is the primary wastewater generated during unconventional natural gas production. Similar to the fracturing process itself, when the flowback fluid enters groundwater aquifers, it interacts with the aquifer matrix through water–rock reactions such as mineral dissolution/precipitation and ion exchange. Therefore, taking flowback fluid as an example, this study comprehensively considers the water–rock interactions occurring after its entry into the aquifer. By coupling a series of hydrogeochemical models, including a mixing model, mineral dissolution–precipitation model, ion exchange model, and regional sensitivity evaluation model, sensitivity indicators of potential groundwater contamination in the Sulige Gas Field were identified. The identified sensitivity indicators were subsequently used to establish recognition curves for flowback fluid-induced groundwater contamination, serving as tools for the identification, assessment, and prediction of groundwater pollution in the study area. Ultimately, a comprehensive model was developed to identify, assess, and predict potential groundwater contamination in hydraulic fracturing zones (Fig. 8.6).

(1) **Conservative Mixing Model**

The mixing model requires significant end-member differences between the chemical compositions of flowback fluid and unpolluted groundwater. The mixing process between flowback fluid and groundwater was computed through the end-member mixing Eqs. (8.1) and (8.2), with simulations performed using the *Mix* module of

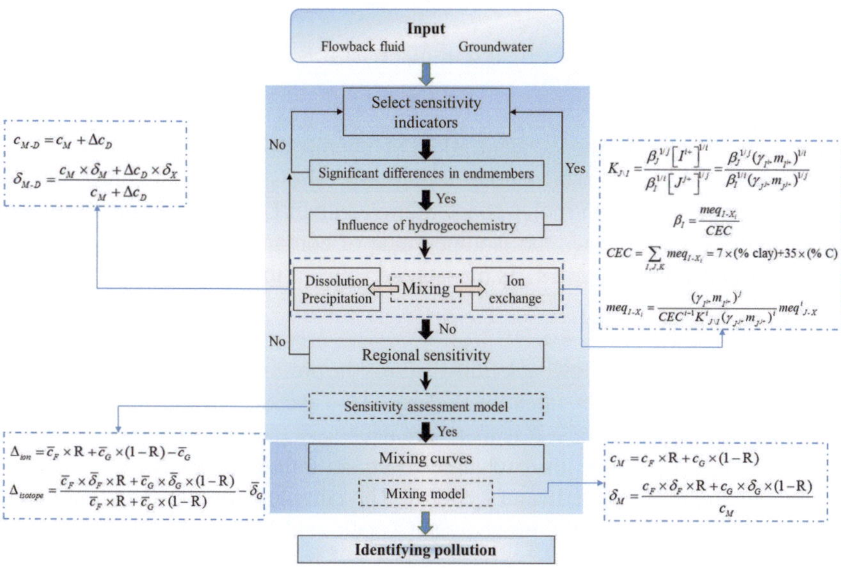

Fig. 8.6 Flowchart for the construction of groundwater pollution identification model

PHREEQC. The model simulated eleven mixing gradients between flowback fluid and groundwater (0, 0.01, 0.1, 1, 5, 10, 30, 50, 70, 90, and 100%), neglecting the influence of hydrogeochemical reactions on the mixing behavior.

$$c_M = c_F \times R + c_G \times (1 - R) \tag{8.1}$$

$$\delta_M = \frac{c_F \times \delta_F \times R + c_G \times \delta_G \times (1 - R)}{c_M} \tag{8.2}$$

Here, c_F, c_G, and c_M denote the concentrations of chemical components in the flowback fluid, groundwater, and their mixture (mmol/L), respectively; δ_F, δ_G, and δ_M correspond to their isotopic values; and R represents the mixing proportion (%).

(2) **Mineral Dissolution–Precipitation Model**

This model assumes that upon the entry of flowback fluid into the aquifer and subsequent mixing with groundwater, changes in pH and ion concentrations can induce dissolution of aquifer minerals or precipitation within the mixed solution. Following these reactions, the mixture and aquifer medium achieve a new water–rock equilibrium. In this study, the flowback fluid and groundwater were first mixed at preset ratios, followed by interaction with aquifer minerals to achieve new equilibrium conditions. The saturation index (SI) was utilize to determine the dissolution or precipitation status and to quantify ion and isotope transfer between the mixture and aquifer matrix (Eqs. 8.3 and 8.4).

$$c_{M-D} = c_M + \Delta c_D \tag{8.3}$$

$$\delta_{M-D} = \frac{c_M \times \delta_M + \Delta c_D \times \delta_X}{c_M + \Delta c_D} \tag{8.4}$$

where $c_{M\text{-}D}$ represents the concentration of chemical components (mmol/L) in contaminated groundwater after considering mineral dissolution and precipitation; Δc_D denotes the change in molar concentration (mmol/L) caused by dissolution or precipitation; $\delta_{M\text{-}D}$ indicates the isotope value of contaminated groundwater after dissolution/precipitation; and δ_X is the isotope value of the minerals involved.

(3) **Ion Exchange Model**

As the infiltration of flowback fluid disturbs the pre-existing water–rock equilibrium within the aquifer, it readily undergoes ion exchange with aquifer minerals, especially clay minerals. It is therefore necessary to establish an ion exchange model to assess how ion exchange reactions influence sensitivity indicators in contaminated groundwater systems. The complete ion exchange modeling process consists of three procedures (Appelo and Postma 2005; Zhang et al. 2021):

1) Establish the ion balance model between the unpolluted aquifer medium and groundwater based on the mass conservation law:

$$K_{J\backslash I} = \frac{\beta_J^{1/j}\left[I^{i+}\right]^{1/i}}{\beta_I^{1/i}\left[J^{j+}\right]^{1/j}} = \frac{\beta_J^{1/j}(\gamma_{I^{i+}} m_{I^{i+}})^{1/i}}{\beta_I^{1/i}(\gamma_{J^{j+}} m_{J^{j+}})^{1/j}} \tag{8.5}$$

$$\beta_I = \frac{meq_{I-X_i}}{\text{CEC}} \tag{8.6}$$

$$\text{CEC} = \sum_{I,J,K} meq_{I-X_i} = 7 \times (\%\ \text{clay}) + 35 \times (\%\ \text{C}) \tag{8.7}$$

$$meq_{I-X_i} = \frac{(\gamma_{I^{i+}} m_{I^{i+}})^j}{\text{CEC}^{i-1} K_{J\backslash I}^i (\gamma_{J^{j+}} m_{J^{j+}})^i} meq_{J-X}^i \tag{8.8}$$

where $K_{J/I}$ is the ion exchange coefficient (following the Gaines-Thomas convention); β_I (β_J) represents the equivalent fraction; i (j) denotes the charge number of ion I (J); $[I^{i+}]$ ($[J^{j+}]$) is the ion activity, equal to the product of the activity coefficient (γ) and molar concentration (m); $meq_{I\text{-}Xi}$ ($meq_{J\text{-}Xi}$) is the ion exchange capacity of the aquifer medium for ion I (J) in meq/kg; and CEC denotes the cation exchange capacity in meq/kg, which can be measured in the laboratory or estimated using Eq. (8.7) (Gaines and Thomas 1953; Breeuwsma et al. 1986; Zheng et al. 2021; He et al. 2022).

2) Apply the previously established conservative mixing model (Eqs. 8.1 and 8.2) to simulate the mixing between flowback fluid and groundwater.
3) Establish a new water–rock equilibrium between contaminated groundwater and the aquifer medium:

$$K_{J\backslash I} = \frac{\left(\frac{meq_{J-X} + umeq_{exchange}}{\text{CEC}}\right)^{1/j} (\gamma_{M-I}(m_{M-I} + \Delta m_I))^{1/i}}{\left(\frac{meq_{I-X} - umeq_{exchange}}{\text{CEC}}\right)^{1/i} (\gamma_{M-J}(m_{M-J} - \Delta m_J))^{1/j}} \tag{8.9}$$

where u is the gravimetric water content (%); Δm is the amount of exchanged ions in contaminated groundwater (mmol); $meq_{exchange}$ is the equivalent ion exchange quantity ($i(j) \times \Delta m$); $\gamma_{M\text{-}I}$ ($\gamma_{M\text{-}J}$) and $m_{M\text{-}I}$ ($m_{M\text{-}J}$) represent the activity coefficients and molar concentrations of ions I (J) in contaminated groundwater, respectively (Zheng et al. 2021; He et al. 2022).

The ion exchange quantity (Δc_C) generated from the flowback fluid-groundwater mixture interacting with the aquifer matrix is incorporated into the mixing model to form the basic ion exchange equation:

$$c_{M-C} = c_M + \Delta c_C \tag{8.10}$$

$$\delta_{M-C} = \frac{c_M \times \delta_M + \Delta c_C \times \delta_Y}{c_M + \Delta c_C} \tag{8.11}$$

In this equation, $c_{M\text{-}C}$ denotes the chemical concentration in contaminated groundwater after ion exchange (mmol/L); $\delta_{M\text{-}C}$ is the isotopic value after ion exchange; and δ_Y represents the isotopic value of the exchanged chemical components in the aquifer medium.

(4) **Regional Sensitivity Assessment Model**

The regional sensitivity assessment model assumes that the concentrations of chemical species and isotopic values in groundwater are randomly distributed within a certain range under natural conditions. When the solute concentration in groundwater follows a normal distribution, 99.7% of random values fall within the mean ± 3σ (σ: standard deviation) range (Huang et al. 2020; Li et al. 2020). If a fraction R of flowback water is mixed into the aquifer groundwater, the chemical and isotopic composition will change by Δ. When hydrogeochemical processes exert minimal influence on the mixing process, the following equation can be applied to evaluate regional sensitivity (Li et al. 2020):

$$\Delta_{ion} = \overline{c}_F \times R + \overline{c}_G \times (1 - R) - \overline{c}_G \tag{8.12}$$

$$\Delta_{isotope} = \frac{\overline{c}_F \times \overline{\delta}_F \times R + \overline{c}_G \times \overline{\delta}_G \times (1 - R)}{\overline{c}_F \times R + \overline{c}_G \times (1 - R)} - \overline{\delta}_G \tag{8.13}$$

Here, Δ_{ion} and $\Delta_{isotope}$ represent the mean ± 3σ of groundwater chemical and isotopic parameters, respectively; $\overline{c}_F$ and $\overline{c}_G$ denote the mean chemical levels in flowback water and groundwater, while $\overline{\delta}_F$ and $\overline{\delta}_G$ represent their respective mean isotopic compositions. R denotes the threshold proportion (%) at which the tracer can identify groundwater contamination by fracturing flowback fluid.

8.3.3 *Identification of Groundwater Pollution*

Strontium isotopes are recognized as highly sensitive indicators for monitoring and identifying environmental pollution. Strontium (Sr) is a rare element naturally occurring in four isotopic forms: ^{84}Sr, ^{86}Sr, ^{87}Sr, and ^{88}Sr (Capo et al. 2014). Among these, ^{84}Sr, ^{86}Sr, and ^{88}Sr are stable isotopes, whereas ^{87}Sr is produced by the radioactive decay of ^{87}Rb (Ladegaard-Pedersen et al. 2020). Because Sr isotopic ratios ($^{87}Sr/^{86}Sr$) are not affected by biogeochemical processes, they are widely employed as geochemical tracers for studying water–rock interactions, fluid mixing, and contaminant source identification (Capo et al. 2014; Ladegaard-Pedersen et al. 2020; Li et al. 2020). Xie et al. (2013) used $^{87}Sr/^{86}Sr$ ratios to characterize groundwater flow paths and associated water–rock interactions in the Datong Basin. Chapman et al.

(2012) applied the same isotope ratio to study produced water from the Marcellus shale, demonstrating its potential to trace groundwater and surface water contamination from shale gas development. Bagheri et al. (2014) used $^{87}Sr/^{86}Sr$ to identify salinity sources in formation water. $^{87}Sr/^{86}Sr$ ratio has thus been extensively applied in tracing the origins of flowback water and monitoring groundwater contamination (Owen et al. 2020; Zheng et al. 2017). Kohl et al. (2014) and Goldberg and Griffith (2014) found that monitoring $^{87}Sr/^{86}Sr$ in produced and drinking waters can provide early warnings of groundwater pollution. However, its applicability for identifying groundwater contamination in the Sulige Gas Field of the Ordos Basin requires further validation.

Based on the analytical results of fracturing flowback fluid and the established groundwater pollution identification model, the $^{87}Sr/^{86}Sr$ ratio was employed as a tracer to identify groundwater contamination in the widely distributed Cretaceous aquifers of the northern Desert Plateau area of the Sulige Gas Field. The characteristic sensitivity of $^{87}Sr/^{86}Sr$ in identifying groundwater contamination in the Sulige region was examined from three perspectives.

(1) Distinct differences in $^{87}Sr/^{86}Sr$ between flowback fluid and groundwater: In the study area, the $^{87}Sr/^{86}Sr$ ratio of groundwater generally falls below 0.712 (Su et al. 2011; Rao et al. 2015), whereas that of fracturing flowback and formation water in the Sulige region typically exceeds 0.714 (Dou et al. 2010; He et al. 2022). Therefore, a pronounced isotopic contrast exists between the two fluids, allowing the use of end-member mixing models to quantify the mixing ratio and assess the degree of groundwater contamination.

(2) Identification period and stability: Compared with conservative ions (Cl^-) and stable isotopes of hydrogen and oxygen (δ^2H, $\delta^{18}O$), the $^{87}Sr/^{86}Sr$ ratio in flowback fluid increases rapidly after the completion of hydraulic fracturing, exceeding 0.714 within 2–4 h (Fig. 4.13). In contrast, Cl^-, δ^2H, and $\delta^{18}O$ rise more gradually, indicating a delayed detection period (Figs. 4.8 and 4.12). Moreover, Cl^-, δ^2H, and $\delta^{18}O$ exhibit considerable spatial variability, whereas $^{87}Sr/^{86}Sr$ remains stable in both flowback and groundwater end-members across the Sulige region (Dou et al. 2010; He et al. 2022). Thus, $^{87}Sr/^{86}Sr$ demonstrates superior temporal and spatial sensitivity for contamination identification.

(3) Influence of hydrogeochemical processes: Similar to the entry of fracturing fluid into low-permeability reservoirs, once flowback fluid infiltrates the aquifer, it disrupts the pre-existing water–rock equilibrium, inducing a series of hydrogeochemical reactions that alter ionic concentrations and isotopic compositions in contaminated groundwater. For strontium, mineral dissolution–precipitation and cation exchange are the dominant processes.

1) Mineral dissolution and precipitation: The aquifer matrix in the study area mainly comprises quartz, feldspar, calcite, and clay minerals (illite, chlorite, and illite–smectite interlayers). Quartz, feldspar, and clay minerals exhibit extremely low dissolution and precipitation rates under natural conditions, thus having negligible short-term effects on groundwater chemistry (Appelo and Postma 2005). Previous studies show that at 25 °C and pH = 5, the average dissolution

times of 1 mm thick quartz, K-feldspar, and mica crystals are 34 million, 0.52 million, and 2.7 million years, respectively (Lasaga 1984). Modeling results further indicate that K-feldspar undergoes no significant dissolution during mixing (Table 8.2). Because the flowback fluid is weakly acidic while the groundwater is slightly alkaline, the intrusion of flowback fluid may induce carbonate mineral dissolution, potentially affecting the $^{87}Sr/^{86}Sr$ ensitivity.

Calcite is the dominant carbonate mineral in the aquifer matrix, accounting for 2–21% of the total mineral composition (Table 3.2). Hydrochemical parameters of flowback fluid and groundwater were input into a mixing model, with average temperatures set at 15 °C and 30 °C, respectively. In Eq. (8.4), δ_X represents the $^{87}Sr/^{86}Sr$ ratio in calcite from the aquifer matrix, valued at 0.710148 (Su et al. 2011). As soluble Sr in the aquifer mainly resides in calcite, the amount of dissolved Sr (Δc_D) was estimated from the Ca released during calcite dissolution (Eq. 8.14):

$$\Delta c_D = \frac{m_{\mathrm{Sr}}}{m_{\mathrm{Ca}}} \times \Delta \mathrm{Ca} \tag{8.14}$$

where m_{Sr} and m_{Ca} represent the Sr and Ca levels in calcite, with an $m_{\mathrm{Sr}}/m_{\mathrm{Ca}}$ ratio of approximately 0.5×10^{-3} (Stoll and Schrag 2001; Ando et al. 2006).

Modeling results indicate that when 0.1–10% of flowback fluid mixes with groundwater, calcite dissolution occurs in the aquifer matrix; however, the maximum change in $^{87}Sr/^{86}Sr$ ratio is only $\Delta\varepsilon_{\mathrm{Sr}} = 0.11$ (Table 8.2). The SI values of Sr-bearing minerals (celestite, strontianite) remain far below 0 throughout the mixing process, indicating undersaturation and thus no precipitation. Therefore, mineral dissolution–precipitation has negligible influence on $^{87}Sr/^{86}Sr$ and does not compromise its sensitivity in tracing contamination.

2) Cation exchange effects: Under steady hydrochemical conditions, ion exchange between groundwater and the aquifer matrix remains in dynamic equilibrium. When exhibits significant compositional shifts (e.g., due to contamination or acidification), cation exchange readjusts ion distributions between solid and liquid phases to reach a new equilibrium (Appelo and Postma 2005). Owing to their surface charge properties, clay minerals play a dominant role in this process; their content in the aquifer matrix is 8–33% (Table 3.2). Thus, intrusion of flowback fluid disrupts this equilibrium, prompting migration of Sr and its isotopes between the contaminated groundwater and solid matrix (Owen et al. 2020; Zheng et al. 2017). Because Na^+ is the predominant cation in flowback fluid, the model primarily considers exchange between Na^+ and Sr^{2+}. The cation-exchange coefficients ($K_{\mathrm{Na\backslash Sr}}$, $K_{\mathrm{Na\backslash K}}$, $K_{\mathrm{Na\backslash Ca}}$, $K_{\mathrm{Na\backslash Mg}}$) are 0.4, 0.25, 0.5, and 0.6, respectively (Appelo and Postma 2005; Zheng et al. 2017, 2021). The CEC, based on the average clay content, is 100.31 meq/kg. δ_Y is 0.714145 (Su et al. 2011), and u is set at 10% (Hou et al. 2017; Zheng et al. 2021).

Simulation results indicate that when the mixing ratio is below 1%, Sr^{2+} in the solution tends to be adsorbed by the aquifer medium; conversely, Sr^{2+} in the solid phase is released into the solution as the mixing ratio increases. The amount of Sr^{2+}

Table 8.2 Sensitivity evaluation of $^{87}Sr/^{86}Sr$ at different mixing ratios (R)

R	0%	0.01%	0.10%	1%	5%	10%	30%	50%	70%	90%	100%
c_M (mmol/L)	7.35E−03	7.43E−03	8.15E−03	1.54E−02	4.75E−02	8.76E−02	2.48E−01	4.09E−01	5.69E−01	7.29E−01	8.10E−01
δ_M	0.710893	0.710934	0.711266	0.712873	0.714098	0.714367	0.714573	0.714617	0.714636	0.714647	0.714651
ε_{SrM}	24.31	24.89	29.57	52.22	69.51	73.3	76.2	76.83	77.1	77.25	77.3
$SI_{Strontianite}$	− 0.81	− 0.81	− 0.81	− 0.87	− 0.91	− 0.86	− 0.64	− 0.48	− 0.35	− 0.25	− 0.2
$SI_{Celestite}$	− 1.7	− 1.7	− 1.66	− 1.42	− 1.05	− 0.88	− 0.66	− 0.58	− 0.53	− 0.5	− 0.48
$SI_{Calcite}$	0.48	0.48	0.45	0.21	− 0.05	− 0.07	0.1	0.27	0.41	0.53	0.59
Δc_D (mmol/L)	0	0	2.50E−06	2.60E−05	9.40E−05	1.20E−04	0	0	0	0	0
$\delta_{M\text{-}D}$	0.710893	0.710934	0.711266	0.712868	0.714091	0.714361	0.714573	0.714617	0.714636	0.714647	0.714651
$\varepsilon_{SrM\text{-}D}$	24.31	24.89	29.57	52.16	69.4	73.22	76.2	76.83	77.1	77.25	77.3
$\Delta_{Calcite}$ (mmol/L)	–	0	5.00E−03	5.20E−02	1.88E−01	2.40E−01	0	0	0	0	0
$\Delta_{K\text{-}feldspar}$ (mmol/L)	–	−1.00E−04	−9.10E−04	−5.03E−03	−9.22E−03	−1.05E−02	−1.06E−02	−9.24E−03	−7.41E−03	−5.34E−03	−4.23E−03
Δc_C (mmol/L)	0	−2.65E−05	−2.24E−04	−1.30E−03	1.39E−02	7.09E−02	3.71E−01	5.31E−01	6.03E−01	6.40E−01	6.53E−01
$\delta_{M\text{-}C}$	0.710893	0.710922	0.711185	0.712754	0.714110	0.714270	0.714319	0.714353	0.714386	0.714415	0.714427
$\varepsilon_{SrM\text{-}C}$	24.31	24.73	28.43	50.55	69.67	71.93	72.62	73.1	73.56	73.97	74.15

released from the solid phase increases significantly with increasing mixing ratio. However, the maximum increase in $^{87}Sr/^{86}Sr$ caused by cation exchange is only $\Delta\varepsilon_{Sr} = 3.73$ (Table 8.2). Although the influence of cation exchange on $^{87}Sr/^{86}Sr$ is greater than that of mineral dissolution and precipitation, its overall impact remains below 5%, which is insufficient to affect the sensitivity of $^{87}Sr/^{86}Sr$ as a tracer.

Based on the discussions above, $^{87}Sr/^{86}Sr$ exhibits distinct sensitivity in identifying contamination of Cretaceous groundwater by fracturing flowback fluids in the Sulige area. It can therefore serve as a reliable indicator for assessing groundwater pollution induced by fracturing flowback fluids in this region. According to the regional sensitivity evaluation model, in the northern Desert Plateau area, groundwater contamination by as little as 0.49–2.15% of fracturing flowback fluid can already be detected through changes in $^{87}Sr/^{86}Sr$. In contrast, when Cl^- is used as a tracer, the corresponding detection threshold is 0.76–3.70%. Although Cl^- also shows relatively good sensitivity, its concentration exhibits considerable variability in the flowback fluid end-member. The average Cl^- concentration in the flowback fluid is 7130 ± 4675 mg/L, with a standard deviation accounting for 66% of the mean, whereas the average $^{87}Sr/^{86}Sr$ ratio is 0.714651 ± 0.0004, with a standard deviation of only 0.06% of the mean value. Therefore, for the entire Sulige hydraulic fracturing zone, $^{87}Sr/^{86}Sr$ serves as a more precise and timely indicator for identifying groundwater contamination in aquifers. In contrast, when Cl^- is used as a tracer, contamination might have already occurred even though the Cl^- signal fails to reveal it accurately.

Flowback fluid, composed of mixed fracturing fluid and formation brine, contains numerous organic and inorganic compounds, which can irreversibly contaminate groundwater. However, organic pollutants are generally present at low concentrations, or their detection is complex and costly, preventing timely monitoring of contamination (Chapman et al. 2012; Li et al. 2020). Strontium is not a major groundwater contaminant, but strontium and its isotopes are characteristic tracers of oilfield waters (McDevitt et al. 2021). In the Sulige Gas Field, $^{87}Sr/^{86}Sr$ ratios are highly sensitive for identifying groundwater contamination, providing early warning for unconventional oil and gas hydraulic fracturing areas. Compared to conventional aqueous components, $^{87}Sr/^{86}Sr$ measurement is more technically demanding and costly, whereas Sr^{2+} and Cl^- are simpler and cheaper to analyze. Therefore, using the groundwater pollution identification model, mixing curves of $^{87}Sr/^{86}Sr$ versus Sr^{2+} and Cl^- were established for the Sulige region (Fig. 8.7). Further, precise identification of groundwater contamination can be achieved by extensive Sr^{2+} and Cl^- measurements combined with limited $^{87}Sr/^{86}Sr$ tests. These curves can be used not only for detecting potential groundwater contamination but also for assessing pollution severity and forecasting. Additionally, if the $^{87}Sr/^{86}Sr$ ratio of groundwater in the Sulige Gas Field exceeds 0.713, the water has likely been affected by hydraulic fracturing activities.

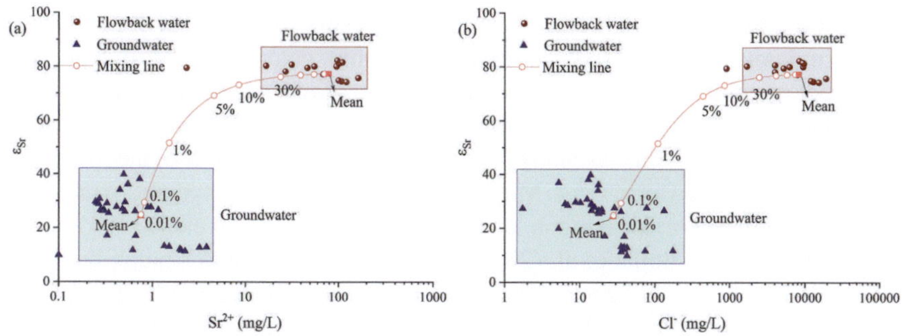

Fig. 8.7 Strontium isotope mixing pathway of ε_{Sr} versus Sr (**a**) and ε_{Sr} versus Cl (**b**) under varying contamination gradients

8.4 Prevention and Control of Groundwater Pollution

Prolonged energy exploitation such as coal, petroleum, and natural gas has led to the formation of extensive, localized groundwater contamination zones in the Ordos and Northern Shaanxi energy bases (Hou et al. 2017; Zhang et al. 2023). Given the expected increase in hydraulic fracturing activities in the Sulige Gas Field, it is crucial to implement proactive measures to prevent and control groundwater pollution in this region, given the high frequency of hydraulic fracturing operations and the associated contamination risks. On the one hand, the management and reuse of flowback water should be strengthened to convert potential pollution sources into valuable resources. On the other hand, it is equally important to strengthen groundwater monitoring by establishing a targeted monitoring network and water quality baselines to facilitate early prevention, rapid detection, and timely remediation of contamination.

Previous studies have investigated groundwater in the Sulige region (Hou et al. 2008; Su et al. 2011), but mainly focused on recharge-runoff-discharge conditions, hydrochemical characteristics, and evolution, with limited attention to hydraulic fracturing activities. Current groundwater monitoring relies mainly on conventional parameters, which are insufficiently sensitive to detect the influence of unconventional oil and gas development, thus limiting timely identification and assessment of hydraulic fracturing impacts on groundwater environments. In light of the projected expansion of hydraulic fracturing in the region, it is imperative to refine and expand both groundwater monitoring parameters and the monitoring network. Groundwater level and quality monitoring points should be strategically installed within each block of the hydraulic fracturing zone. Based on aquifer characteristics at different stratigraphic levels (Salawusu Formation, Cretaceous Huanhe Formation, Luohe Formation), multi-level monitoring wells should be installed to establish a regional groundwater monitoring network for the Sulige hydraulic fracturing area.

Furthermore, it is essential to promptly establish groundwater quality baselines in the study area. Unlike conventional background values, groundwater quality baselines reflect the natural or pre-disturbance water quality status of an assessment area before contamination or ecological damage occurs (Bondu et al. 2021). Due to long-term human activities (e.g., agriculture, mining), determining pristine groundwater background values is difficult. Establishing a groundwater baseline can better reflect the impacts of hydraulic fracturing on groundwater quality. Rhodes and Horton (2015) established baseline groundwater quality for domestic wells in the Marcellus Shale region, Pennsylvania, USA. Microearthquakes induced by hydraulic fracturing may have an impact on local radon levels (Botha et al. 2019; Xu et al. 2019; Daraktchieva et al. 2021). Botha et al. (2019) investigated groundwater radon baselines prior to shale gas development and fracturing in the Karoo Basin, South Africa. Similarly, Daraktchieva et al. (2021) studied radon baselines in the context of shale gas development in North Yorkshire, England. Li et al. (2020) established groundwater baseline values in the Xishui shale gas area, Sichuan Basin, China, using hydrochemical isotope data. Montcoudiol et al. (2019) reported results from a 6-month baseline groundwater monitoring program near an active shale gas pad in northern Poland. A comprehensive baseline assessment documents natural salinity sources, geological and legacy contamination, and anthropogenic influences such as land use activities unrelated to shale gas development. Based on baseline sampling, parameter selection, and threshold definition, optimal baseline values were determined (Bondu et al. 2021). Fluids associated with unconventional oil and gas fracturing contain major and trace contaminants, including Cl, Na, Ba, Sr, B, Mn, As, Cd, Cr, Ni, Pb, Sb, and Se. Additionally, such fluids display unique isotopic fingerprints, including $\delta^{34}S\text{-}SO_4$, $\delta^{18}O\text{-}SO_4$, $\delta^{2}H\text{-}\delta^{18}O$, $\delta^{15}N\text{-}NO_3$, $\delta^{18}O\text{-}NO_3$, $\delta^{15}N\text{-}NH_4$, $\delta^{13}C$-DIC, ^{14}C-DIC, $\delta^{7}Li$, $\delta^{11}B$, and $^{87}Sr/^{86}Sr$, and even radionuclides like ^{226}Ra, ^{228}Ra, ^{222}Rn and ^{238}U. These provide multiple pathways for selecting baseline indicators and defining threshold levels (Bondu et al. 2021). Given the high sensitivity of Sr isotopes in detecting groundwater contamination, establishing Sr and $^{87}Sr/^{86}Sr$ baselines in the study area is essential for rapid and effective monitoring of fracturing impacts and supporting groundwater contamination assessment and prediction.

For the Sulige Gas Field, it is necessary to establish a flowback water database to support fracturing operations and improve the accuracy of groundwater contamination identification. Given the high mobility of gases, the likelihood of methane intrusion into aquifers is generally higher than that of fracturing fluid or flowback water leakage (Moritz et al. 2015; Li et al. 2020). In North American shale gas regions, multiple documented incidents of methane intrusion into groundwater aquifers have been reported (Harkness et al. 2017; Bondu et al. 2021). Therefore, establishing methane baseline values is necessary to monitor its variation in groundwater alongside hydrochemical indicators.

Leakage is a major pathway for groundwater contamination from fracturing activities, often affecting soil before groundwater (He et al. 2022; Zhang et al. 2024). Recognizing the potential effects of flowback water contamination on soil nitrogen cycling, Yang et al. (2024) examined how major flowback constituents influence nitrogen mineralization dynamics. Ma et al. (2019) further elucidated the migration mechanisms of organic compounds from flowback water in the vadose zone under the influence of environmental variables, particularly recharge rates. Mallants et al. (2022) simulated the attenuation of chemical substances from flowback fluids along soil-groundwater pathways following a hypothetical leakage event. In addition, previous studies have suggested that fracturing flowback fluids may indirectly contaminate groundwater by remobilizing pre-existing colloidal pollutants (Sang et al. 2014). *Final Report of the Scientific Inquiry into Hydraulic Fracturing in the Northern Territory* (Pepper et al. 2018) reported that the likelihood of small-volume spills occurring at onshore natural gas operation sites is "moderate" (i.e., possible). Therefore, the soil and groundwater contamination induced by frequent hydraulic fracturing activities deserves considerable attention. Elucidating the migration and transformation processes of typical contaminants from flowback fluids within soils and groundwater systems is essential for preventing environmental pollution in hydraulic fracturing areas.

References

Ando A, Kawahata H, Kakegawa T (2006) Sr/Ca ratios as indicators of varying modes of pelagic carbonate diagenesis in the ooze, chalk and limestone realms. Sed Geol 191:37–53. https://doi.org/10.1016/j.sedgeo.2006.01.003

Appelo CAJ, Postma D (2005) Geochemistry, groundwater and pollution, 2nd edn. Balkema Publishers, Amsterdam

Bagheri R, Nadri A, Raeisi E et al (2014) Hydrochemical and isotopic ($\delta^{18}O$, $\delta^{2}H$, $^{87}Sr/^{86}Sr$, $\delta^{37}Cl$ and $\delta^{81}Br$) evidence for the origin of saline formation water in a gas reservoir. Chem Geol 384:62–75. https://doi.org/10.1016/j.chemgeo.2014.06.017

Bondu R, Kloppmann W, Naumenko-D'ezes MO et al (2021) Potential impacts of shale gas development on inorganic groundwater chemistry: implications for environmental baseline assessment in shallow aquifers. Environ Sci Technol 55(14):9657–9671. https://doi.org/10.1021/acs.est.1c01172

Botha R, Lindsay R, Newman RT et al (2019) Radon in groundwater baseline study prior to unconventional shale gas development and hydraulic fracturing in the Karoo Basin (South Africa). Appl Radiat Isot 147:7–13. https://doi.org/10.1016/j.apradiso.2019.02.006

Breeuwsma A, Wösten JHM, Vleeshouwer JJ et al (1986) Derivation of land qualities to assess environmental problems from soil surveys. Soil Sci Soc Am J 50:186–190. https://doi.org/10.2136/sssaj1986.03615995005000010035x

Capo RC, Stewart BW, Rowan EL et al (2014) The strontium isotopic evolution of Marcellus Formation produced waters, southwestern Pennsylvania. Int J Coal Geol 126:57–63. https://doi.org/10.1016/j.coal.2013.12.010

Chapman EC, Capo RC, Stewart BW et al (2012) Geochemical and strontium isotope characterization of produced waters from Marcellus shale natural gas extraction. Environ Sci Technol 46(6):3545–3553. https://doi.org/10.1021/es204005g

Dai J (2022) The significance of giant gas fields with annual output of more than 10 billion cubic meters in the development of natural gas industry in China. Natural Gas and Oil 40(4):1–15 (in Chinese with English abstract)

Daraktchieva Z, Wasikiewicz JM, Howarth CB, Miller CA (2021) Study of baseline radon levels in the context of a shale gas development. Sci Total Environ 753:141952. https://doi.org/10.1016/j.scitotenv.2020.141952

Davies RJ, Mathias SA, Moss J et al (2012) Hydraulic fractures: how far can they go? Mar Pet Geol 37:1–6. https://doi.org/10.1016/j.marpetgeo.2012.04.001

Dou W, Liu X, Wang T (2010) The origin of formation water and the regularity of gas and water distribution for the Sulige gas field, Ordos Basin. Acta Petrol Sin 31(5):767–773 (in Chinese with English abstract)

Fei S, Yu H, Chen C et al (2022) Key technologies to development of tight sandstone gas reservoirs by horizontal wells: a case study on the upper Paleozoic gas reservoirs in Sulige Gasfield. J xi'an Shiyou Univ 37(4):26–35 (in Chinese with English abstract)

Gaines GL, Thomas HC (1953) Adsorption studies on clay minerals. II. A formulation of the thermodynamics of exchange adsorption. J Chem Phys 21:714. https://doi.org/10.1063/1.1698996

Goldberg RB, Griffith EM (2014) Strontium isotopes as a potential fingerprint of total dissolved solids associated with hydraulic fracturing activities in the Barnett Shale, Texas. Environ Geosci 24(4):151–165. https://doi.org/10.1306/eg.06191716501

Harkness JS, Darrah TH, Warner NR et al (2017) The geochemistry of naturally occurring methane and saline groundwater in an area of unconventional shale gas development. Geochim Cosmochim Acta 208:302–334. https://doi.org/10.1016/j.gca.2017.03.039

He X, Li P, Shi H et al (2022) Identifying strontium sources of flowback fluid and groundwater pollution using $^{87}Sr/^{86}Sr$ and geochemical model in Sulige gasfield, China. Chemosphere 135594. https://doi.org/10.1016/j.chemosphere.2022.135594

He X (2023) Mechanism of chemical interactions between water and rocks during hydraulic fracturing in tight sandstone reservoirs. Ph.D. thesis, Chang'an University (in Chinese with English abstract)

Hou G, Zhang M, Liu F et al (2008) Groundwater exploration in the Ordos Basin. Geological Publishing House, Beijing (in Chinese)

Hou G, Zhao Z et al (2017) Groundwater and ecological environment of the Ordos Energy Base. Geological Publishing House, Beijing (in Chinese)

Huang T, Pang Z, Li Z et al (2020) A framework to determine sensitive inorganic monitoring indicators for tracing groundwater contamination by produced formation water from shale gas development in the Fuling Gasfield, SW China. J Hydrol 581:124403. https://doi.org/10.1016/j.jhydrol.2019.124403

Jia A, Wei Y, Guo Z et al (2022) Development status and prospect of tight sandstone gas in China. Nat Gas Ind 42(1):83–92 (in Chinese with English abstract)

Kohl CAK, Capo RC, Stewart BW et al (2014) Strontium isotopes test long-term zonal isolation of injected and Marcellus Formation water after hydraulic fracturing. Environ Sci Technol 48:9867–9873. https://doi.org/10.1021/es501099k

Ladegaard-Pedersen P, Achilleos M, Dörflinger G et al (2020) A strontium isotope baseline of Cyprus. Assessing the use of soil leachates, plants, groundwater and surface water as proxies for the local range of bioavailable strontium isotope composition. Sci Total Environ 708:134714. https://doi.org/10.1016/j.scitotenv.2019.134714

Lasaga AC (1984) Chemical kinetics of water-rock interactions. J Geophys Res Solid Earth 89:4009–4025. https://doi.org/10.1029/JB089iB06p04009

Li L (2021) Development of natural gas industry in China: review and prospect. Nat Gas Ind 41(8):1–11 (in Chinese with English abstract)

Li Z, Huang T, Ma B et al (2020) Baseline groundwater quality before shale gas development in Xishui, Southwest China: analyses of hydrochemistry and multiple environmental isotopes (^{2}H, ^{18}O, ^{13}C, $^{87}Sr/^{86}Sr$, ^{11}B, and noble gas isotopes). Water 12(6):1741. https://doi.org/10.3390/w12061741

Li Q, Ma L, Liu T (2022) Transformation among precipitation, surface water, groundwater, and mine water in the Hailiutu River Basin under mining activity. J Arid Land 14(6):620–636. https://doi.org/10.1007/s40333-022-0020-1

Ma L, Hurtado A, Eguilior S, Borrajo JFL (2019) Forecasting concentrations of organic chemicals in the vadose zone caused by spills of hydraulic fracturing wastewater. Sci Total Environ 696:133911. https://doi.org/10.1016/j.scitotenv.2019.133911

Mallants D, Kirby J, Golding L et al (2022) Modelling the attenuation of flowback chemicals for a soil-groundwater pathway from a hypothetical spill accident. Sci Total Environ 806:150686. https://doi.org/10.1016/j.scitotenv.2021.150686

McDevitt B, Geeza TJ, Gillikin DP et al (2021) Freshwater mussel soft tissue incorporates strontium isotopic signatures of oil and gas produced water. ACS ES&T Water 1:2046–2056. https://doi.org/10.1021/acsestwater.1c00135

Montcoudiol N, Banks D, Isherwood C et al (2019) Baseline groundwater monitoring for shale gas extraction: definition of baseline conditions and recommendations from a real site (Wysin, Northern Poland). Acta Geophys 67:365–384. https://doi.org/10.1007/s11600-019-00254-w

Moritz A, Hélie JF, Pinti DL et al (2015) Methane baseline concentrations and sources in shallow aquifers from the shale gas-prone region of the St. Lawrence Lowlands (Quebec, Canada). Environ Sci Technol 49(7):4765–4771. https://doi.org/10.1021/acs.est.5b00443

Owen J, Bustin RM, Bustin AMM (2020) Insights from mixing calculations and geochemical modeling of Montney Formation post hydraulic fracturing flowback water chemistry. J Petrol Sci Eng 195:107589. https://doi.org/10.1016/j.petrol.2020.107589

Pepper R, Anderson A, Ashworth P et al (2018) Final report of the scientific inquiry into hydraulic fracturing in the Northern Territory. https://frackinginquiry.nt.gov.au

Rao W, Jin K, Jiang S et al (2015) Chemical and strontium isotopic characteristics of shallow groundwater in the Ordos Desert Plateau, North China: implications for the dissolved Sr source and water-rock interactions. Geochemistry 75:365–374. https://doi.org/10.1016/j.chemer.2015.07.003

Rhodes AL, Horton NJ (2015) Establishing baseline water quality for household wells within the Marcellus Shale gas region, Susquehanna County, Pennsylvania, U.S.A. Appl Geochem 60:14–28. https://doi.org/10.1016/j.apgeochem.2015.03.004

Sang W, Stoof CR, Zhang W et al (2014) Effect of hydrofracking fluid on colloid transport in the unsaturated zone. Environ Sci Technol 48:8266–8274. https://doi.org/10.1021/es501441e

Stoll HM, Schrag DP (2001) Sr/Ca variations in Cretaceous carbonates: relation to productivity and sea level changes. Palaeogeogr Palaeoclimatol Palaeoecol 168:311–336. https://doi.org/10.1016/S0031-0182(01)00205-X

Su X, Wu C, Dong W, Hou G (2011) Strontium isotope evolution mechanism of the cretaceous groundwater in Ordos desert plateau. J Chengdu Univ Technol 38(3):348–358 (in Chinese with English abstract)

Wang L, Dong Y, Zhang Q et al (2021) Effect of upward migration of hydraulic fracturing fluid on shallow groundwater. Safety Environ Eng 28(3):198–205 (in Chinese with English abstract)

Xie X, Wang Y, Ellis A et al (2013) Delineation of groundwater flow paths using hydrochemical and strontium isotope composition: a case study in high arsenic aquifer systems of the Datong basin, northern China. J Hydrol 476:87–96. https://doi.org/10.1016/j.jhydrol.2012.10.016

Xu Y, Sajja M, Kumar A (2019) Impact of the hydraulic fracturing on indoor radon concentrations in Ohio: a multilevel modelling approach. Front Public Health 7:76. https://doi.org/10.3389/fpubh.2019.00076

Yang Q, Wang L, Ma H et al (2016) Hydrochemical characterization and pollution sources identification of groundwater in Salawusu aquifer system of Ordos Basin, China. Environ Pollut 216:340–349. https://doi.org/10.1016/j.envpol.2016.05.076

Yang R, Hou B, Zhang L (2024) Dissolved organic compounds in shale gas extraction flowback water as principal disturbance factors of soil nitrogen dynamics. Sci Total Environ 908:168197. https://doi.org/10.1016/j.scitotenv.2023.168197

Zhang J, Fan Q, Wang Y et al (2019) Mixed large well pattern development technology of tight sandstone gas in Sulige Gas Field, Ordos Basin. Xinjiang Pet Geol 40(6):714–719 (in Chinese with English abstract)

Zhang Y, He L, Xue G et al (2021) Research and application of produced fluid transportation technology in Sulige Gas Field. Inner Mongolia Petrochem Ind 47(2):63–66 (in Chinese)

Zhang H, Han X, Wang G et al (2023) Spatial distribution and driving factors of groundwater chemistry and pollution in an oil production region in the Northwest China. Sci Total Environ 875:162635. https://doi.org/10.1016/j.scitotenv.2023.162635

Zhang H, Han X, Wang G et al (2024) Hydrogeochemical and isotopic evidences of the underlying produced water intrusion into shallow groundwater in an oil production area, Northwest China. Sci Total Environ 916:170242. https://doi.org/10.1016/j.scitotenv.2024.170242

Zheng Z, Zhang H, Chen Z et al (2017) Hydrogeochemical and isotopic indicators of hydraulic fracturing flowback fluids in shallow groundwater and stream water, derived from Dameigou shale gas extraction in the Northern Qaidam Basin. Environ Sci Technol 51(11):5889–5898. https://doi.org/10.1021/acs.est.6b04269

Zheng Z, Cui X, Zhu P et al (2021) Sensitivity assessment of strontium isotope as indicator of polluted groundwater for hydraulic fracturing flowback fluids produced in the Dameigou Shale of Qaidam Basin. J Groundw Sci Eng 9(2):93–101. https://doi.org/10.19637/j.cnki.2305-7068.2021.02.001

Chapter 9
Flowback Water Management and Resources Utilization

The application of hydraulic fracturing technology has long been controversial, with continuous competition between its economic benefits and environmental risks. Environmental challenges, particularly water resource depletion and pollution, remain at the core of this controversy. Achieving green, cost-effective, and sustainable management of flowback fluids have always been a critical and challenging issue in the development of unconventional oil and gas resources. A variety of coexisting resources are associated with oil and gas production, such as sandstone-type uranium deposits, geothermal resources, rock salt, potassium-rich brine, lead–zinc and precious metal ores, hydrogen sulfide, carbon dioxide, helium, mercury in natural gas, and trace elements occurring in crude oil and bitumen. For unconventional oil and gas, flowback water should not be viewed solely as a pollutant-laden industrial effluent but also as a valuable coexisting resource with distinctive utilization potential. Through scientific management and systematic resource utilization, pollution risks of flowback water can be effectively mitigated while achieving efficient resource recovery and reducing production costs. This chapter explores strategies for managing and valorizing flowback water from unconventional reservoirs, with particular emphasis on its lithium recovery potential, offering new perspectives on the green and sustainable use of co-produced resources.

9.1 Flowback Water Management

Water management plays a crucial role during drilling, fracturing, and completion processes. In unconventional oil and gas production, up to 15% of the total cost is related to water management, with flowback water treatment being the most critical component (Gregory et al. 2011). The flowback process of gas wells in the Sulige Gas Field is complex and exhibits three stages: initial liquid, gas–liquid mixture, and

X. He and P. Li, *Geochemical Processes in Tight Gas Hydraulic Fracturing Stimulation*, Environmental Earth Sciences, https://doi.org/10.1007/978-3-032-15180-3_9

liquid–gas mixture. Flowback processes involve larger volumes and complex pollutant compositions, necessitating multiple tailored treatment techniques. The primary challenges in treating flowback fluids lie in their extreme salinity and chemically complex organic content, which significantly elevate treatment costs. Currently, deep well reinjection, reuse, and post-treatment discharge are the predominant strategies for managing flowback fluids worldwide (Zhong et al. 2021). These approaches universally necessitate sequential pretreatment, including filtration, hardness reduction, and salt removal through integrated physicochemical and biological interventions (Jimenez et al. 2018). Flowback fluid management offers dual operational modalities: decentralized on-site treatment or offsite transport to dedicated facilities for consolidated processing (Shi et al. 2023). In recent years, due to the relatively high transportation costs and the increased risk of leakage during vehicle or pipeline transport, on-site treatment has attracted widespread attention (Chang et al. 2020).

9.1.1 Flowback Water Storage

Based on the understanding of the geochemical characteristics and variation patterns, flowback water can be stored and treated at diverse stages. Analyzing temporal variations in chemical and isotopic compositions divides the flowback process into three stages (Fig. 9.1a and b). During the first stage, chemical and isotopic compositions show rapid increase around 7 days post-flowback. TDS values in flowback water from four horizontal wells similarly increase approximately 7 days post-flowback (Fig. 9.1d). Chemical compositions in the first stage are complex due to mixing and water–rock interactions. Over 60% of the flowback water returns to the surface within a week (Fig. 9.2). A cumulative volume of 1597 m^3 of flowback water was collected over 33 days of sampling in the horizontal well, with a flowback rate of 63.85% (Fig. 9.2a). In contrast, the directional well accumulated a volume of 563 m^3 of flowback water over 6.7 days of sampling, which increased to 637 m^3 after four months (Fig. 9.2b). Conversely, the horizontal well had recovered 1100 m^3 of flowback water within the first week of flowback, and this figure only increased to 1790 m^3 after 50 days. Hence, efficient treatment of flowback water during this stage is crucial for reuse in subsequent fracturing fluid preparation. The second stage, roughly 7–14 days after the flowback, chemical compositions fluctuates wildly due to inhomogeneous mixing. In the third stage (after 14 days), flowback water primarily consists of high-salinity formation water, marking the gas production stage. Metal ions and elements like Li, Sr, Ba, Br are abundant in gas reservoir formation water (Liu et al. 2023; Zhou et al. 2022). Flowback water in this stage offers potential sources of recyclable chemical elements. Hence, desalting and recycling of these elements become the focus of this stage.

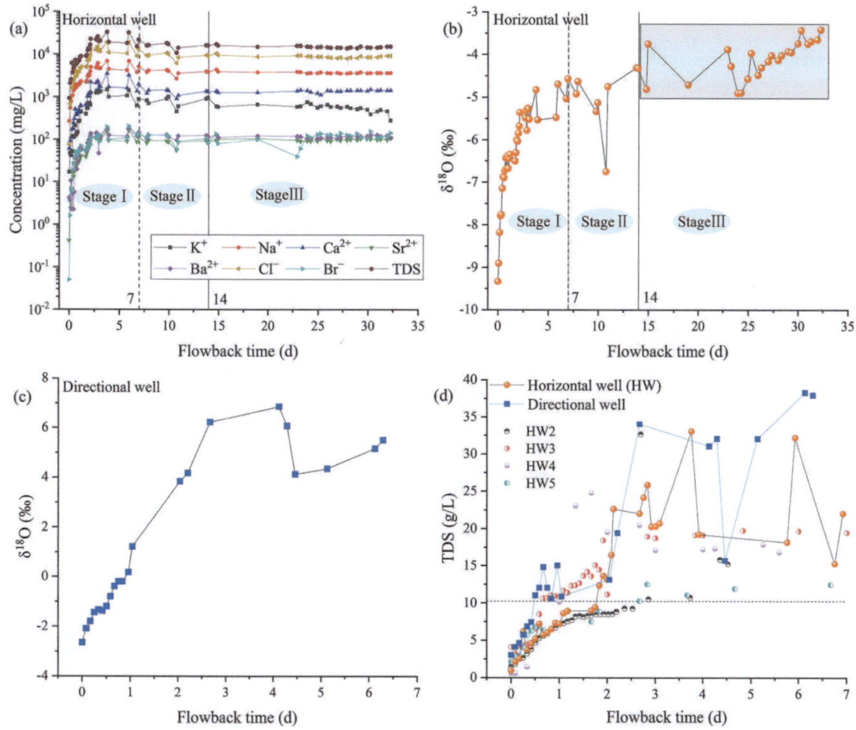

Fig. 9.1 Temporal variations in chemical compositions of flowback water and flowback stages: **a** and **b** horizontal well; **c** directional well; **d** TDS variations in six fractured wells

9.1.2 Treatment and Reuse

Hydraulic fracturing consumes a significant amount of freshwater while producing several times its volume of wastewater with complex chemical compositions, posing substantial challenges for water treatment. At present, treatment methods for fracturing flowback water are generally classified into four categories: physical, chemical, physicochemical, and biological approaches. Physical methods mainly target suspended contaminants in flowback fluids, including gravity separation, centrifugation, and filtration. The chemical method involves adding specific chemical agents into the flowback water; after a certain reaction time, pollutants decompose or precipitate under the action of reagents. Typical chemical treatment methods include coagulation-sedimentation, neutralization, oxidation–reduction, and micro-electrolysis. Physicochemical methods include air flotation and adsorption. The biological method creates favorable environmental conditions for microbial growth and reproduction, accelerating microbial proliferation. Microorganisms metabolize and oxidatively decompose organic matter, converting it into stable inorganic substances. Due to the complex and variable composition of flowback fluids, a

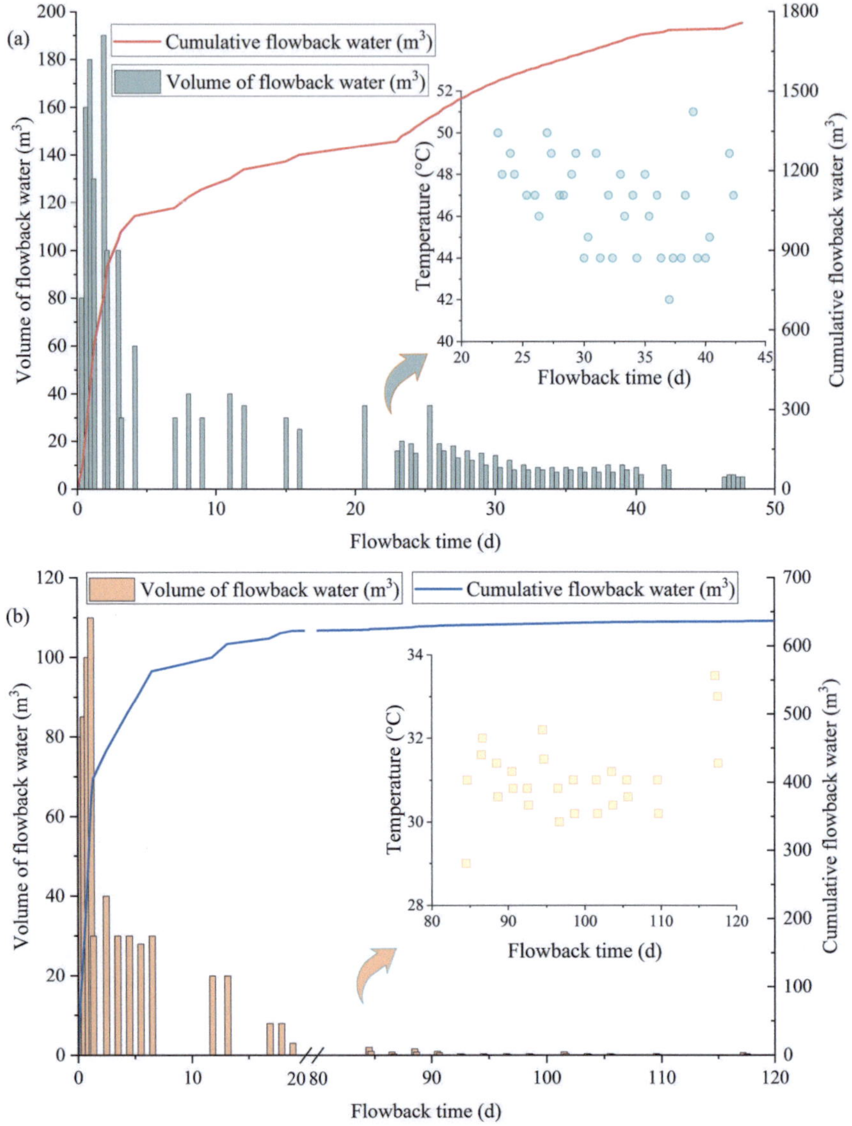

Fig. 9.2 Evolution of flowback water volume in the horizontal well (**a**) and the directional well (**b**)

single treatment approach is often insufficient to meet discharge standards; therefore, integrated processes combining physical, chemical, and biological methods are frequently employed (Silva et al. 2017; Shi et al. 2023).

The repurposing of hydraulic fracturing flowback fluids holds strategic value for hydrocarbon sector advancement, especially in the water-scarce regions such as the Sulige Gas Field. Reusing treated flowback water to prepare new fracturing fluids is regarded as the most effective and sustainable treatment strategy. Recycling flowback water for new fracturing operations reduces both treatment expenses and freshwater demand. Hydraulic fracturing operations induce substantial leaching of metallic constituents from reservoir formations. However, multivalent cations in particular can strongly affect the performance of fracturing fluids. Among them, Ca^{2+}, Mg^{2+}, and Fe are the most concerning in fracturing fluid preparation. Studies establish threshold concentrations making up guar-based fracturing fluid by 275 mg/L (Ca^{2+}), 250 mg/L (Mg^{2+}), and 200 mg/L (Fe) (Ke et al. 2020). Crosslinked fluid systems typically require higher water quality than non-crosslinked fluids (Table 9.1).

Flowback water from the Sulige Gas Field exhibits elevated Ca^{2+} levels varying from 60.1 to 6212 mg/L, which can adversely affect fracturing fluid performance and thus require calcium removal. Mg^{2+} and Fe levels are relatively low, varying from 6.08 to 395 mg/L and 0.39 to 204.24 mg/L, respectively, satisfying the technical specifications for hydraulic fracturing fluid formulation (Table 9.1). For anions, SO_4^{2-} concentrations in flowback fluids from six wells range from 24.0 to 2690 mg/L (median: 120 mg/L), with only one sample exceeding 1000 mg/L. Therefore, SO_4^{2-} in Paleozoic tight sandstone flowback water does not pose a constraint on water reuse. The HCO_3^- concentration ranges from 0 to 3966 mg/L, with an average of 984 mg/L (median 894 mg/L), and only about 10% of samples exceed 1500 mg/L. When mixed with flowback fluids of lower HCO_3^- levels, HCO_3^- generally poses little threat to reuse. Moreover, elevated salinity in flowback water can affect fracturing fluid performance. Therefore, certain desalination treatments are required for high-salinity flowback water (Zhong et al. 2021; Shi et al. 2023). Flowback water desalination can be effectively achieved through dilution techniques. Using flowback water instead of freshwater for fracturing substantially cuts operational costs. However, challenges and limitations remain, such as the generally inferior performance of fracturing fluids prepared with flowback water and their relatively higher costs compared to freshwater (Li et al. 2016).

Table 9.1 Water quality standard for returned water to prepare fracturing fluid

Parameter	Unit	Threshold	
		Crosslinked fluid	Non-crosslinked fluid
pH		6.0 ~ 7.5	6.0 ~ 7.5
$Ca^{2+} + Mg^{2+}$	mg/L	< 150	< 1000
Fe	mg/L	< 20	< 200
Al	mg/L	< 20	< 20
SO_4^{2-}	mg/L	< 1000	< 2000
$CO_3^{2-} + HCO_3^-$	mg/L	< 1500	< 1500
Bacteria	Number/mL	< 100	< 1000
Organic impurities	%	< 0.2	< 0.2

9.2 Resources Utilization

9.2.1 Chemical Elements Extraction

The extraction of high-value metal elements from hydraulic fracturing flowback water has been highly anticipated as a promising recovery route. Flowback water from the Sulige Gas Field is typically enriched in metal ions and elements, including Sr, Ba, Br, and Li. During the critical phase of global energy transition and the implementation of the "dual-carbon" strategy, lithium, known as the "white petroleum of the twenty-first century" and a core strategic resource for the new energy industry, has seen a rapidly escalating supply–demand gap. According to the latest forecast by the International Energy Agency (IEA), the global lithium supply–demand gap may reach as high as 50% by 2030, highlighting the urgent need to explore new lithium extraction pathways (International Energy Agency 2021; Busch et al. 2025). Notably, beyond traditional salt lakes and pegmatite-type lithium deposits, lithium resources contained in associated brines such as oilfield water are emerging as a new international focus for lithium development due to their widespread distribution, large reserves, and environmentally friendly extraction characteristics (Yang et al. 2024).

Globally, lithium resources are primarily hosted in brine-type deposits (65%) and hard-rock deposits (U. S Geological Survey 2024). Brine-type deposits include salt lake brines, geothermal fluids, and oilfield waters. At present, high-grade salt lake lithium resources are predominantly concentrated in the South American "Lithium Triangle" (Chile, Argentina, and Bolivia). Representative examples include Chile's Salar de Atacama (Li: 900–7000 mg/L, mean 1400 mg/L), Argentina's Hombre Muerto Salt Lake (190–900 mg/L, mean 521 mg/L), and Bolivia's Salar de Uyuni (80–1150 mg/L, mean 321 mg/L) (Munk et al. 2025). Oilfield waters are increasingly recognized as promising new carriers for lithium resource recovery. Our project team has systematically summarized the distribution of lithium concentrations in produced waters from major hydrocarbon basins worldwide (Fig. 9.3). Results show that lithium concentrations reach up to 1890 mg/L in the Qaidam Basin oilfield waters, while the peak lithium concentration in unconventional flowback fluids occurs in shale gas reservoirs (633.93 mg/L). Further statistical analysis reveals that the average lithium concentration in shale gas flowback water (562 samples) is 50.71 mg/L (median 34.95 mg/L); in coalbed methane produced water (565 samples), it is much lower (mean 0.48 mg/L); while in tight gas flowback water (219 samples), lithium concentrations exhibit strong heterogeneity, ranging from 0.01–204 mg/L with a mean of 12.91 mg/L (Table 9.2).

With the rapid advancement of unconventional oil and gas development, the production of oilfield water has shown an exponential increase. It is projected that by 2030, the global output of produced water from oil and gas fields will reach 49.9–358.5 billion m^3, with China's production estimated at 50–73 million m^3 (Zhao et al. 2023). A recent study reported that the total volume of flowback water generated by a single hydraulically fractured well throughout its production cycle can reach

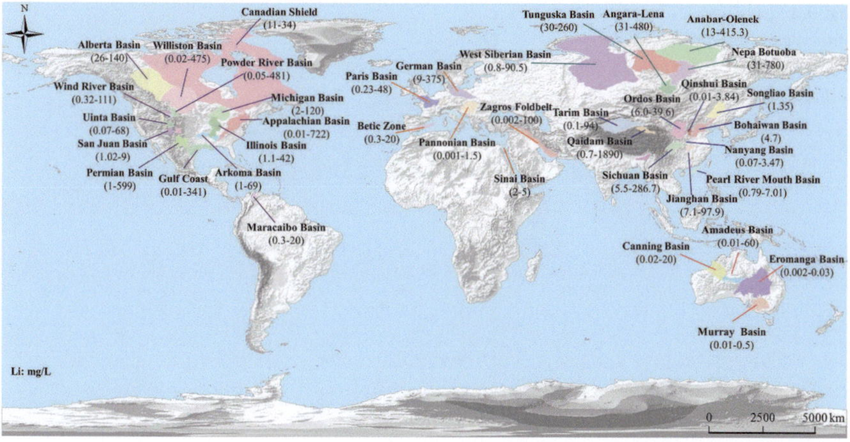

Fig. 9.3 The distribution of lithium concentrations in oilfield waters from major hydrocarbon basins worldwide (mg/L)

Table 9.2 Statistical results of lithium concentration in different types of brine (mg/L)

Type		Max	Mean	Median	Number of Samples
Salt lake water		27,560	253.96	30.43	1435
Geothermal water		1200	9.14	0.45	9587
Conventional oilfield water		1890	23.68	5.00	6905
Unconventional oilfield water	Shale gas	633.93	50.71	34.95	562
	Tight gas	204	12.91	5.4	219
	Coalbed methane	36	0.48	0.20	565

up to 2.5 times that of a conventional well. From 2005 to 2019, North American hydraulic fracturing activities produced roughly 314 billion tons of saline wastewater, and this amount is projected to rise dramatically to 1.7 trillion tons by 2050 (Zolfaghari et al. 2024). In the Sulige Gas Field, multi-layer and multi-stage fracturing technologies, along with the adoption of volumetric fracturing concepts, have led to a steady increase in the amount of fracturing fluid used per well. The average fracturing fluid consumption is 757 m^3 per vertical well and up to 4042 m^3 per horizontal well. Between 2018 and 2020, the total fracturing fluid consumption in the Sulige Gas Field reached about 2.69 million m^3, which is 2.5 times that used between 2015 and 2017. Meanwhile, to maintain stable production, the number of new wells drilled in the gas field increased nearly fourfold between 2015 and 2022. By 2020, the produced water volume had exceeded 2 million m^3, and this figure continues to rise annually with the increasing number and scale of fractured wells (He 2023). The immense volume of produced water, coupled with its considerable lithium enrichment potential, makes lithium recovery from unconventional oil and gas flowback water a strategic alternative for mitigating lithium resource supply–demand gap.

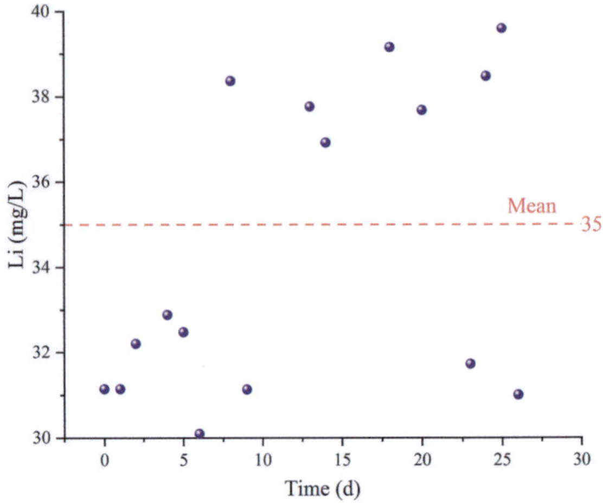

Fig. 9.4 Lithium concentration variation in produced water from a well in the Sulige Gas Field

The produced water from the Sulige Gas Field contains a high concentration of lithium. Continuous monitoring of the lithium concentration in produced water from a well in the Sulige Gas Field over nearly one month showed that the concentration remained stable between 30.09 and 39 mg/L, with an average of 35 mg/L (Fig. 9.4), exceeding China's threshold grade for lithium resource development (25 mg/L). Lithium extraction experiments in the Sulige Gas Field demonstrated superior performance of manganese-based adsorbents over aluminum-based ones. Deep purification of the desorbed solution through nanofiltration, reverse osmosis, electrodialysis, crystallization, and washing produced lithium carbonate with 99.1% purity. Based on the annual volume and average lithium concentration of flowback water, the estimated annual lithium metal discharge from the Sulige Gas Field is approximately 100 tons.

Despite the extensive reserves and widespread geographical distribution of flowback water that indicate considerable lithium recovery potential, the relatively low lithium concentration and highly variable chemical composition pose substantial technological challenges in the extraction process. Compared with the mature lithium extraction technologies developed for salt lake brines, the study of lithium recovery from oilfield brines is still at an early stage, with relatively few research outcomes reported. As two major types of liquid lithium mineral resources, oilfield brines and salt lake brines share significant physicochemical similarities, which provide valuable insights for technological development. The lithium concentration in both oilfield brines and salt lake brines shows a positive correlation with boron (B) and TDS (Fig. 9.5). Salt lake brines are characterized by high Mg and B concentrations. The high Mg/Li ratio constitutes the main constraint factor in the lithium extraction from salt lake brines (Nikkhah et al. 2024). Conversely, oilfield brines exhibit high Ca but low Mg concentrations (Fig. 9.5) and are enriched in organic substance (Zhang et al. 2025; He et al. 2026). Consequently, the lithium extraction techniques designed

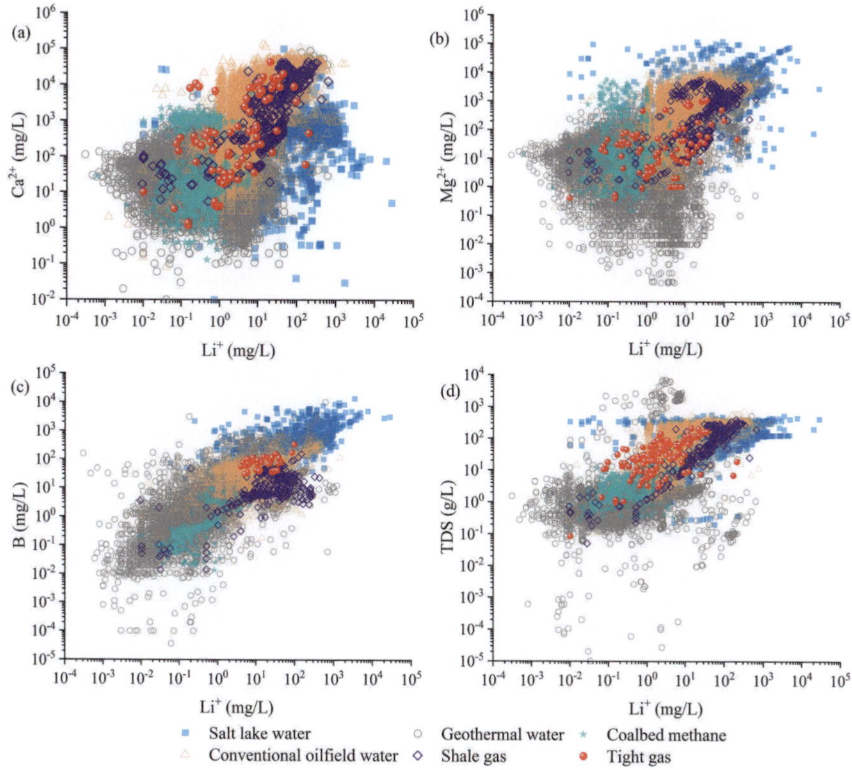

Fig. 9.5 Relationship between Li^+ and Ca^{2+} (**a**), Mg^{2+} (**b**), B (**c**), and TDS (**d**) in different types of water

for salt lake brines cannot be directly applied to oilfield brines, but their relatively mature systems can serve as a basis for targeted optimization.

At present, lithium extraction mainly focuses on precipitation, adsorption, membrane separation, and solvent extraction methods (Nikkhah et al. 2024; Yang et al. 2024; Zhang et al. 2025). However, these technologies face multiple challenges in engineering applications, with existing processes showing limited compatibility with the unique operational conditions of oilfields. Research trends are shifting from single techniques to multi-process coupling approaches. The evolution of direct lithium extraction (DLE) technology has progressed from single-process approaches to integrated systems coupling multiple techniques, primarily based on the synergistic application of solvent extraction, adsorption, ion exchange, and membrane separation (Almousa et al. 2025). This advancement has greatly enhanced extraction efficiency and adaptability, achieving lithium recovery rates typically above 90% (Zhang et al. 2025). The process generally involves: (1) extracting brine from the reservoir; (2) employing coupled technologies such as precipitation, adsorption, solvent extraction, and membrane separation to selectively recover lithium; (3) refining the lithium-enriched brine into lithium products; and (4) reinjecting the depleted brine into the original reservoir. However, the industrial application of DLE technologies is hindered by various technical and commercialization barriers. Technically, the insufficient adaptability to diverse feedstocks poses the foremost challenge, as the heterogeneity of lithium resources prevents the development of a universal DLE process framework (Yang et al. 2024; Zhang et al. 2025).

In order to mitigate the projected gap in future lithium supply, the exploitation of unconventional lithium resources has become a growing global priority in recent years, leading to the initiation of multiple oilfield brine lithium extraction projects (Table 9.3). ExxonMobil, a leading international oil company, has announced its foray into lithium extraction and plans to construct the Evergreen Project in Arkansas, USA. The project will have an initial production capacity of 10,000 t/a of lithium carbonate, with production expected to begin in 2027 and a projected long-term capacity of 100,000 t/a. Representative oilfield brine lithium extraction projects include the Lanxess and South West Arkansas projects developed by Standard Lithium in Arkansas, USA. The former has entered the Definitive Feasibility Study stage, while the latter is currently undergoing Preliminary Feasibility Study and Front-End Engineering Design (FEED), with production expected in 2026 and 2027, respectively. In Canada, E3 Lithium has established seven subprojects in Alberta, with three core projects containing an inferred resource of 7 million tons of lithium carbonate equivalent (LCE) and lithium concentrations of 53–75 mg/L. The company is partnering with Imperial Oil to develop a lithium hydroxide project with an annual capacity of 20,000 tons. In China, five oilfield brine lithium extraction projects have been deployed across Qinghai, Sichuan, and Hubei, including one in the Jianghan Oilfield, two in the Southwest Oilfield, and two in the Qinghai Oilfield. In 2019, Sinopec launched a pilot project in the Jianghan Oilfield for lithium extraction from associated brines using adsorption technology, successfully producing battery-grade lithium carbonate in 2023. The project is designed for an annual output of 30 tons of battery-grade lithium carbonate, employing a hybrid "adsorption-membrane" process with a lithium recovery rate of at least 85% (Table 9.4). CNPC has established four pilot projects for lithium extraction from oilfield brines. The Longwangmiao Gas Field project commenced production at the end of 2022 and successfully passed performance acceptance testing in July 2023. The pilot project has a scale of 50 t/a industrial-grade lithium carbonate. After 72 h of continuous operation, the produced lithium carbonate achieved a purity of 99.2%, with an average daily output of 135 kg. The process includes pretreatment via "air flotation oil removal + air stripping desulfurization + catalytic oxidation + flocculation-sedimentation + two-stage filtration" followed by a lithium extraction system integrating "adsorption, membrane separation, and precipitation" achieving an overall lithium recovery rate of $\geq 70\%$ and an adsorption recovery rate of $\geq 80\%$ (Table 9.4; Zhu et al. 2023; Zhang et al. 2025).

Effective lithium extraction from oil/gas field water necessitates preliminary treatment to eliminate H_2S residuals, colloidal suspensions, and organic contaminants. The process faces challenges such as high chemical reagent consumption during brine pretreatment and recovery, as well as interference from concentrated impurity ions during lithium extraction. Overall, lithium extraction from low-level oilfield waters

Table 9.3 Comparison and applicability of different lithium extraction technologies

Technology	Operation	Cost	Efficiency	Purity	Advantage	Applicability
Precipitation	Li precipitated on addition of precipitants (e.g., phosphate, carbonate)	Low	Low-medium	Low-medium	Operational simplicity, mature technology, energy-efficient performance	Low Mg/Li and high Li concentration brines
Adsorption	Li adsorbed onto the adsorbents (e.g., Al-based, Mn-based, Ti-based adsorbents)	Medium	Medium	Medium	High adsorption capacity and selectivity, simple operation process	High Mg/Li and low Li concentration brines
Solvent extraction	Li extracted by organic phase (neutral phosphorus-based, diketone extractants	High	High	High	High selectivity, simple operation process	High Mg/Li and low Li concentration brines
Membrane technology	Li rejected by membrane (e.g., nanofiltration, reverse osmosis)	High	High	High	Obtain high-concentration lithium, environmental friendly	High-medium Mg/Li and high Li concentration brines
Direct lithium extraction (DLE)	Synergistically combining adsorption, solvent extraction, and membrane separation	Very high	High	Very high	Environmental friendly	High Mg/Li and complex brines

Table 9.4 Representative lithium extraction projects from oil/gas field produced water (OGPW) worldwide (Zhang et al. 2025)

Country	Area	Company	Project name	Average Li concentration (mg/L)	Absorbent type	Extraction technology	Product	Capacity (t/a)	Year of production	Recovery rate
USA	Arkansas	Standard Lithium	Demonstration Proj	237	Ti-based	LiSTR + membrane + precipitation	Li_2CO_3	OGPW: 2.5 m^3/h	2020	DLE ≥ 90%
					Al-based	LSS + membrane + precipitation	Li_2CO_3		2022	DLE ≥ 95%
	Arkansas	Standard Lithium	Lanxess Proj	208	Al-based	Adsorption + membrane separation + precipitation	Li_2CO_3	5400	2026	DLE ≥ 90%
	Arkansas	Standard Lithium	Southwest Arkansas Proj. (SWA)	437	Al-based	Adsorption + membrane separation + precipitation + evaporation	$LiOH{\cdot}H_2O$	30,000	2027	DLE ≥ 90%
	Utah	Anson Resources	Paradox Lithium Proj	123	Al-based	Adsorption + membrane separation + precipitation	Li_2CO_3	13,074	2026	DLE ≥ 91.5%
	Taxes	Volt Lithium	Permian Basin Proj	30–55	Al-based	Adsorption + membrane	Concentrated LiCl solution	OGPW: 1590 m^3/d	2024	DLE ≥ 99%

(continued)

Table 9.4 (continued)

Country	Area	Company	Project name	Average Li concentration (mg/L)	Absorbent type	Extraction technology	Product	Capacity (t/a)	Year of production	Recovery rate
	Louisiana	International Battery Metals (IBAT)	IBAT Modular Direct Extraction pilot Proj	300	/	Adsorption + membrane	Concentrated LiCl solution	5738	2022	DLE: 81%; OAR: 69%
Canada	Alberta	E3 Lithium	Clearwater Proj	75 ± 5	Al-based	Adsorption + membrane separation + precipitation	$LiOH{\cdot}H_2O$	32,250 (25,850)	2027	OAR ≥ 75%
	Saskatchewan	Arizona Lithium	Prairie I Proj	101	Al-based	Adsorption + membrane separation + precipitation	Li_2CO_3	150 (6000)	2025	DLE ≥ 90%
China	Qinghai	PetroChina	Jianbei pilot Proj		Al-based	Adsorption + membrane separation + precipitation	Li_2CO_3 + Li_2PO_4	100 + 8	2023	OAR ≥ 70%
	Qinghai	PetroChina	Nanyishan pilot Proj		Al-based	Adsorption + membrane separation	Li_2CO_3 + KCl	400 + 3000		
	Sichuan	PetroChina	Longwangmiao pilot Proj	25.5–73.3	Al-based	Adsorption + membrane separation + softening + evaporation + precipitation	Li_2CO_3	50	2023	OAR ≥ 70%

(continued)

Table 9.4 (continued)

Country	Area	Company	Project name	Average Li concentration (mg/L)	Absorbent type	Extraction technology	Product	Capacity (t/a)	Year of production	Recovery rate
	Sichuan	PetroChina	Weiyuan pilot Proj		Al-based	Adsorption + NF/RO + ion exchange + crystallization	Li_2CO_3 + KCl + NaBr	100	2024	OAR ≥ 70%
	Hubei	Sinopec	Jianghan pilot Proj		Mn-based	Adsorption + membrane separation + precipitation	Li_2CO_3	30	2022	DLE ≥ 85%

Note: *LiSTR* lithium stirred tank reactors, *LSS* lithium selective sorption, *NF* nanofiltration, *RO* reverse osmosis
(25,850): Expected production capacity
DLE ≥: Recovery rate in DLE stage; OAR ≥: Overall recovery rate

(Li < 100 mg/L) is technically feasible, and purification and precipitation technologies developed for salt lake brines can be applied in oilfield water regions (Zhang et al. 2025). From a technical perspective, future improvements in lithium recovery should focus on three aspects: (1) Strengthen the investigation of oilfield water resources and leverage oil and gas exploration technologies to quantify lithium reserves in oilfield waters; (2) Accelerate the advancement of DLE technologies to enhance the selectivity, anti-fouling capability, and longevity of adsorbents, extractants, and membranes; (3) Promote the coupling of different lithium extraction methods, integrating lithium recovery processes with oilfield water treatment technologies (Liu et al. 2023; Zhu et al. 2023; Almousa et al. 2025; Zhang et al. 2025). Additionally, the comprehensive utilization of geothermal energy, waste heat, solar energy, and other valuable elements (e.g., Br, I, Rb, Cs, Sr, U) should be considered to achieve multi-resource integration and reduce the cost of lithium recovery from oilfield waters.

9.2.2 Utilization of Geothermal Energy

The majority of China's oilfields are now in the middle to late stages of exploitation. High water cut renders these oilfields practically "geothermal water reservoirs" with persistently high development costs. The integrated exploitation of oilfield geothermal energy or co-production of "oil and heat" along with comprehensive and cascade utilization of geothermal resources, will be key to enhancing oilfield efficiency (Wang et al. 2024). Fracturing flowback water is an important geothermal resources. Capturing geothermal energy from flowback water has garnered increasing attention (Kiaghadi et al. 2017; Grauberger et al. 2023). Formation water in tight gas reservoirs is more enriched compared to shale reservoirs (Wu et al. 2022). For instance, in the Sulige Gas Field, the number of productive wells with a high water cut is rising. The mean liquid–gas ratio (the amount of water produced from the formation per 10^4 m^3 of gas produced) has increased from 0.5 $m^3/10^4$ m^3 to 0.75 $m^3/10^4$ m^3. In this study, the mean temperature of flowback water from horizontal and directional wells were 47 °C and 31 °C, respectively (Fig. 9.2). In comparison with the directional well (90 °C), the horizontal has a higher initial formation temperature (104–109 °C) and richer formation water (Fig. 9.2). Additionally, the geothermal gradient in the Sulige gas field is approximately 3 °C/100 m (Fig. 9.6). It suggests that formation temperatures in reservoirs located at depths greater than 3000 m are typically higher than 100 °C. Thus, horizontal well in reservoirs deeper than 3000 m may be prime locations (sweet spots) for geothermal utilization from flowback water (Fig. 9.6). Facing the dilemma of rapid decline of tight gas production, capturing geothermal energy from tight gas wells could extend their economic life. Additionally, the hydraulic fracturing process for deep dry hot rock geothermal resources encounters similar environmental and engineering challenges as those associated with unconventional hydrocarbon extraction (Jiang et al. 2024). The expertise and technical advancements gained through geochemical processes during hydraulic

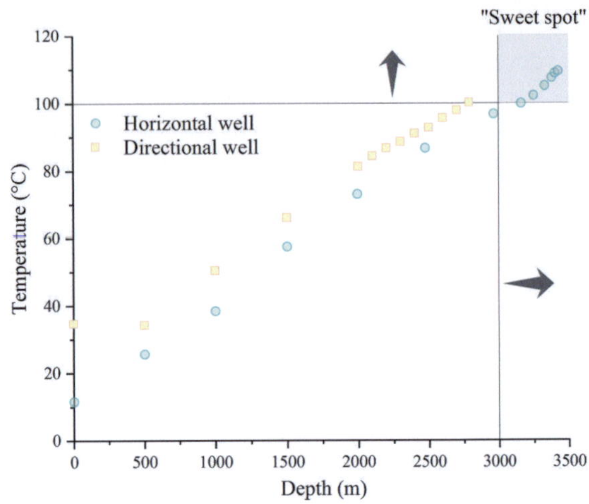

Fig. 9.6 Relationship between formation temperature and buried depth in the study area

fracturing in unconventional hydrocarbon extraction, such as optimizing fracturing fluid selection, mitigating formation damage, and addressing environmental impacts, offer valuable insights for geothermal development. Applying these experiences knowledge can reduce costs and risks associated with geothermal projects, thereby enhancing their economic viability and feasibility.

9.2.3 Diversified Reuse

In addition to reuse for fracturing fluid preparation, recovery of valuable chemical elements, and utilization of geothermal energy, flowback water has various potential applications after treatment, such as road de-icing, landscape irrigation, soil remediation, and power generation (Zhong et al. 2021; Shi et al. 2023). Owing to its elevated salt content, processed flowback water is suitable for road de-icing purposes. Additionally, flowback water contains various plant growth-promoting nutrients such as K and N, suggesting that, drawing on experience from papermaking wastewater reuse, low-salinity treated flowback water could be further explored for irrigation and soil improvement (Shi et al. 2023). Nevertheless, during flowback water reuse, it is critical to prevent any new environmental pollution, including impacts on groundwater, surface water, and soil. Given the fragile ecosystem in the Sulige region, beyond the aforementioned resource utilization pathways, further exploration of innovative approaches for using flowback and produced water to enhance the local ecological environment is warranted (He 2023; Shi et al. 2023).

References

Almousa M, Lim YH, AlMubaidin M, Alshami A (2025) Comparative feasibility of lithium extraction technologies in U.S. oilfields. Desalin Water Treatment 322:101128. https://doi.org/10.1016/j.dwt.2025.101128

Busch P, Chen Y, Ogbonna P et al (2025) Effects of demand and recycling on the when and where of lithium extraction. Nat Sustain 8:773–783. https://doi.org/10.1038/s41893-025-01561-5

Chang H, Liu S, Tong T et al (2020) On-site treatment of shale gas flowback and produced water in Sichuan Basin by fertilizer drawn forward osmosis for irrigation. Environ Sci Technol 54(17):10926–10935. https://doi.org/10.1021/acs.est.0c03243

Grauberger BM, Cole GM, Robbins CA et al (2023) Viability of waste heat capture, storage, and transportation for decentralized flowback and produced water treatment. Appl Energy 330:120342. https://doi.org/10.1016/j.apenergy.2022.120342

Gregory KB, Vidic RD, Dzombak DA (2011) Water management challenges associated with the production of shale gas by hydraulic fracturing. Elements 7:181–186. https://doi.org/10.2113/gselements.7.3.181

He X, Li P, Qian H et al (2026) Biogeochemistry during hydraulic fracturing: a critical review of reservoirs, fluids, processes, and implications. Geoenergy Sci Eng 256:214143. https://doi.org/10.1016/j.geoen.2025.214143

He X (2023) Mechanism of chemical interactions between water and rocks during hydraulic fracturing in tight sandstone reservoirs. Ph.D thesis, Chang'an University (in Chinese with English abstract)

International Energy Agency (2021) The role of critical minerals in clean energy transitions. https://www.iea.org/reports/the-role-of-critical-minerals-in-clean-energy-transitions

Jiang O, Cao L, Zhu W, Zheng X (2024) Review of reservoir damage mechanisms induced by working fluids and the design principles of reservoir protection fluids: from oil–gas reservoirs to geothermal reservoirs. Energies 17:4895. https://doi.org/10.3390/en17194895

Jimenez S, Micó MM, Arnaldos M et al (2018) State of the art of produced water treatment. Chemosphere 192:186–208. https://doi.org/10.1016/j.chemosphere.2017.10.139

Ke C, Peng L, Li X et al (2020) Study on reuse of fracturing flowback fluids in Sulige gasfield. Oilfield Chem 37(3):409–414 (in Chinese)

Kiaghadi A, Sobel RS, Rifai HS (2017) Modeling geothermal energy efficiency from abandoned oil and gas wells to desalinate produced water. Desalination 414:51–62. https://doi.org/10.1016/j.desal.2017.03.024

Li L, Al-Muntasheri GA, Liang F (2016) A review of crosslinked fracturing fluids prepared with produced water. Petroleum 2(4):313–323. https://doi.org/10.1016/j.petlm.2016.10.001

Liu Q, Yang P, Tu W et al (2023) Lithium recovery from oil and gas produced water: opportunities, challenges, and future outlook. J Water Process Eng 55:104148. https://doi.org/10.1016/j.jwpe.2023.104148

Munk LA, Boutt D, Butler K et al (2025) Lithium brines: origin, characteristics, and global distribution. Econ Geol 120(3):575–597. https://doi.org/10.5382/econgeo.5134

Nikkhah H, Ipekçi D, Xiang W et al (2024) Challenges and opportunities of recovering lithium from seawater, produced water, geothermal brines, and salt lakes using conventional and emerging technologies. Chem Eng J 498:155349. https://doi.org/10.1016/j.cej.2024.155349

Shi H, He X, Zhou C et al (2023) Hydrochemistry, sources and management of fracturing flowback fluid in tight sandstone gasfield in Sulige Gasfield (China). Arch Environ Contam Toxicol 84:284–298. https://doi.org/10.1007/s00244-023-00983-6

Silva TLS, Morales-Torres S, Castro-Silva S et al (2017) An overview on exploration and environmental impact of unconventional gas sources and treatment options for produced water. J Environ Manage 200:511–529. https://doi.org/10.1016/j.jenvman.2017.06.002

U. S Geological Survey (2024) Mineral commodity summaries. U. S. Geological Survey, Reston

Wang S, Shi Y, Fang C et al (2024) Status and development trends of geothermal development and utilization in oilfields of China. Lithol Reserv 36(2):23–32 (in Chinese with English abstract)

Wu W, Zhao J, Wei X et al (2022) Evaluation of gas-rich "sweet-spot" and controlling factors of gas–water distribution in tight sandstone gas provinces: an example from the Permian He8 Member in Sulige Gas Province, central Ordos Basin, Northern China. J Asian Earth Sci 227:105098. https://doi.org/10.1016/j.jseaes.2022.105098

Yang S, Wang Y, Pan H et al (2024) Lithium extraction from low-quality brines. Nature 636:309. https://doi.org/10.1038/s41586-024-08117-1

Zhang C, Liu B, He S et al (2025) Associated lithium extraction from oil/gas field produced water: resources, technologies, and practices. J Water Process Eng 77:108356. https://doi.org/10.1016/j.jwpe.2025.108356

Zhao Z, Zhou X, Wang T et al (2023) Feasibility, technical status and prospects of lithium recovery from produced water in oil and gas fields. J Environ Eng Technol 13(4):1434–1443 (in Chinese with English abstract)

Zhong C, Zolfaghari A, Hou D et al (2021) Comparison of the hydraulic fracturing water cycle in China and North America: a critical review. Environ Sci Technol 55:7167–7185. https://doi.org/10.1021/acs.est.0c06119

Zhou X, Ji Z, Wang J et al (2022) Highly efficient recovery of bromine from shale gas wastewater by selective electrochemical oxidation. J Environ Chem Eng 10:107946. https://doi.org/10.1016/j.jece.2022.107946

Zhu R, Cao J, Liu T et al (2023) Research progress of lithium extraction technology and industrialization of unconventional brines in global. Inorg Chem Ind 55(11):1–11 (in Chinese with English abstract)

Zolfaghari A, Gehman J, Kondash AJ et al (2024) Wastewater production footprint of conventional and unconventional oil and gas wells in North America. Nat Water 2:749–757. https://doi.org/10.1038/s44221-024-00286-7

Chapter 10
Summary and Outlooks

This chapter synthesizes the core discoveries of this book. It also identifies critical knowledge gaps in three emerging fields: biogeochemistry during hydraulic fracturing, lithium recovery potential from flowback water, and contaminant dynamics in soil-groundwater system. Microfluidics and in-situ core recovery technology will offer unprecedented insights into intricate fluid-mineral-microbe interplay during hydraulic fracturing. Research on lithium enrichment patterns in fracturing flowback fluids and the development of multidimensional lithium resource assessment models will critically advance unconventional brine lithium extraction. Deciphering the migration and transport dynamics of characteristic contaminants across soil-groundwater systems is fundamental for preventing environmental pollution in hydraulic fracturing zones.

10.1 Summary

This study focuses on the Sulige Gas Field, China's largest onshore integrated natural gas field. Through field-scale hydraulic fracturing experiments and laboratory water–rock interaction tests, combined with hydrochemical, isotopic, X-ray fluorescence (XRF), X-ray diffraction (XRD), scanning electron microscopy (SEM), and energy dispersive spectrum (EDS) analyses, the geochemical processes involved in hydraulic fracturing of tight sandstone reservoirs were characterized at multiple scales. Furthermore, the impacts and damage mechanisms of water–rock reactions on tight reservoirs, potential groundwater contamination induced by large-scale fracturing activities, and strategies for flowback water management and resource utilization were systematically analyzed. The main conclusions are as follows:

(1) The tight sandstone reservoir characteristics in the Sulige area were comprehensively determined. The tight sandstones of the He 8 Member in Sulige are mainly lithic sandstone and lithic quartz sandstone, characterized by high quartz

X. He and P. Li, *Geochemical Processes in Tight Gas Hydraulic Fracturing Stimulation*, Environmental Earth Sciences, https://doi.org/10.1007/978-3-032-15180-3_10

(45.9–81.6%) and clay minerals (16.5–47.4%), and low carbonate and feldspar contents. The clay minerals consist primarily of illite (3.9–27.3%), kaolinite, and chlorite (10.9–31.8%), with no smectite detected. Compaction, cementation, and dissolution during diagenesis led to further densification of the reservoir sandstone, with the destruction of primary intergranular pores and the development of secondary intragranular and intercrystalline pores. The porosity of the tight sandstone ranges from 7.7 to 12.6%, and the permeability from 0.16 to 1.42 mD, indicating a typical low-permeability tight reservoir.

(2) The hydrochemical characteristics, solute origins, and water quality properties of the tight sandstone flowback fluids were clarified. Cl^- and Na^+ are the dominant ions in the flowback fluids from the Sulige Gas Field, which are classified as high-salinity Cl – Na type waters. Controlled by the mixing process between fracturing fluid and formation water, the solute concentrations in flowback fluids generally increase over time, while individual components exhibit distinct variation patterns due to water–rock interactions. Flowback water is a chemically complex mixture of fracturing fluid and formation brine that has undergone extensive water–rock reactions, showing poor water quality and high pollution potential. Major contamination indicators include COD_{Mn}, NH_4-N, Cl^-, Na^+, TDS, Ba^{2+}, Fe, TH, B, Al, Mn, and pH.

(3) Geochemical processes during tight gas hydraulic fracturing were identified. Upon injection, the fracturing fluid reacts with reservoir rocks, causing dissolution of soluble minerals, including halite, calcite, and detrital cements. Simultaneous ion exchange between the rock matrix and fluid occurs for Ca^{2+}, Sr^{2+}, Na^+, and B, with as much as 60% of Sr^{2+} (22 mg/L) transferred from solution to the solid phase. In the presence of oxygen and oxidants, pyrite oxidation occurs within the reservoir, increasing SO_4^{2-} and Fe concentrations in the flowback fluid, with Fe rising by up to 40 mg/L. Mixing between the fracturing fluid and formation water enriched in Ca^{2+}, Ba^{2+}, and Al often induces secondary mineral precipitation such as calcite, barite, and gibbsite. These interrelated water–rock chemical processes jointly influence the geochemical composition of flowback fluids and alter the physical properties of tight reservoirs.

(4) Clarified the impacts and damage mechanisms of water–rock chemical interactions on tight reservoirs. The injection of low-salinity fracturing fluids enhances mineral dissolution and corrosion, thereby increasing porosity. During the fracturing process of Well *S1*, the theoretical dissolution of calcite reached 985 kg (in 2481 m^3 of fracturing fluid), corresponding to an increase of approximately 0.36 m^3 in pore volume. However, water–rock interactions can induce swelling of illite, transforming platy crystals into fibrous forms that disperse and migrate before aggregating to form bundle-shaped pore throats. In alkaline fracturing environments, kaolinite readily fragments and disperses, causing pore-blocking damage during migration. Simulation results indicate up to 11.7 mg/L of barite and 2.5 mg/L of gibbsite may precipitate during hydraulic fracturing. Clay mineral instability and inorganic scaling can block pores and microfractures, ultimately reducing reservoir permeability.

(5) Developed a model to identify potential groundwater contamination in hydraulic fracturing zones. In view of the high pollution potential of flowback fluids and the high vulnerability of groundwater in the Sulige Gas Field, a model was developed to identify groundwater contamination caused by flowback fluids. A strontium isotope-based groundwater contamination identification curve for the Sulige region was established. The findings reveal that in the northern desert plateau of Sulige, even a minor mixture of 0.49%-2.15% of flowback fluid with groundwater can be detected through changes in the $^{87}Sr/^{86}Sr$ ratio. With the anticipated intensification of hydraulic fracturing operations, strengthening groundwater protection in the Sulige area is essential. This requires both the proper treatment and recycling of flowback fluids and the enhancement of groundwater monitoring to establish water quality baselines and a hydrochemical database of flowback fluids. The methodologies and results of this study can be extended to other unconventional oil and gas basins, providing valuable support for the green and sustainable development of unconventional oil and gas.

(6) Flowback water management and resource utilization were discussed. For unconventional oil and gas systems, flowback water should not merely be regarded as highly polluting industrial wastewater. As an important co-produced resource, it possesses unique resource attributes. Through scientific management and systematic resource utilization, pollution risks can be effectively mitigated while promoting efficient resource recovery and reducing production costs. Flowback water from tight sandstone reservoirs, which contained high-temperature formation water, is a potential geothermal resource. Based on geochemical insights, flowback water can be stored in multiple stages, reused for fracturing fluid, and recycled chemical elements, achieving sustainable treatment and utilization. Given its lithium extraction potential, it is essential to strengthen the assessment of lithium resource reserves and develop efficient lithium extraction technologies.

10.2 Outlooks

Through a combination of field-scale hydraulic fracturing tests and laboratory experiments, this study explored the geochemical processes in tight sandstone reservoirs and achieved several meaningful interim findings. Nevertheless, due to the large scale and high complexity of hydraulic fracturing operations, this research inevitably has some limitations, and numerous aspects of water–rock interactions in tight reservoirs remain to be further investigated. The heterogeneity of low-permeability reservoirs and the chemical complexity of fracturing fluids make the fracturing fluid-tight sandstone interactions highly intricate. Different gas wells, or even different gas-bearing intervals within the same well, often require distinct fracturing techniques and fluid systems. In such processes, physical, chemical, and biological interactions are intricately interwoven and coupled. A comprehensive understanding of the geochemical processes and their effects on reservoirs during hydraulic fracturing requires simultaneous consideration of chemical field variations, hydrodynamic changes, and the influence of microbial activities. The present study primarily examines the chemical aspects of water–rock interactions in tight sandstone reservoirs and does not include the potential role of microbial processes. Future research on water–rock interactions during hydraulic fracturing should involve collaborative in-situ studies between hydrogeologists and petroleum development institutions. Particular attention should be paid to the effects of formation brine and microbial activities on geochemical processes. In view of the potential water resource and environmental risks associated with increasingly frequent hydraulic fracturing in the Sulige area, groundwater monitoring in fracturing zones should be strengthened to prevent groundwater and soil contamination possibly induced by large-scale fracturing operations. Furthermore, the treatment and resource utilization of large volumes of wastewater generated by hydraulic fracturing in the study area should be prioritized in future research and management strategies.

10.2.1 Biogeochemistry During Hydraulic Fracturing

Microorganisms are essential participants across the full lifecycle of unconventional hydrocarbon production, affecting oil/gas generation, energy recovery efficiency, the quality of flowback fluids, and wastewater treatment (He et al. 2026). With the growing understanding of deep subsurface microbiology, microbial activities have received increasing attention. In unfractured reservoirs, researchers have identified various microorganisms, such as methanogenic archaea, sulfate-reducing bacteria, and iron-reducing bacteria (Daly et al. 2016). Moreover, flowback waters host abundant halotolerant and thermophilic microbial consortia (Cluff et al. 2014; Daly et al. 2016). Hydraulic fracturing creates fracture networks, enhances fluid mobility, and introduces chemical additives and surface microorganisms, thereby providing favorable physical and chemical conditions for microbial metabolism in the reservoir. These processes foster the establishment of new subsurface microbial ecosystems. In turn, microorganisms mediate a range of chemical reactions that profoundly influence the fracturing process and the evolution of flowback fluids (Hernandez-Becerra et al. 2023; He et al. 2026). Thus, microorganisms represent critical agents governing the hydraulic fracturing process.

Examination of microbial community dynamics in flowback fluids has emerged as a key method for elucidating the biochemical mechanisms occurring during hydraulic fracturing. Studies have revealed that microbial communities in shale flowback fluids are predominantly composed of halophilic and thermophilic species, such as *Halanaerobium*, *Marinobacter*, *Methanohalophilus*, and *Candidatus* Frackibacter. Moreover, the microbial community structure varies markedly across different stages of flowback (Cluff et al. 2014; Daly et al. 2016). Cliffe et al. (2022) reported that guar gum serves as a metabolic substrate for thiosulfate-reducing microorganisms, facilitating the production of hazardous H_2S, thereby adversely affecting shale gas production. Platt et al. (2023) found through coal-microbe interaction experiments that the addition of algal enhancers can effectively stimulate microbial methanogenesis in coal seams. Furthermore, research conducted by Borton et al. (2018) revealed that amino acid co-fermentation (Stickland reaction) is a key metabolic pathway in shale-microbe interactions. By rationally engineering amino acid-based microbial metabolic networks, it is possible to enhance biomethane yields and mitigate contaminant generation. Integrating field hydraulic fracturing experiments with laboratory simulations, and employing innovative techniques including tracer tracing, isotopic fingerprinting, microbial omics, and microfluidic chip analysis, enables the in-depth elucidation of biogeochemical processes occurring during hydraulic fracturing.

(1) **Hydraulic fracturing tracer test**

Considering the complexity of the hydraulic fracturing process, a tracer test is designed to effectively identify the sources of solutes in flowback fluids, characterize the mixing process between fracturing fluid and formation water, and delineate the propagation of hydraulic fractures. Various tracers are introduced into the fracturing fluids at different stages of the operation (Fig. 10.1). Currently, the most widely used tracers include chemical, isotopic, and trace-element tracers (Table 10.1). Rare earth element tracers, characterized by high sensitivity, minimal background interference, and cost-effectiveness, can serve as suitable alternative tracers.

Before initiating the fracturing, samples of the fracturing fluid and preparation water are collected. After the completion of hydraulic fracturing, flowback water sampling is commenced without delay. In the early stage of flowback (pure liquid

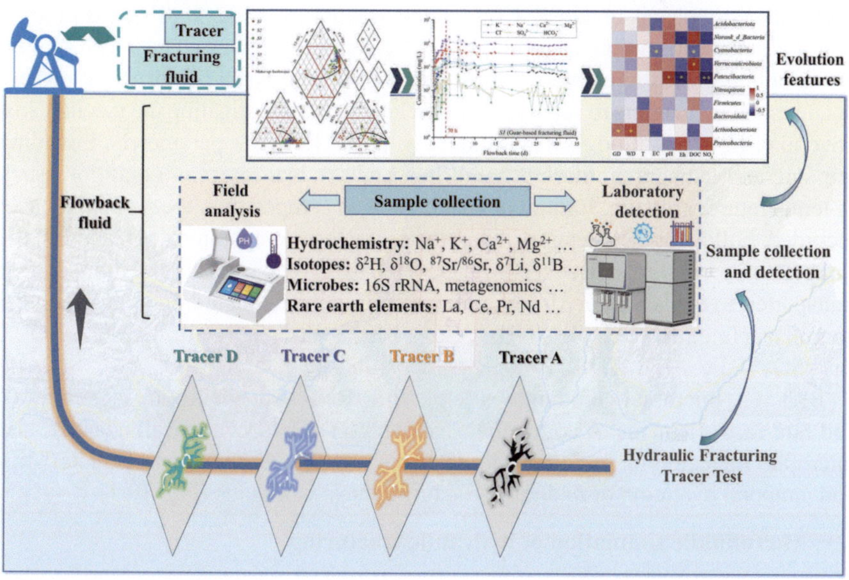

Fig. 10.1 Schematic diagram of hydraulic fracturing tracer test

Table 10.1 Comparison and applicability of different tracers

Type	Tracer	Advantage	Disadvantage	Detection method
Chemical tracers	Soluble inorganic salts, dyes, halogenated hydrocarbons, alcohols	Diversified, easily detected	Low precision, large usage, high cost, environmental hazard	Spectrophotometer, chromatographic analyzer
Stable isotopes	$\delta^{12}C$, $\delta^{13}C$, $\delta^{18}O$, $\delta^{15}N$	Low dosage, high precision	Few types, high cost, complex testing process	Isotope ratio mass spectrometer
Radioisotopes	Tritium	Low dosage, low cost, high precision	High environmental and health risks, high cost	Liquid scintillation counter
Trace substance tracers	Rare earth elements, fluorescent substances, trace ions	Diversified, easily detected, low cost, high precision	Imperfect explanation method	Inductively coupled plasma mass spectrometry

phase), samples can be taken every half hour. With time, sampling intervals are progressively extended and dynamically adjusted according to sample conditions and operational needs to ensure extensive sampling during the pure liquid stage, while still obtaining representative samples during the gas–liquid and liquid–gas mixed flow stages. Ultimately, a high-frequency, long-duration sequence of flowback water samples is obtained. Continuous monitoring of the well's fluid production is carried out to collect formation water samples at later stages. Throughout the sampling process, cumulative flowback volume and visible color variations are documented in real time to supply critical field data for investigating the mechanisms governing flowback fluid formation and evolution. Relevant parameters, including pressure and temperature during fracturing, and original reservoir conditions such as temperature, pressure, formation water, and gas composition, are simultaneously recorded. Following field testing of parameters such as temperature, pH, and Eh, the samples are processed and preserved as per analytical requirements and immediately transported to the laboratory for comprehensive analyses, including hydrochemical parameters (e.g., K^+, Li^+, Na^+, NH_4^+, Ca^{2+}, Mg^{2+}, Ba^{2+}, Sr^{2+}, Cl^-, HCO_3^-, NO_3^-, Br^-, SO_4^{2-}, Fe, Al, B, TDS), isotopic compositions (e.g., δ^2H, $\delta^{18}O$, $^{87}Sr/^{86}Sr$, δ^7Li, $\delta^{11}B$, $\delta^{13}C$), microbial communities (e.g., 16S rRNA sequencing, metagenomics), and rare earth elements. A comprehensive analysis of hydrochemical, isotopic, and microbial indicators is performed to determine the biogeochemical characteristics and temporal evolution of hydraulic fracturing flowback fluids (Fig. 10.1).

(2) **Microfluidic simulation of hydraulic fracturing**

Microfluidic platforms leverage compact design and superior performance to revolutionize biomedical-to-energy applications, while their integration of spectroscopic, elemental, and fluorescent analytics offers unprecedented insights into intricate fluid-mineral-microbe interplay during hydraulic fracturing. Based on insights gained from field fracturing tests regarding the geochemical features and functional microbial communities of flowback fluids, a laboratory microfluidic simulation of hydraulic fracturing is subsequently designed. Core samples from unfractured gas reservoirs were obtained and analyzed for elemental composition (XRF), mineralogy (XRD), and microstructural features (SEM). These analyses determined the fundamental characteristics of the tight sandstone, including mineral composition and pore distribution, providing essential baseline data for subsequent microfluidic fracturing simulations. To realistically simulate in-situ reservoir environment such as temperature and pressure, microfluidic fracturing simulations involving fracturing fluid-tight sandstone interactions were carried out using a high-temperature, high-pressure reaction system. The experimental procedures were designed as follows:

Microfluidic Chip Design: Following the approaches of Li et al. (2020), Fouke et al. (2022), and Zhou et al. (2022), the microfluidic chip was designed in a Y-shaped configuration, featuring a main flow channel (approximately 30 mm in length and 0.25 mm in width) and two fluid inlets (~5 mm long, 0.25 mm wide). Rock samples were sliced into rectangular thin sections (~35 mm × 10 mm × 5 mm thick) and further cut into one triangular piece (top of the "Y") and two trapezoidal pieces (sides of the "Y"). The three rock pieces were temporarily separated with 0.25 mm paper spacers, glued onto a glass slide in a Y-shaped configuration, and after spacer removal, polished to around 0.5 mm thickness. Following channel cleaning, a second perforated glass slide was bonded with adhesive to form the experimental microfluidic chip (Fig. 10.2). Multiple microfluidic chips were fabricated following the above procedure for subsequent experiments.

Experimental Apparatus: The experimental microfluidic platform comprised a microfluidic simulation unit, microflow injection system, control and data acquisition

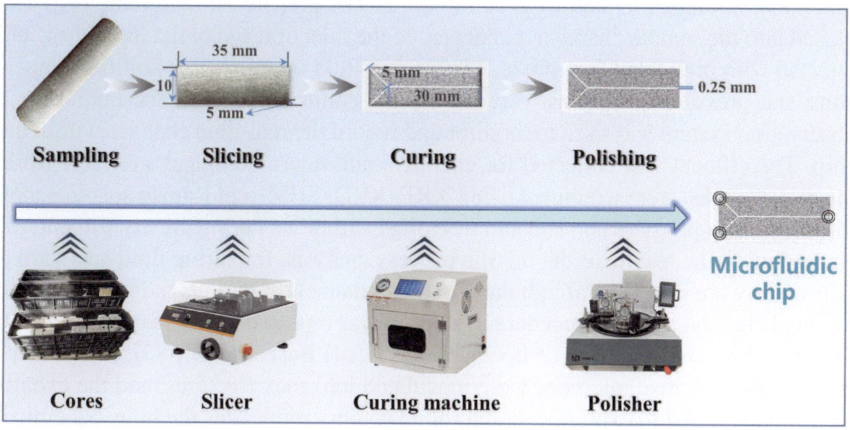

Fig. 10.2 Manufacturing process of microfluidic chips

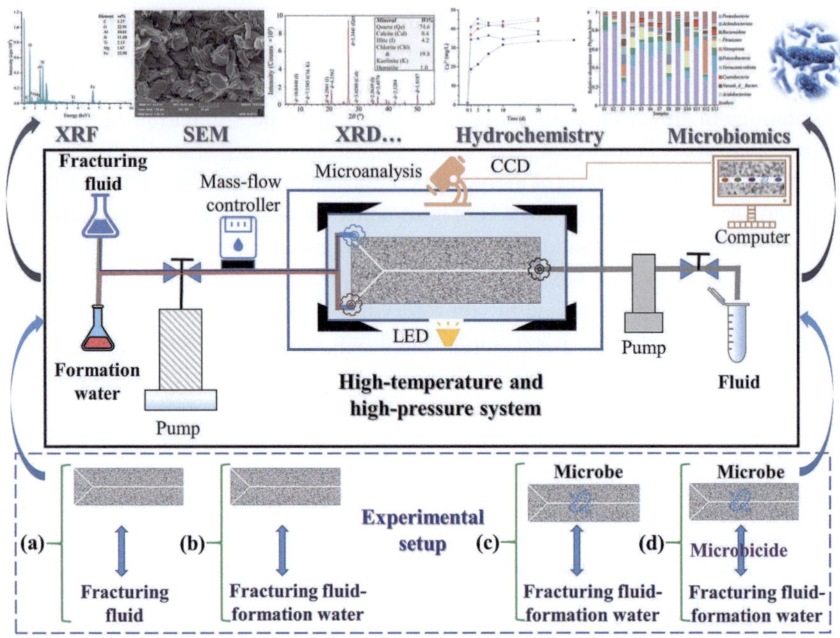

Fig. 10.3 Schematic diagram of the microfluidic simulation of hydraulic fracturing

system, microscopic imaging module, high-temperature and high-pressure system, and auxiliary peripherals (Fig. 10.3).

Experimental Procedures: (1) Based on the chemical composition and microbial community analyses of flowback fluids and formation water, synthetic solutions mimicking formation water were prepared in the laboratory, and representative functional microbial strains were cultured. (2) The prepared microfluidic chip was placed into the sample chamber, connected to the injection and outlet interfaces, and injected with pretreated (gel-broken) fracturing fluid to ensure flowability. Temperature and pressure conditions were set via the control software. The microscopic observation system was used to monitor and record the real-time changes within the chip. The effluent was collected for chemical and microbiological analyses, while the rock samples were examined using XRF, XRD, SEM, and Raman spectroscopy to characterize physicochemical and microbial variations during the experiment. (3) According to the results of the mixing process analysis, fracturing fluid and formation water were injected through the two inlet channels at various ratios to simulate mixing behavior, and the procedure in step (2) was repeated to elucidate its effect on the formation and evolution of flowback fluids. (4) Based on step (3), fluorescent-tagged representative microbes were inoculated into rock fractures, and the experiment was repeated to explore how microbial activity modulates the biogeochemical evolution of flowback fluids. (5) Following step (4), biocides were incorporated into the fracturing fluid, and the tests were repeated to assess their inhibitory effects on microbial activity and corresponding impacts on the evolution of flowback fluids (Fig. 10.3). A deionized water blank control group was set up concurrently.

By integrating pre- and post-experimental analyses of aqueous, mineral, and microbial components with visualized microscopic images recorded during the tests, the study compared how the mixing of fracturing fluid and formation water influences water–rock interactions and microbial behavior, thereby elucidating how different mixing ratios regulate reaction kinetics and microbial metabolic activity. Using microbial abundance and metabolic product data, the impact of microbial activity on geochemical reactions was assessed, highlighting its essential role in the evolution and formation of fracturing flowback fluids. Fluorescence labeling enabled real-time tracking of the dynamic interplay between biocides and microorganisms, revealing the inhibition efficiency of biocides on functional microbial populations and their consequent effects on the biogeochemical evolution of flowback fluids. Using microscopic analytical techniques such as SEM and Raman spectroscopy, changes in the chemical composition and microstructure of clay minerals before and after biogeochemical interactions were compared, revealing the impact of microbial activity on clay minerals. Integrating all analytical results, the synergistic processes among fluids, minerals, and microorganisms were systematically elucidated, providing the understanding of the origin and evolution of flowback fluids, as well as essential parameters for subsequent model development.

(3) **In-situ field experiment**

In-situ experiments on water–rock interactions, studies, particularly through pre- and post-test core/fluid sampling, represent the most direct and reliable approach for geochemical processes validation (Maity and Ciezobka 2019; He et al. 2026). Nevertheless, the implementation of in-situ experiments is constrained by reservoir depth constraints, economic feasibility, and post-stimulation core retrieval difficulties (He et al. 2026). Two studies on core recovery were conducted in the Wolfcamp-Eagle Ford shale projects in Texas, USA (Jew et al. 2022) and Sinopec's recent retrieval of 1,122.96 m cores from six stimulated wells in the Fuling Gas Field, China (Xia and Luo 2024). These preserved cores enable fracture network quantification (density, dimensions, aperture) for stimulation efficacy assessment, with future analyses anticipated to decode subsurface geochemical process dynamics during hydraulic fracturing (He et al. 2026).

10.2.2 Lithium Enrichment and Resource Assessment

(1) **Lithium enrichment of flowback water**

Influenced by water–rock interactions between fracturing fluids and formation work, as well as mixing with formation water, large amounts of lithium are present in fracturing flowback fluids (Fig. 10.4). Can these vast volumes of flowback water serve

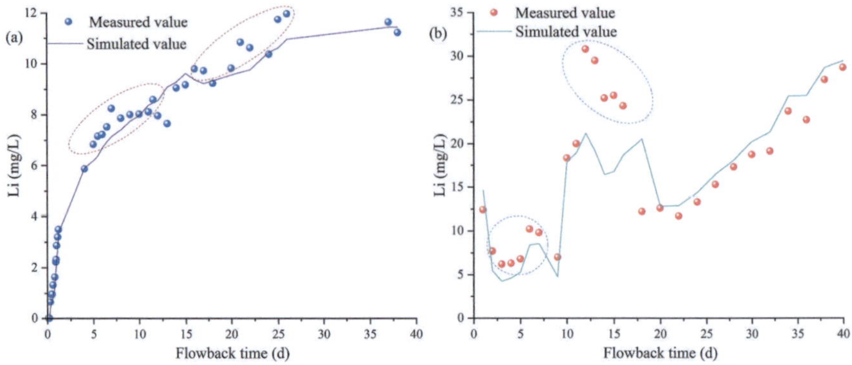

Fig. 10.4 Evolution of measured (Point) and simulated (Line, modeling results of fracturing fluid and formation water mixing) Li concentrations of the flowback water from shale reservoirs (**a** Huang et al. 2022; **b** Ni et al. 2024)

as a potential lithium resource? At present, systematic studies on lithium occurrence characteristics, controlling factors, enrichment-migration mechanisms under fluid-mineral coupling, and multi-dimensional resource evaluation are still lacking, which restricts the development of lithium extraction from unconventional oil and gas flowback water.

With the recent surge in unconventional lithium extraction, increasing attention has been paid to the lithium enrichment mechanisms in salt lake brines and geothermal waters. The formation of salt-lake-type lithium deposits is governed by complex geological-climatic coupling processes: lithium derived from rock weathering in the drainage basin is transported by surface runoff into closed basins, where it becomes progressively concentrated under intense evaporation, eventually forming high-lithium brines (López Steinmetz and Salvi 2021; Munk et al. 2025). Key controlling factors include (1) source rock properties (e.g., Li-rich granites or volcanic rocks), (2) hydrogeological conditions (degree of basin closure, water residence time, recharge intensity), and (3) climatic characteristics such as evaporation rate, precipitation, and wind speed (López Steinmetz and Salvi 2021; Zhang et al. 2025). Lithium enrichment in high-temperature geothermal systems is primarily controlled by deep-seated geological processes. The lithium source mechanisms can be summarized as: (1) dehydration reactions in subduction zones lead to the initial enrichment of lithium in crustal materials; (2) partial melting of the mantle wedge transfers incompatible elements, such as lithium, into the melt; (3) magmatic differentiation promotes the progressive concentration of lithium in the residual melt; and (4) later hydrothermal circulation further extracts lithium from the enclosing rocks (Liu et al. 2025).

Compared with salt lake and geothermal systems, research on lithium enrichment mechanisms in oil and gas field waters is relatively limited, with current understanding primarily derived from studies in shale gas reservoirs (Phan et al. 2016;

Huang et al. 2022; Lee et al. 2024). Recent findings suggest that in shale reservoirs, lithium mainly resides in the interlayers of clay minerals, with a minor portion bound to organic matter (Lee et al. 2024). Phan et al. (2016) applied a sequential extraction technique to quantify lithium distribution in shale minerals. Their results revealed that structural lithium within clay minerals constitutes 75–91% of the total lithium content, lithium in carbonate matrices contributes less than 3%, and about 20% of lithium is associated with oxidizable organic matter and sulfide phases (Phan et al. 2016). During hydraulic fracturing, lithium release from clay mineral interlayers is primarily controlled by ion exchange reactions (notably the substitution of Li^+ by Ca^{2+}) and by organic–inorganic interactions (Phan et al. 2020; Lee et al. 2024). Isotopic geochemical analyses (δ^7Li, δ^{11}B, δ^{138}Ba) indicate that both shale thermal maturity and the degree of water–rock interaction collectively regulate lithium concentrations in produced waters (McDevitt et al. 2024). Laboratory experiments have demonstrated that fracturing fluid-shale interactions can raise lithium concentrations in flowback fluids by about 20% (Huang et al. 2022), a trend that has been corroborated by field observations from shale formations in both the Appalachian and Sichuan basins (Fig. 10.4) (Phan et al. 2016; Huang et al. 2022; Ni et al. 2024). During unconventional oil and gas development, hydraulic fracturing alters the reservoir's temperature–pressure-chemical conditions, significantly influencing lithium occurrence and migration. Specifically: (1) high-pressure fluids promote lithium desorption from clay interlayers; (2) fracture network expansion increases water–rock reaction interfaces; and (3) redox changes during fracturing trigger secondary mineral precipitation or dissolution. To better understand these mechanisms, comprehensive investigations on fluid-mineral coupling processes are required.

Field-scale hydraulic fracturing tests should be implemented to collect high-resolution, long-duration flowback water samples (Fig. 10.1). Through detailed geochemical analyses, the hydrochemical characteristics of the flowback fluids can be established, and the behavior of lithium ions, isotopic compositions, and associated trace elements can be used to delineate lithium's geochemical occurrence and controlling processes. Analytical approaches such as hydrochemical modeling, multivariate statistical analysis, and isotopic fingerprinting will further clarify lithium sources and dominant controls. Complementary laboratory-scale water–rock interaction experiments will quantify the abundance of soluble, exchangeable, and carbonate-bound lithium in rock. By integrating field and laboratory findings, the enrichment and activation mechanisms of lithium under coupled fluid-mineral processes can be systematically elucidated.

(2) **Lithium resource assessment**

Building a rigorous and scientific evaluation framework, underpinned by a detailed understanding of lithium distribution patterns and enrichment mechanisms, is essential to enable the efficient exploitation of lithium resources from lithium-rich brines. In recent years, preliminary advances have been achieved in the assessment of emerging lithium-bearing systems, including geothermal and oilfield waters. Wei et al. (2024) conducted a comprehensive analysis of hydrochemical datasets from

1989 geothermal sites in 30 provinces of China, providing the first quantitative estimate of the nation's geothermal lithium potential. The study suggests that China's geothermal waters discharge approximately 3233 tons of metallic lithium annually. Using long-term monitoring data from 368 production wells in the Weiyuan and Changning gas fields of the Sichuan Basin, Sang et al. (2023) verified that over 800 tons of lithium carbonate could be recovered annually from produced water, highlighting the industrial potential of lithium extraction from oilfield brines. Recent exploration in the Western Canada Sedimentary Basin identified an inferred lithium resource of 75.9 million tons, an indicated resource of 14.9 million tons, and a measured resource of 7.2 million tons of lithium (Bishop and Robbins 2025). These studies confirm the industrial value of oilfield brines as emerging lithium resource carriers. However, conventional estimation approaches based solely on the "concentrations multiply volumes" method are overly simplistic and inadequate for guiding large-scale resource development. Frontier studies are advancing toward intelligent, multi-parameter, multi-dimensional evaluation frameworks, mainly in two aspects: (1) Geochemistry-geostatistics integration, Peng et al. (2025) developed a data-driven model using cutoff-based thresholds, integrating spatial coordinates, stratigraphic information, and key geochemical parameters (e.g., TDS, Na, K, Mg, Ca) to successfully predict lithium enrichment zones in Devonian brines of the Alberta Basin, filling lithium data gaps (Fig. 10.5). (2) Machine-learning prediction, Knierim et al. (2024) developed a random forest model that integrates 15 geological, petrophysical, and thermal parameters to map lithium concentration in the Smackover Formation (Arkansas), estimating lithium resources of 5.1–19 million tons, equivalent to 35–136% of current U.S. resources (Fig. 10.5). Collectively, these pioneering studies have expanded the capability for lithium resource delineation and quantification into regions and stratigraphic intervals with sparse lithium data, achieving spatial prediction and assessment of lithium endowment.

Nevertheless, lithium resource assessments must go beyond spatial targeting to integrate extraction costs and recovery efficiencies for a more realistic evaluation of economic viability. Lithium extraction from oilfield brines is still in the pilot testing phase. Current studies largely remain at the static concentration level and lack systematic integration of three critical dimensions: (1) Geological framework (e.g., clay mineral composition); (2) Engineering parameters (e.g., fracturing fluid composition, flowback regime); and (3) Economic constraints (e.g., water treatment cost, market price of lithium). With the deep integration of big data and artificial intelligence, machine learning is driving a methodological revolution in geosciences. Next-generation evaluation frameworks leverage deep learning algorithms to uncover hidden relationships within geological and engineering datasets, enabling simultaneous analysis of lithologic, geochemical, and production dynamic data. This approach offers a new pathway toward a unified "geological potential-engineering feasibility-economic viability" evaluation model. Therefore, building on insights into lithium enrichment mechanisms in flowback waters, multi-source datasets encompassing geological attributes (formation lithology, mineral composition), geochemical parameters (ion concentrations, isotopic ratios), engineering metrics (fracturing scale, flowback rate), and economic indicators (extraction cost,

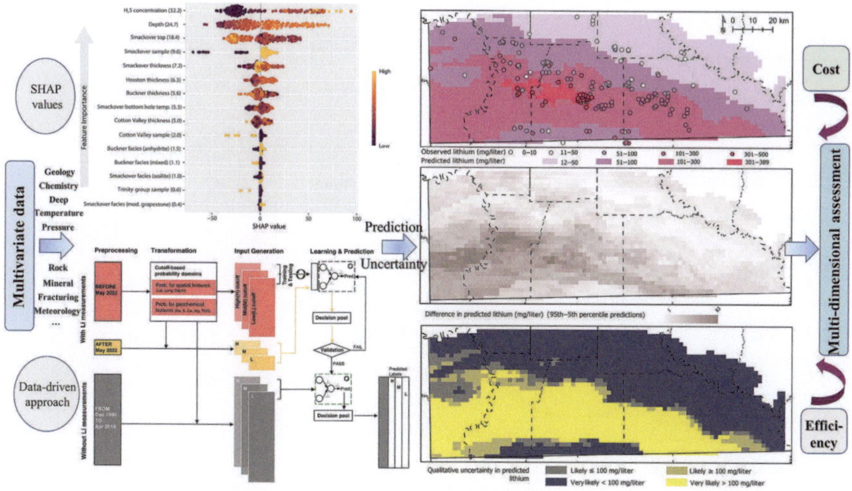

Fig. 10.5 Conceptual framework for constructing a multi-dimensional evaluation system of lithium resources in flowback water (Modified from Knierim et al. 2024 and Peng et al. 2025)

recovery efficiency) can be integrated using machine learning algorithms to enable accurate prediction and visualization of lithium resource potential (Fig. 10.5).

10.2.3 Migration and Transformation of Typical Pollutants in Soil and Groundwater

The potential contamination of soil and groundwater by flowback fluid through multiple pathways has consistently attracted public attention. Once the flowback fluid enters subsurface environments, how do pollutants migrate and transform, and how can contamination be detected quickly? These are crucial challenges for regions engaged in unconventional hydraulic fracturing. Understanding the migration and transformation processes of typical pollutants in soil and groundwater is thus fundamental for preventing environmental pollution in hydraulic fracturing zones. After identifying the major pollutants in flowback fluid, soil samples from hydraulic fracturing areas are collected for laboratory column and sandbox experiments. These experiments generalize the structural and functional complexity of aquifer systems, emphasizing pollutant migration and transformation across the vadose and saturated zones under fluctuating water table conditions. The setups are shown in Figs. 10.6 and 10.7. Pollutant concentrations and forms in soil and groundwater are analyzed before and after the experiments. Combined with geochemical modeling, the migration and transformation patterns of pollutants in multi-media soil-groundwater systems will be systematically examined.

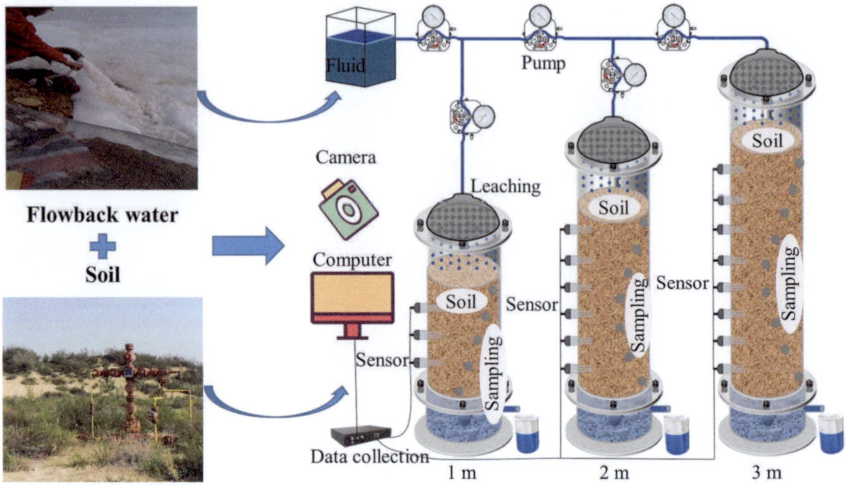

Fig. 10.6 Experimental simulation of pollutants migration and transformation within soil

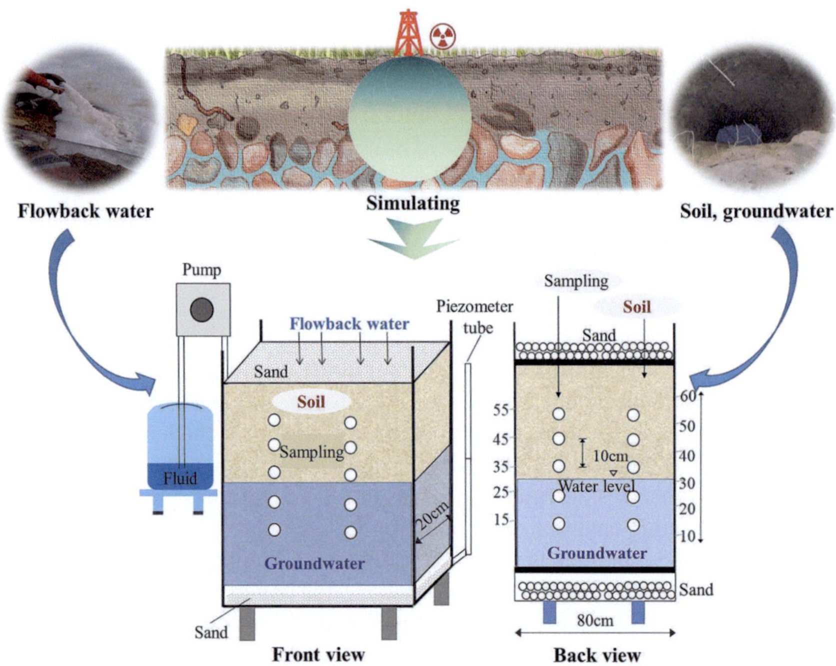

Fig. 10.7 Experimental simulation of pollutants migration and transformation across the soil-groundwater system

(1) **Pollutant migration and transformation within soil**

To investigate the migration and transformation behavior of pollutants during flowback fluid infiltration, soil column experiments are designed in the laboratory. Mixed flowback fluid from different production stages will be used as the water source. The columns will be packed with representative soil types from the study area (sandy soil and loess), as shown in Fig. 10.6. Depending on soil properties, the column length will range from 1–3 m, with sampling ports installed at various vertical intervals, and the column diameter set at 20–30 cm. Flowback fluid will be gradually introduced from the top under a controlled hydraulic head, and infiltration processes will be monitored using a camera and sensors. When infiltration fronts reach each sampling port, soil samples will be collected. Once effluent appears at the outlet, samples of the leachate and upper soil layers will be taken for analysis. The chemical composition and pollutant content of both pre- and post-experimental samples (liquid and soil) will be determined.

(2) **Pollutant migration and transformation across the soil-groundwater system**

The soil-groundwater migration experiment uses a transparent acrylic tank (80 × 90 × 20 cm) with an open top covered by perforated plastic film to reduce evaporation. Two vertical rows of five sampling ports each are placed at depths of 15, 25, 35, 45, and 55 cm from the bottom. The lower part of the tank is packed with quartz sand and groundwater to simulate an aquifer, while the upper section contains soil. The base is connected to a piezometer and a Mariotte bottle via a three-way valve, with a peristaltic pump controlling the water level. A 5-cm quartz sand layer (2–3 mm particle size) is placed at both the top and bottom to maintain the water table (Fig. 10.7). Flowback fluid containing a color tracer is sprayed onto the soil surface. Once the flowback fluid infiltrates and reaches the simulated aquifer, soil samples are taken from three sampling ports in one column and groundwater samples from two ports in the other. When the infiltrated flowback reaches the aquifer base, additional samples are collected from all five ports for analysis of typical pollutants in soil and groundwater before and after the experiment.

References

Bishop BA, Robbins LJ (2025) Overview of brine-hosted lithium mineral resources, distribution, and genesis in the Western Canada Sedimentary Basin. Canad J Mineral Petrol.https://doi.org/10.3749/2400035

Borton MA, Hoyt DW, Roux S et al (2018) Coupled laboratory and field investigations resolve microbial interactions that underpin persistence in hydraulically fractured shales. PNAS 115(28):E6585–E6594. https://doi.org/10.1073/pnas.1800155115

Cliffe L, Hernandez-Becerra N, Boothman C et al (2022) Guar gum stimulates biogenic sulfide production in microbial communities derived from UK fractured shale production fluids. Microbiol Spectrum 10(6):e03640-e3722. https://doi.org/10.1128/spectrum.03640-22

Cluff MA, Hartsock A, MacRae JD et al (2014) Temporal changes in microbial ecology and geochemistry in produced water from hydraulically fractured Marcellus Shale gas wells. Environ Sci Technol 48:6508–6517. https://doi.org/10.1021/es501173p

Daly RA, Borton MA, Wilkins MJ et al (2016) Microbial metabolisms in a 2.5-km-deep ecosystem created by hydraulic fracturing in shales. Natural Microbiol 1(10):16146. https://doi.org/10.1038/nmicrobiol.2016.146

Fouke BW, Bhattacharjee AS, Fried GA et al (2022) Sulfate-reducing bacteria streamers and iron sulfides abruptly occlude porosity and increase hydraulic resistance in proppant-filled shale fractures. AAPG Bull 106:179–208. https://doi.org/10.1306/07132120124

He X, Li P, Qian H et al (2026) Biogeochemistry during hydraulic fracturing: a critical review of reservoirs, fluids, processes, and implications. Geoenergy Sci Eng 256:214143. https://doi.org/10.1016/j.geoen.2025.214143

Hernandez-Becerra N, Clife L, Xiu W et al (2023) New microbiological insights from the Bowland shale highlight heterogeneity of the hydraulically fractured shale microbiome. Environ Microbiome 18:14. https://doi.org/10.1186/s40793-023-00465-1

Huang T, Li Z, Long Y et al (2022) Role of desorption-adsorption and ion exchange in isotopic and chemical (Li, B, and Sr) evolution of water following water-rock interaction. J Hydrol 610:127800. https://doi.org/10.1016/j.jhydrol.2022.127800

Jew AD, Druhan JL, Ihme M et al (2022) Chemical and reactive transport processes associated with hydraulic fracturing of unconventional oil/gas shales. Chem Rev 122(9):9198–9263. https://doi.org/10.1021/acs.chemrev.1c00504

Knierim KJ, Blondes MS, Masterson A et al (2024) Evaluation of the lithium resource in the Smackover formation brines of southern Arkansas using machine learning. Sci Adv 10:eadp8149. https://doi.org/10.1126/sciadv.adp8149

Lee KJ, You J, Gao Y et al (2024) Release, transport, and accumulation of lithium in shale brines. Fuel 356:129629. https://doi.org/10.1016/j.fuel.2023.129629

Li Q, Jew AD, Brown GE et al (2020) Reactive transport modeling of shale−fluid interactions after imbibition of fracturing fluids. Energy Fuels 34:5511–5523. https://doi.org/10.1021/acs.energyfuels.9b04542

Liu M, Kong Y, Guo Q (2025) Sources and enrichment mechanisms of lithium, rubidium, and cesium in waters of magmatic-hydrothermal systems. Earth Sci Rev 270:105241. https://doi.org/10.1016/j.earscirev.2025.105241

López Steinmetz RL, Salvi S (2021) Brine grades in Andean salars: when basin size matters A review of the lithium triangle. Earth Sci Rev 217:103615. https://doi.org/10.1016/j.earscirev.2021.103615

Maity D, Ciezobka J (2019) An interpretation of proppant transport within the stimulated rock volume at the hydraulic-fracturing test site in the Permian Basin. SPE Reservoir Eval Eng 22(2):477–491. https://doi.org/10.2118/194496-PA

McDevitt B, Tasker TL, Coyte R et al (2024) Utica/point pleasant brine isotopic compositions (δ^7Li, δ^{11}B, δ^{138}Ba) elucidate mechanisms of lithium enrichment in the Appalachian Basin. Sci Total Environ 947:15174588. https://doi.org/10.1016/j.scitotenv.2024.174588

Munk LA, Boutt D, Butler K et al (2025) Lithium brines: origin, characteristics, and global distribution. Econ Geol 120(3):575–597. https://doi.org/10.5382/econgeo.5134

Ni Y, Liu S, Li L et al (2024) Geochemical characteristics of shale gas hydraulic fracturing flowback/produced water in Zheng'an block, Northern Guizhou Province, China. J Natural Gas Geosci 9:27–37. https://doi.org/10.1016/j.jnggs.2024.01.001

Peng X, Chen Z, Jiang C et al (2025) A data-driven approach for exploring unconventional lithium resources in Devonian sedimentary brines, Alberta, Canada. Nat Resour Res 34:1885–1900. https://doi.org/10.1007/s11053-025-10461-6

Phan TT, Capo RC, Stewart BW et al (2016) Factors controlling Li concentration and isotopic composition in formation waters and host rocks of Marcellus Shale, Appalachian Basin. Chem Geol 420:162–179. https://doi.org/10.1016/j.chemgeo.2015.11.003

Phan TT, Hakala JA, Sharma S (2020) Application of isotopic and geochemical signals in unconventional oil and gas reservoir produced waters toward characterizing in situ geochemical fluid-shale reactions. Sci Total Environ 714:136867. https://doi.org/10.1016/j.scitotenv.2020.136867

Platt GA, Davis KJ, Schweitzer HD et al (2023) Algal amendment enhances biogenic methane production from coals of different thermal maturity. Front Microbiol 14:1097500. https://doi.org/10.3389/fmicb.2023.1097500

Sang S, Hu C, Cen Y et al (2023) Evaluation and development prospect of lithium resources associated with produced water of shale gas in southern Sichuan. Chem Eng Oil Gas 52(3):41–45 (in Chinese with English abstract)

Wei S, Zhang W, Fu Y et al (2024) Distribution characteristics and resource potential evaluation of lithium in geothermal water in China. Geol China 51(5):1527–1553 (in Chinese with English abstract)

Xia M, Luo H (2024) Sinopec leads the world in post-compression coring technology. China Petrochem News 2024–06–03(007). https://doi.org/10.28130/n.cnki.ncshb.2024.000690 (In Chinese)

Zhang X, Miao W, Han G et al (2025) Genesis and resource of lithium brines in the Qaidam Basin of North Qinghai-Xizang plateau: an overview. Econ Geol 120(5):1089–1111. https://doi.org/10.5382/econgeo.5164

Zhou L, Michelson K, Fried G et al (2022) Rates of sulfate reduction by microbial biofilms that form on shale fracture walls within a microfluidic testbed. ACS ES&T Eng 2:1619–1631. https://doi.org/10.1021/acsestengg.2c00048

MIX
Papier aus verantwortungsvollen Quellen
Paper from responsible sources
FSC® C105338
FSC
www.fsc.org

If you have any concerns about our products,
you can contact us on
ProductSafety@springernature.com

In case Publisher is established outside the EU,
the EU authorized representative is:
Springer Nature Customer Service Center GmbH
Europaplatz 3, 69115 Heidelberg, Germany

Printed by Libri Plureos GmbH
in Hamburg, Germany